TRIGONOMETRY
for College Students

TRIGONOMETRY
for College Students

Nancy Myers

Bunker Hill Community College

D. VAN NOSTRAND COMPANY

New York   Cincinnati   Toronto   London   Melbourne

Cover photograph by Stan Wakefield

D. Van Nostrand Company Regional Offices:
New York   Cincinnati

D. Van Nostrand Company International Offices:
London   Toronto   Melbourne

Copyright © 1981 by Litton Educational Publishing, Inc.

Library of Congress Catalog Card Number: 80-53133
ISBN: 0-442-31179-6

Published by D. Van Nostrand Company
135 West 50th Street, New York, N.Y. 10020

10   9   8   7   6   5   4   3   2   1

Preface

TRIGONOMETRY FOR COLLEGE STUDENTS is an introductory book in the sense that it assumes no previous knowledge of trigonometry. It assumes that students are familiar with the algebra of linear and quadratic expressions and equations.

The book has several unique features. We review algebraic concepts as they are needed, rather than in a general review at the beginning. We have covered a broad range of topics, each including many worked examples and detailed explanations. Complicated proofs are left to the end of units, where they do not interrupt the flow of the text.

By experimenting with several approaches to trigonometry, we have found that the student without previous experience in the subject understands best when we begin with right triangle trigonometry. We proceed to general angles, oblique triangles, and then analytic trigonometry. We first use basic algebra and then apply concepts of functions, quadratics, and other algebraic concepts to the trigonometric functions.

We have chosen the unit form of organization because it allows the student to master a single concept in each section and a limited number of related concepts in each unit. Specific Objectives are stated at the beginning of each unit. Each section number and topic within the unit correspond to an objective number. A Self-Test completes each unit. The self-test items may be in random order, but the answers are also keyed to the objective numbers. Thus a student can refer by number from a self-test item to the relevant objective and corresponding section. This organization within each unit, along with Cumulative Review units spaced strategically throughout the book, makes the book suitable for self-paced courses and other types of self-instruction as well as for traditional courses.

The unit form of organization also allows the instructor flexibility in choosing topics. The book falls generally into three parts. The first part, Units 1 to 7, introduces the basic techniques of right triangle trigonometry, angles in a coordinate system, and oblique triangle trigonometry, with applications. The student is then ready to move on to analytic trigonometry, including radian measure, the trigonometric ratios as functions and their graphs, and finally to identities and equations involving trigonometric functions in Units 8 to 16. Units 1 to 16 may be considered a core course in trigonometry at the intermediate level. The third part, Units 17 to 21, explores more sophisticated topics in trigonometry, along with the related topics of logarithmic functions and complex numbers.

The answers to all Exercises and Self-Tests (except proofs of identities) are given at the back of the book. A Test Manual, available from the publisher, contains four quiz forms for each unit, a pretest, and two forms of a final exam.

As the text developed, detailed comments were provided by Albert Liberi, Westchester Community College; Richard Semmler, Northern Virginia Community College; and Ara Sullenberger, Tarrant County Junior College. Other reviewers included Arnold Knuppel, San Jacinto College, and Ronald Schryer, Orange Coast College.

At Bunker Hill Community College, the mathematics department, with Chairpersons Joan McGowan and then Dorothy Ryan, provided help and encouragement. In particular, Robert (Ted) Carlson checked all the worked examples and exercise answers. Maria Trenga, a former student, typed the manuscript. And scores of students helped with their comments and questions.

Nancy Myers

Contents

TRIGONOMETRY
for College Students

Unit 1

The Trigonometric Ratios

INTRODUCTION

The word "trigonometry" comes from the Greek words meaning "measurement of triangles." The trigonometric ratios are relationships among the angles and sides of right triangles. In this unit you will learn the names and definitions of the six trigonometric ratios. Also, you will learn how to find values for the trigonometric ratios using special triangles and using tables of values.

OBJECTIVES

When you have finished this unit you should be able to:

1. Label the opposite side, the adjacent side, and the hypotenuse of a right triangle where the right angle and an acute angle are indicated.
2. State the six trigonometric ratios in terms of the opposite side, the adjacent side, and the hypotenuse.
3. Use special triangles to find values of trigonometric ratios of special angles.
4. Use a table of trigonometric ratios to find values of trigonometric ratios of acute angles given in degrees and tens of minutes, and to find an acute angle to the nearest degrees and tens of minutes given a value of a trigonometric ratio.
5. Use linear interpolation to find values of trigonometric ratios of acute angles given in degrees and minutes, and to find an acute angle to the nearest degrees and minutes given a value of a trigonometric ratio.

Section 1.1	Right Triangles

Historically, right triangle trigonometry was the first type of trigonometric study. Isolated uses of the trigonometric ratios for right triangles appear to have been developed by the Babylonians, and may also have been known to the Egyptians, earlier than 1600 B.C. Right triangle trigonometry as we know it was begun by Greek mathematicians and astronomers living in Alexandria between about 200 B.C. and 300 A.D. One of the most famous of these astronomers was Claudius Ptolemy (about 150 A.D.), whose theory that the sun and planets revolve about the earth was believed for many centuries.

Recall that a **right triangle** is a triangle with one **right angle**. You may recall from other courses that a unit of measure for angles is the **degree**. There are 360° (360 degrees) in a circle. This measure can be traced through the Greeks of Alexandria back to the ancient Babylonians. A circle can be divided equally into four right angles. Therefore, each right angle measures 90°.

Right triangle trigonometry is based on assumptions of Euclidean geometry. In Euclidean geometry, the sum of the angles of a triangle is exactly equal to the measure of two right angles, or 180°. Thus it is impossible to have a triangle with more than one right angle. If there were two right angles in a triangle, the two right angles would equal 180°, and there would be no degrees left for the third angle of the triangle. Thus the right angle of a right triangle is uniquely determined, meaning there is just one right angle, and the other two angles of a right triangle are each less than 90°.

Since the right angle of a right triangle is uniquely determined, the side opposite the right angle is also uniquely determined. This side is called the **hypotenuse** of the right triangle. The word "hypotenuse" comes from Greek words meaning "to stretch under," indicating the side of the right triangle which stretches across the right angle. The hypotenuse is the longest of the three sides of the right triangle because it stretches across the 90° angle, which is the largest angle:

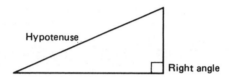

Any angle which is more than 0° but less than 90° is called an **acute** angle. In a right triangle, the two angles other than the right angle are acute angles. The two acute angles are not uniquely determined. However, the right angle is 90°, and the sum of all the angles is 180°. Therefore, the sum of the two acute angles is 90°. If the acute angles are equal, then each is 45°. If they are not equal, then one is more than 45° and the other is less.

In general, suppose θ (theta) is either one of the acute angles of a right triangle. (The Greek letter θ is often used to indicate an unknown angle, just as the letter x is often used in algebra to indicate an unknown number.) The side of the triangle across from θ is called the **opposite side**. The side of the triangle which with the hypotenuse forms θ is called the **adjacent side**.

EXAMPLE 1.1.   Label the opposite side, the adjacent side, and the hypotenuse of this triangle, where the right angle and an acute angle are indicated:

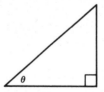

Solution.   First locate the hypotenuse, which is opposite the right angle. Then label the side which is not part of θ as the opposite side. The remaining side is the adjacent side, and with the hypotenuse forms θ:

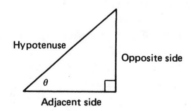

Exercise 1.1

Label the opposite side, adjacent side, and hypotenuse for each of these triangles, where the right angle and an acute angle are indicated:

1.

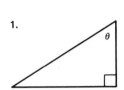

2.

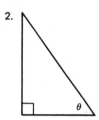

3.

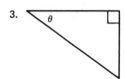

4.

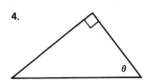

Section	
1.2	Definition of the Trigonometric Ratios

A **ratio** is a quotient of two numbers. A **proportion** is two equal ratios. For example,

$$\frac{1}{2} = \frac{2}{4}$$

is a proportion. A basic theorem of Euclidean geometry is: If a line is drawn across a triangle parallel to one side of the triangle, the line cuts the other two sides proportionally. Consider the triangle ABC, with $B'C'$ parallel to BC:

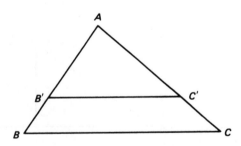

(B' is read "B prime" and C' is read "C prime.") $B'C'$ cuts sides AB and AC proportionally. That is,

$$\frac{AB'}{B'B} = \frac{AC'}{C'C}.$$

We may also say

$$\frac{AB'}{AC'} = \frac{B'B}{C'C}.$$

Suppose two right triangles have an acute angle θ which is the same in each triangle:

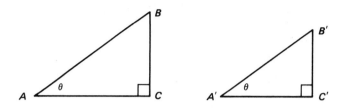

Regardless of the sizes of the triangles, when an acute angle θ is the same in each triangle, then the sides of the triangles are proportional. To show that the sides BC and AB in one triangle are proportional to the sides $B'C'$ and $A'B'$ in the other, we place the triangles so that the points B and A' coincide and the sides AB and $A'B'$ form a line:

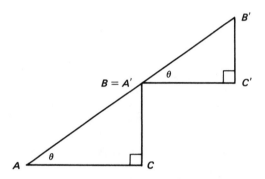

Then, we extend the sides AC and $B'C'$ until they meet in a right angle at a point D. The result is a triangle $AB'D$, where $A'C'$ is parallel to AD:

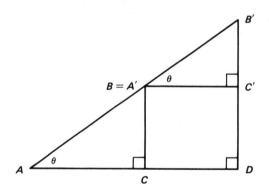

The sides of triangle $AB'D$ are cut proportionally by $A'C'$. Therefore,

$$\frac{B'C'}{A'B'} = \frac{C'D}{AA'}.$$

But $C'D$ is the same length as BC, and A' is the same point as B, so

$$\frac{B'C'}{A'B'} = \frac{BC}{AB}.$$

This proportion is true for any right triangle $A'B'C'$ compared to ABC when the acute angle θ is the same in each triangle. Thus the ratio of the sides BC to AB is constant for a given acute angle θ, regardless of the size of the right triangle:

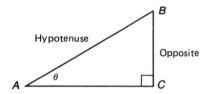

Observe that BC is the side opposite θ and AB is the hypotenuse. The fact that the ratio of these sides is constant leads to one of the fundamental definitions of trigonometry.

Definition: If θ is an acute angle of a right triangle, the **sine** of θ is the ratio of the side opposite θ and the hypotenuse. We abbreviate this definition:

$$\sin \theta = \frac{\text{opposite side}}{\text{hypotenuse}}.$$

The word "sine" comes from a Latin word meaning "fold of a garment." It is also connected with a word meaning "chord of a circle." These connections will become clear when we reach Unit 9.

Using similar methods, we can show that the ratios of the side adjacent to θ and the hypotenuse, and the side opposite θ and the side adjacent to θ, are also each constant. Therefore, we have two other fundamental ratios.

Definition: If θ is an acute angle of a right triangle, the **cosine** of θ is the ratio of the side adjacent to θ and the hypotenuse.
Also, the **tangent** of θ is the ratio of the side opposite θ and the side adjacent to θ.
We abbreviate these definitions:

$$\cos \theta = \frac{\text{adjacent side}}{\text{hypotenuse}}$$

$$\tan \theta = \frac{\text{opposite side}}{\text{adjacent side}}.$$

We will find that the **reciprocals** of the fundamental ratios are also very useful.

Definition:
The **cotangent** of θ is the reciprocal of the tangent of θ.
The **secant** of θ is the reciprocal of the cosine of θ.
The **cosecant** of θ is the reciprocal of the sine of θ.
We abbreviate these definitions:

$$\cot \theta = \frac{1}{\tan \theta}$$

$$\sec \theta = \frac{1}{\cos \theta}$$

$$\csc \theta = \frac{1}{\sin \theta}.$$

Observe that the cosecant of θ is abbreviated "csc θ" to distinguish it from the cosine of θ. In some books, the cotangent of θ is also abbreviated "ctn θ."

We may write the reciprocals in terms of the side opposite θ, the side adjacent to θ, and the hypotenuse. Since

$$\sin \theta = \frac{\text{opposite side}}{\text{hypotenuse}},$$

we have

$$\csc \theta = \frac{1}{\sin \theta} = \frac{\text{hypotenuse}}{\text{opposite side}}.$$

Similarly, since

$$\cos \theta = \frac{\text{adjacent side}}{\text{hypotenuse}},$$

$$\sec \theta = \frac{1}{\cos \theta} = \frac{\text{hypotenuse}}{\text{adjacent side}};$$

and since

$$\tan \theta = \frac{\text{opposite side}}{\text{adjacent side}},$$

$$\cot \theta = \frac{1}{\tan \theta} = \frac{\text{adjacent side}}{\text{opposite side}}.$$

These ratios are the six **trigonometric ratios.** We may summarize them:

$$\sin \theta = \frac{\text{opposite side}}{\text{hypotenuse}}$$

$$\cos \theta = \frac{\text{adjacent side}}{\text{hypotenuse}}$$

$$\tan \theta = \frac{\text{opposite side}}{\text{adjacent side}}$$

$$\cot \theta = \frac{\text{adjacent side}}{\text{opposite side}}$$

$$\sec \theta = \frac{\text{hypotenuse}}{\text{adjacent side}}$$

$$\csc \theta = \frac{\text{hypotenuse}}{\text{opposite side}}$$

The six trigonometric ratios are easiest to remember if you always list them in the order shown. Then observe that the first and last are reciprocals, the second and fifth are reciprocals, and the two middle are reciprocals.

Exercise 1.2

Use opposite side, adjacent side, and hypotenuse to complete each statement:

1. $\sin \theta =$ 2. $\cot \theta =$ 3. $\sec \theta =$

4. $\tan \theta =$ 5. $\csc \theta =$ 6. $\cos \theta =$

(Check the preceding pages to see how you did.)

Section
1.3

The Special Triangles

You should recall from previous courses that all right triangles have a special property called the Pythagorean theorem. The Pythagorean theorem states that, if the sides of a right triangle are labeled a, b, and c, with c the hypotenuse, then

$$c^2 = a^2 + b^2.$$

Two right triangles have further special properties. These right triangles are called the **special triangles**. The first special triangle has both acute angles 45°. The second special triangle has one acute angle 30° and one acute angle 60°. The 45°, 30°, and 60° acute angles are called the **special angles**.

First, we consider the special triangle with both acute angles 45°. This triangle is **isosceles**; that is, sides a and b are equal. Suppose sides a and b each are equal to 1:

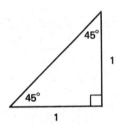

We use the Pythagorean theorem to find the hypotenuse c:

$$c^2 = a^2 + b^2$$

$$c^2 = 1^2 + 1^2$$

$$c^2 = 2$$

$$c = \sqrt{2}.$$

This special triangle is called the **45–45–90 triangle** for its angles, or the **1–1–$\sqrt{2}$ triangle** for its sides.

Using either of the 45° angles, we can find the values of the six trigonometric ratios for $\theta = 45°$. Suppose we use the 45° angle at the lower left:

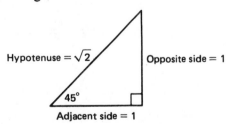

Then,

$$\sin 45° = \frac{\text{opposite side}}{\text{hypotenuse}} = \frac{1}{\sqrt{2}}$$

$$\cos 45° = \frac{\text{adjacent side}}{\text{hypotenuse}} = \frac{1}{\sqrt{2}}$$

$$\tan 45° = \frac{\text{opposite side}}{\text{adjacent side}} = \frac{1}{1} = 1$$

$$\cot 45° = \frac{\text{adjacent side}}{\text{opposite side}} = \frac{1}{1} = 1$$

$$\sec 45° = \frac{\text{hypotenuse}}{\text{adjacent side}} = \frac{\sqrt{2}}{1} = \sqrt{2}$$

$$\csc 45° = \frac{\text{hypotenuse}}{\text{opposite side}} = \frac{\sqrt{2}}{1} = \sqrt{2}.$$

Recall from Section 1.2 that, if the acute angles are 45°, we may use sides a and b of any size without changing the values of the trigonometric ratios. However, it is usual to take $a = 1$ and $b = 1$. Furthermore, we may rationalize the denominator in the first two ratios:

$$\frac{1}{\sqrt{2}} = \frac{1 \cdot \sqrt{2}}{\sqrt{2} \cdot \sqrt{2}} = \frac{\sqrt{2}}{2}.$$

Thus the first two ratios are often written in the form

$$\sin 45° = \frac{\sqrt{2}}{2}$$

and

$$\cos 45° = \frac{\sqrt{2}}{2}.$$

We will usually use the direct form, rather than the rationalized form, throughout this book.

Now, we consider the special triangle with one acute angle 30° and one acute angle 60°. Suppose the shortest side a, opposite the 30° angle, is 1. Then, suppose a triangle which is the mirror image of the original is placed below, with side b common to both triangles. The two triangles make an equiangular triangle; that is, all the angles are 60°. All the sides also are equal, and are 2:

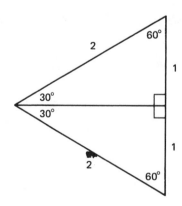

Since, in the original triangle, $a = 1$ and $c = 2$, we find b using the Pythagorean theorem:

$$c^2 = a^2 + b^2$$

$$2^2 = 1^2 + b^2$$

$$4 = 1 + b^2$$

$$b^2 = 3$$

$$b = \sqrt{3} .$$

This special triangle is called the **30–60–90 triangle** for its angles, or the **1–$\sqrt{3}$ –2 triangle** for its sides.

Using the 30° angle, we can find the values of the six trigonometric ratios for $\theta = 30°$:

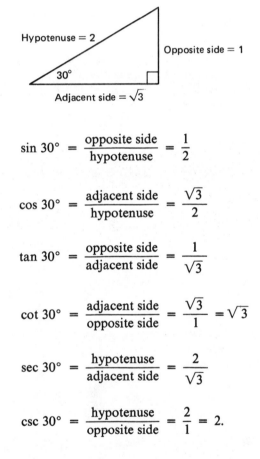

$$\sin 30° = \frac{\text{opposite side}}{\text{hypotenuse}} = \frac{1}{2}$$

$$\cos 30° = \frac{\text{adjacent side}}{\text{hypotenuse}} = \frac{\sqrt{3}}{2}$$

$$\tan 30° = \frac{\text{opposite side}}{\text{adjacent side}} = \frac{1}{\sqrt{3}}$$

$$\cot 30° = \frac{\text{adjacent side}}{\text{opposite side}} = \frac{\sqrt{3}}{1} = \sqrt{3}$$

$$\sec 30° = \frac{\text{hypotenuse}}{\text{adjacent side}} = \frac{2}{\sqrt{3}}$$

$$\csc 30° = \frac{\text{hypotenuse}}{\text{opposite side}} = \frac{2}{1} = 2.$$

Alternatively, using the 60° angle, we can find the values of the six trigonometric ratios for $\theta = 60°$:

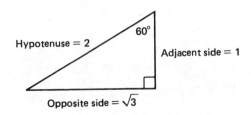

$$\sin 60° = \frac{\text{opposite side}}{\text{hypotenuse}} = \frac{\sqrt{3}}{2}$$

$$\cos 60° = \frac{\text{adjacent side}}{\text{hypotenuse}} = \frac{1}{2}$$

$$\tan 60° = \frac{\text{opposite side}}{\text{adjacent side}} = \frac{\sqrt{3}}{1} = \sqrt{3}$$

$$\cot 60° = \frac{\text{adjacent side}}{\text{opposite side}} = \frac{1}{\sqrt{3}}$$

$$\sec 60° = \frac{\text{hypotenuse}}{\text{adjacent side}} = \frac{2}{1} = 2$$

$$\csc 60° = \frac{\text{hypotenuse}}{\text{opposite side}} = \frac{2}{\sqrt{3}}.$$

Again, if the acute angles are 30° and 60°, we may start with a side *a* of any size without changing the values of the trigonometric ratios. Furthermore, tan 30°, sec 30°, cot 60°, and csc 60° may be written in a form where the denominator is rationalized. We will use the direct form throughout this book.

EXAMPLE 1.2.   Find sec 45°.

Solution.   We draw the 45–45–90 triangle and label its sides:

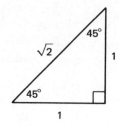

Then, using either of the 45° angles,

$$\sec 45° = \frac{\text{hypotenuse}}{\text{adjacent side}} = \frac{\sqrt{2}}{1} = \sqrt{2}.$$

EXAMPLE 1.3.   Find tan 30°.

Solution.   We draw the 30–60–90 triangle and label its sides:

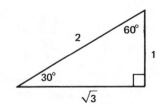

Then, using the 30° angle,

$$\tan 30° = \frac{\text{opposite side}}{\text{adjacent side}} = \frac{1}{\sqrt{3}} .$$

Exercise 1.3

Use special triangles to find the value:

1. sin 30°   2. cos 45°   3. tan 45°   4. sec 60°   5. csc 60°

6. cot 30°   7. csc 45°   8. tan 60°   9. cos 30°   10. cos 60°

Section 1.4   Tables of Trigonometric Ratios

Clearly we will need to use angles other than the special angles. To find the values of the trigonometric ratios for other acute angles we use tables of trigonometric ratios. Table II at the back of this book is such a table. Almost all standard tables of trigonometric ratios are printed in a form similar to Table II.

A part of Table II is reproduced below. The angles are given in **degrees and tens of minutes**. There are 60 minutes (written 60′) in a degree. Thus 1° = 60′. Most tables include a column

Degrees	Radians	Sin	Cos	Tan	Cot	Sec	Csc		
36°00′	.6283	.5878	.8090	.7265	1.376	1.236	1.701	.9425	54°00′
10	312	901	073	310	368	239	695	396	50
20	341			355	360	241	688	367	40
30				.7400	1.351				
	603	111	916	766	288	266	630	105	10
		134	898						
38°00′	.6632	.6157	.7880	.7813	1.280	1.269	1.624	.9076	52°00′
10	661	180	862	860	272	272	618	047	50
20	690	202	844	907	265	275	612	.9018	40
30	.6720	.6225	.7826	.7954	1.257	1.278	1.606	.8988	30
40	749	248	808	.8002	250	281	601	959	20
50	778	271	790	050	242	284	595	930	10
39°00′	.6807	.6293	.7771	.8098	1.235	1.287	1.589	.8901	51°00′
10	836	316	753	146	228	290	583	872	50
20	865	338	735	195	220	293	578	843	40
30	.6894	.6361	.7716	.8243	1.213	1.296	1.572	.8814	30
40	923	383	698	292	206	299	567	785	20
50	952	406	679	342	199	302	561	756	10
40°00′	.6981	.6428	.7660	.8391	1.192	1.305	1.556	.8727	50°00′
10	.7010			441	185	309	550	698	50
20					178	312			
	796	.7009	.7133	844			423	912	
		030	112						
50	825	050	092	942	006	410	418	883	10
45°00′	.7854	.7071	.7071	1.000	1.000	1.414	1.414	.7854	45°00′
		Cos	Sin	Cot	Tan	Csc	Sec	Radians	Degrees

marked "radians," which we will use in Unit 8. Then, the table gives the values of the six trigonometric ratios next to each angle. The values are given as decimal approximations of the ratios to four significant digits.

EXAMPLE 1.4.   Find the value of:

   a. sin 38°   b. sec 38°20′

Solutions.   a. We find 38° in the left-hand column, and then go across to the column marked "sin" at the top. We find

$$\sin 38° = .6157.$$

Remember to be careful to skip the "radians" column.

   b. We find 38°20′ in the left-hand column, and then go across to the column marked "sec" at the top. Observe that only the last three digits are given. We look for the complete number closest above, and see that

$$\sec 38°20′ = 1.275.$$

To find values for angles between 45° and 90°, we consider two angles θ and θ' (theta prime), where $\theta + \theta' = 90°$. Such angles θ and θ' are called **complementary angles**. If θ is one acute angle of a right triangle, then its complementary angle θ' is the other acute angle:

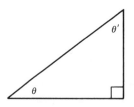

We use the opposite and adjacent sides, first for θ and then for θ', to observe that

$$\sin \theta = \cos \theta' \text{ and } \cos \theta = \sin \theta'$$

$$\tan \theta = \cot \theta' \text{ and } \cot \theta = \tan \theta'$$

$$\sec \theta = \csc \theta' \text{ and } \csc \theta = \sec \theta'.$$

Because of these relationships, the related ratios are called **cofunctions**. Sine and cosine are cofunctions, tangent and cotangent are cofunctions, and secant and cosecant are cofunctions. Clearly, the names "cosine," "cotangent," and "cosecant" are derived from the cofunction relationship. It is a common error to confuse the cofunctions with the reciprocal ratios. Only the tangent and cotangent are both each other's cofunction and each other's reciprocal.

Tables of trigonometric ratios use the cofunction relationship to list values for angles between 45° and 90°. Observe that the angles between 45° and 90° are listed from bottom to top at the right-hand side of the table, and the cofunctions are listed below one another at the bottom of the table.

EXAMPLE 1.5.   Find the value of:

   a. cos 52°   b. csc 51°40′

Solutions.   a. We find 52° in the right-hand column, and then go across to the column marked "cos" at the bottom. We find

$$\cos 52° = .6157.$$

Observe that $38° + 52° = 90°$, and so $\sin 38° = \cos 52°$.

b. We find $51°40'$ in the right-hand column, reading from bottom to top so that $51°40'$ is above $51°$. Then, we go across to the column marked "csc" at the bottom. We find

$$\csc 51°40' = 1.275.$$

Observe that $38°20' + 51°40' = 89°60' = 90°$. Therefore, $\sec 38°20' = \csc 51°40'$. Remember that when you use the *right-hand* column of angles, and the *bottom* ratio names, you must always be sure to read the degrees and tens of minutes *from bottom to top*.

EXAMPLE 1.6.   Find the value of:

a. $\tan 50°50'$ b. $\tan 51°10'$

Solutions.   a. We find $50°50'$ above $50°$ but below $51°$, so

$$\tan 50°50' = 1.228.$$

b. We find $51°10'$ above $51°$, so

$$\tan 51°10' = 1.242.$$

If we are given the value of a trigonometric ratio for an acute angle, we can use the table of trigonometric ratios to find the angle to the nearest degrees and tens of minutes. First, we look for the value going down from the ratio name at the top of the table. If we find the value, we read the angle from the left-hand side of the table. If we do not find the exact value, but do find a smaller value and a larger value, we use the closer value.

If the value is not found using the top ratio names, then we look for the value going up from the ratio names at the bottom of the table. When we find the value, or the closest value, we read the angle from the right-hand side of the table, remembering to read from bottom to top.

EXAMPLE 1.7.   Find θ to the nearest degrees and tens of minutes when:

a. $\sin \theta = .6271$ b. $\sin \theta = .7698$ c. $\cot \theta = 1.269$

Solutions.   a. We look under "sin" at the top of the table. When we reach .6271, we look to the left and find

$$\theta = 38°50'.$$

b. We must go to the ratio names at the bottom of the table. Going up from "sin" at the bottom of the table, we find .7698. Looking to the right, and remembering to read degrees and tens of minutes from bottom to top,

$$\theta = 50°20'.$$

c. We see that the value is between 1.272 and 1.265 under "cot." Since 1.269 is closer to 1.272, we look to the left of 1.272 and find

$$\theta = 38°10',$$

to the nearest degrees and tens of minutes.

Exercise 1.4

Use Table II to find:

1. cos 15°   2. tan 84°   3. sin 32°10′   4. csc 43°50′

5. tan 48°50′   6. cot 86°10′   7. sec 77°50′   8. sec 19°40′

9. cos 7°30′   10. sin 87°50′

Use Table II to find θ to the nearest degrees and tens of minutes:

11. cos θ = .9744   12. tan θ = .8899   13. tan θ = 1.150   14. sec θ = 1.239

15. sin θ = .9261   16. sin θ = .3330   17. cot θ = 1.666   18. csc θ = 1.666

19. tan θ = 1.125   20. cos θ = .1250

Section 1.5	Linear Interpolation

It is possible to use Table II to find values of the trigonometric ratios for angles given in degrees and minutes, and to find angles in degrees and minutes for values of the trigonometric ratios. We use a method called **linear interpolation**.

Suppose we know values at two points. If the two points are sufficiently close together (10′ is sufficiently close), we may approximate the points between by a straight line through the two known points:

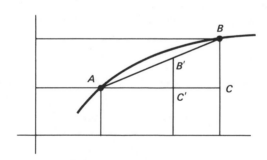

Observe that we have a triangle ABC, with a line $B'C'$ parallel to BC. Then $B'C'$ cuts the sides proportionally.

EXAMPLE 1.8.   Find sin 23°38′.

Solution.   From Table II, we know the two points

$$\sin 23°30' = .3987$$

and

$$\sin 23°40' = .4014.$$

In our diagram, the sides of the triangles are the differences between the points:

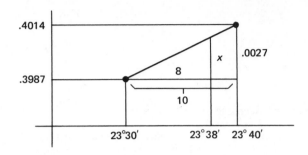

Since the sides of the triangles are proportional,

$$\frac{8}{10} = \frac{x}{.0027}$$

$$x = \frac{8(.0027)}{10}$$

$$= .0022.$$

We see from the diagram that

$$\sin 23°38' = .3987 + x.$$

Therefore,

$$\sin 23°38' = .3987 + .0022$$

$$= .4009.$$

It is usually convenient to replace the diagram by a chart:

$$
10 \left[\; 8 \left[\begin{array}{l} \sin 23°30' \;=.3987 \\ \sin 23°38' \;= \\ \sin 23°40' \;=.4014 \end{array}\right.\!\!\! \Big] x \right] .0027
$$

and proceed as before.

Observe from Table II that the sine, tangent, and secant *increase* as the angles increase. However, the cofunctions *decrease* as the angles increase. When the values decrease, we must subtract x from the first given value.

EXAMPLE 1.9.   Find cos 41°53'.

Solution.   From Table II,

$$
10 \left[\; 3 \left[\begin{array}{l} \cos 41°50' \;=.7451 \\ \cos 41°53' \;= \\ \cos 42°00' \;=.7431 \end{array}\right.\!\!\! \Big] x \right] .0020
$$

The proportion is

$$\frac{3}{10} = \frac{x}{.0020}$$

$$x = \frac{3(.0020)}{10}$$

$$= .0006.$$

Since the cosine decreases,

$$\cos 41°53' = .7451 - .0006$$

$$= .7445.$$

You can tell whether you should add or subtract by remembering that your value should always be between the two given values.

To find the angle given the value of a trigonometric ratio, we find the closest value on each side of the given value. Then, x in the proportion is a number of minutes between $0'$ and $10'$.

EXAMPLE 1.10.   Find θ to the nearest degrees and minutes if $\sec \theta = 1.229$.

Solution.   From Table II, we find the closest values on either side of 1.229 and under "sec." Then, we construct a chart similar to those in the preceding examples:

Since the values increase, we have subtracted the first value from the second and third. The proportion is

$$\frac{x}{10} = \frac{.001}{.003}$$

$$x = \frac{10(.001)}{.003}$$

$$= 3',$$

to the nearest minutes. Therefore,

$$\theta = 35°33'.$$

EXAMPLE 1.11.   Find θ to the nearest degrees and minutes if $\cot \theta = .6720$.

Solution.   Using the bottom heading, and remembering to read the right-hand angles from bottom to top, we have the chart:

Since the values decrease, we have subtracted the second and third values from the first. The proportion is

$$\frac{x}{10} = \frac{.0025}{.0042}$$

$$x = \frac{10(.0025)}{.0042}$$

$$= 6',$$

to the nearest minute. Therefore,

$$\theta = 56°06'.$$

A Note on Calculators

With scientific calculators readily available at reasonable prices, tables of trigonometric ratios and linear interpolation are becoming obsolete. You can find the value of a trigonometric ratio for an angle given in degrees and minutes, or find an angle in degrees and minutes given a value of a trigonometric ratio, using a scientific calculator. For this book, we have used the Texas Instruments SR-30 or Slimline TI-35, but the methods may be adapted to almost any make and model of scientific calculator.

First, you must realize that calculators work in decimals, but degrees and minutes are not given in decimals. For example, 27.10° and 27°10' are not the same. To convert 27°10' to a decimal, divide the number of minutes by 60:

$$10 \div 60 = .16666667.$$

Now, add the number of degrees:

$$.16666667 + 27 = 27.166667.$$

You now have the degrees and minutes as a decimal, and may press the "sin," "cos," or "tan" key. To find the other trigonometric ratios, use the *reciprocals* (remember not to confuse them with the cofunctions) and the "$1/x$" key. For example, since csc θ is the reciprocal of sin θ, to find csc 27.166667° press the "sin" key and then the "$1/x$" key. The result is 2.1901947, or 2.190 to four significant digits.

To find an angle given the value of a trigonometric ratio, you reverse the process. Given a value for sin θ, cos θ, or tan θ, press the key marked "inv," and then the "sin," "cos," or "tan" key. (On some calculators you may have a "second function" and "sin^{-1}," "cos^{-1}," and "tan^{-1}.") The result is an angle given as a decimal. For example, given tan θ = 1.332, enter 1.332, then press the "inv" key and the "tan" key. The result is 53.102583°. To convert this result to degrees and minutes, subtract the degrees:

$$53.102583 - 53 = .10258277.$$

Now, multiply by 60:

$$.10258277 \times 60 = 6.1549664.$$

Therefore, θ = 53°06' to the nearest degrees and minutes, or 53°10' to the nearest degrees and tens of minutes.

Because some calculators carry different numbers of digits, the number of decimal places in your results may differ from ours. Also, the last digit may differ by one unit. For example, our results are from the SR-30. However, the TI-35 gives 53.102583°, then 0.1025828 and, in the last step, 6.1549664 again. Since we are rounding to the nearest minute, or tens of minutes, these differences are insignificant.

If you are given a value for cot θ, sec θ, or csc θ, use the "$1/x$" key and the reciprocal of the ratio. For example, since cot θ is the reciprocal of tan θ, to find θ given cot $\theta = 1.332$, enter 1.332. Then, press the "$1/x$" key, the "inv" key, and finally the "tan" key. The result is 36.897417. Subtracting 36 and multiplying by 60, $\theta = 36°54'$ to the nearest degrees and minutes, or $36°50'$ to the nearest degrees and tens of minutes.

Finally, observe from Table II that the sine and cosine are never more than 1. If you enter a number more than 1, and press the "inv" and "sin" or "cos" keys, your calculator should return an error signal.

Exercise 1.5

Find out from your instructor whether you should use Table II and linear interpolation, or whether you may use a calculator. Find:

1. sin 25°32′	2. tan 36°49′	3. cos 17°18′	4. cos 39°56′
5. csc 43°23′	6. sec 44°02′	7. cot 57°18′	8. cot 69°57′
9. sec 87°39′	10. csc 53°11′		

Find θ to the nearest degrees and minutes:

11. tan $\theta = .4397$	12. sin $\theta = .5230$	13. cos $\theta = .7238$	14. cos $\theta = .9600$
15. sec $\theta = 1.359$	16. csc $\theta = 2.549$	17. cot $\theta = .2912$	18. cot $\theta = .7050$
19. csc $\theta = 1.340$	20. sec $\theta = 10.23$		

Self-test

1. Use Table II and linear interpolation to find θ to the nearest degrees and minutes if tan θ = .3936.

1._____

2. Use Table II to find sec 76°20′.

2._____

3. Label the opposite side, the adjacent side, and the hypotenuse of this triangle, where the right angle and an acute angle are indicated:

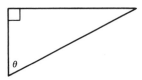

4. Use opposite side, adjacent side, and hypotenuse to complete the statement tan θ =

4._____

5. Use special triangles to find the value of:
   a. sin 45°
   b. cot 60°

5a._____

5b._____

Unit 2

Right Triangle Trigonometry

INTRODUCTION

In this unit you will apply the six trigonometric ratios to solving right triangles; that is, finding the sides and angles of a right triangle. You will then be able to apply the methods for solving right triangles to solving verbal problems where the situation described involves a right triangle.

OBJECTIVES

When you have finished this unit you should be able to:

1. Find any angle or side of a right triangle, given one acute angle and one side, or two sides.
2. Solve verbal problems involving right triangles.

Section 2.1

Solving Right Triangles

We know that one angle of a right triangle is a right angle. Given any side and one of the acute angles, we can find the other two sides and the other acute angle. Given any two sides, we can find the other side and the two acute angles. When the sides of a right triangle are given as a, b, and c, with c the hypotenuse, the acute angles are usually called α (alpha) and β (beta), the first two letters of the Greek alphabet. Throughout this unit, we will always label the side opposite α as a, and the side opposite β as b. The hypotenuse is of course labeled c. A diagram of a right triangle might look like this:

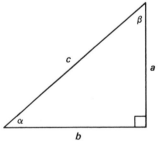

To **solve a right triangle** means to find all of its angles and sides. In order to solve a right triangle, we must be given an acute angle and a side, or two sides.

Suppose we are given a right triangle with angle $\alpha = 30°$, and side $a = 3$. We know that side a is opposite angle α. Also, side b is opposite angle β, and the hypotenuse is c. A diagram of the

triangle is

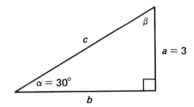

Clearly, $\beta = 60°$, and the triangle is a 30–60–90 triangle. In such a triangle, the hypotenuse c is twice the shortest side, so

$$c = 2a$$
$$= 2(3)$$
$$= 6.$$

Using the Pythagorean theorem,

$$c^2 = a^2 + b^2$$
$$6^2 = 3^2 + b^2$$
$$36 = 9 + b^2$$
$$b^2 = 27$$
$$b = \sqrt{27}$$
$$= \sqrt{9(3)}$$
$$= 3\sqrt{3} .$$

Thus the triangle is solved, and $\beta = 60°$, $b = 3\sqrt{3}$, and $c = 6$.

We can also solve the triangle above using the trigonometric ratios we derived for the 30° angle in Section 1.3. We use the diagram

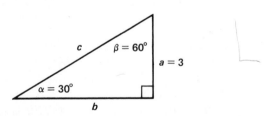

Again, β is clearly 60°. We know side a and angle α, and we know from Section 1.2 that, for any angle θ,

$$\sin \theta = \frac{\text{opposite side}}{\text{hypotenuse}}$$

so

$$\sin 30° = \frac{3}{c}.$$

But, from Section 1.3,

$$\sin 30° = \frac{1}{2}.$$

Therefore,

$$\frac{1}{2} = \frac{3}{c}$$

$$1(c) = 2(3)$$

$$c = 6.$$

Also, from Section 1.2,

$$\tan \theta = \frac{\text{opposite side}}{\text{adjacent side}}$$

$$\tan 30° = \frac{3}{b}.$$

But, from Section 1.3,

$$\tan 30° = \frac{1}{\sqrt{3}}.$$

Therefore,

$$\frac{1}{\sqrt{3}} = \frac{3}{b}$$

$$1(b) = \sqrt{3}\,(3)$$

$$b = 3\sqrt{3}.$$

We have found β, b, and c, and so the triangle is solved using the trigonometric ratios.

When we have a right triangle which is not a special triangle, we must use the trigonometric ratios to solve the triangle. Recall from Section 1.4 that we find approximations of values of the trigonometric ratios in a table such as Table II, to four significant digits.

You may have seen significant digits in other mathematics courses or in science courses. In the applied sciences, the use of significant digits is very precise. You should find out how many significant digits you can use in different situations in applied courses such as physics. For this text, however, we will usually use three significant digits. Three significant digits means we round off to three digits, where the first digit is not zero, regardless of the location of the decimal point. Thus .572, 5.72, 572, and 57,200 all are given to three significant digits. Furthermore, an angle given in degrees and tens of minutes has three significant digits. For example, 57°20′ has three significant digits. We will usually find angles to the nearest degrees and tens of minutes.

EXAMPLE 2.1.   Solve the right triangle with $\alpha = 56°10'$ and $a = 12$.

Solution.   First, we draw a diagram of the triangle, remembering that side a is opposite angle α, side b is opposite angle β, and the hypotenuse is c:

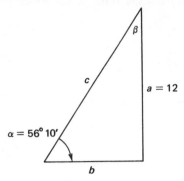

Since $\alpha + \beta = 90°$, we have $\beta = 90° - \alpha$; thus $\beta = 90° - 56°10'$. To perform this subtraction, remember that there are $60'$ in $1°$. We "borrow" $1°$ from $90°$, and write $90°$ as $89°60'$. The subtraction is

$$\begin{array}{r} 89°60' \\ -\ 56°10' \\ \hline 33°50' \end{array}$$,

and so $\beta = 33°50'$.

Now, to find b and c, we use the trigonometric ratios. The acute angle α is given. The side opposite α is a, which is given. The side adjacent to α is b. Recall that, for any angle θ,

$$\tan \theta = \frac{\text{opposite side}}{\text{adjacent side}}$$

and

$$\cot \theta = \frac{\text{adjacent side}}{\text{opposite side}}.$$

Therefore, to find b we may use

$$\tan \alpha = \frac{a}{b}$$

or

$$\cot \alpha = \frac{b}{a}.$$

The computations are usually easier when the unknown side is in the numerator; therefore, we choose

$$\cot \alpha = \frac{b}{a}.$$

Substituting the known values for α and a,

$$\cot 56°10' = \frac{b}{12},$$

and solving for b,

$$b = 12 \cot 56°10'.$$

We use Table II to find the value of cot 56°10′:

$$b = 12(.6703)$$

$$b = 8.04$$

to three significant digits.

There remains to find c, the hypotenuse. Recall that, for any angle θ,

$$\sin \theta = \frac{\text{opposite side}}{\text{hypotenuse}}$$

and

$$\csc \theta = \frac{\text{hypotenuse}}{\text{opposite side}}$$

Since a is the side opposite α and c is the hypotenuse, again using the given angle α and the given side a, we could choose

$$\sin \alpha = \frac{a}{c}$$

or

$$\csc \alpha = \frac{c}{a}.$$

Choosing the ratio which has the unknown side in the numerator, we choose

$$\csc \alpha = \frac{c}{a}.$$

Substituting α and a,

$$\csc 56°10′ = \frac{c}{12},$$

and solving for c,

$$c = 12 \csc 56°10′.$$

Using Table II to find csc 56°10′,

$$c = 12(1.204)$$

$$c = 14.4$$

to three significant digits.

If you are using a scientific calculator, it is particularly convenient to choose a trigonometric ratio so that the unknown side is in the numerator. You can solve for the unknown side in just one simple series of operations. For example, to find c, transform 56°10′ to the decimal form 56.166667, press "sin" and then "$1/x$" to obtain 1.203861. Then, multiply by the denominator 12 to obtain 14.446332, or 14.4 to three significant digits. You can find b similarly, using "tan" and "$1/x$."

The solution of the triangle is $\beta = 33°50′$, $b = 8.04$, and $c = 14.4$. Recall that $\alpha = 56°10′$ and $a = 12$ were given. Observe that the smallest side b is opposite the smaller acute angle β, and the hypotenuse c is the largest side.

To check, we may use the Pythagorean theorem:

$$c^2 = a^2 + b^2$$

$$14.4^2 \overset{?}{\approx} 12^2 + 8.04^2$$

$$207.36 \overset{?}{\approx} 144 + 64.6416$$

$$207 \approx 209$$

where $\approx$ means "approximately equal to." The two sides are not exactly equal because there are approximations in the table of trigonometric ratios, and we have approximated further in rounding off to three significant digits.

EXAMPLE 2.2.   Solve the right triangle with $\beta = 41°20'$ and $c = 5.2$.

Solution.   First we draw a diagram:

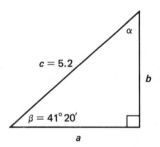

Observe that we could again put β at the top of the triangle. Then b would be at the bottom, opposite β:

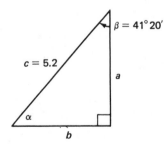

To find α, we write

$$\begin{array}{r} 89°60' \\ -41°20' \\ \hline 48°40' \end{array},$$

and so $\alpha = 48°40'$.

The hypotenuse c is given, the angle β is given, and a is the side adjacent to β. Recall that, for any angle θ,

$$\cos \theta = \frac{\text{adjacent side}}{\text{hypotenuse}}$$

and

$$\sec \theta = \frac{\text{hypotenuse}}{\text{adjacent side}}.$$

We can use

$$\cos \beta = \frac{a}{c}$$

or

$$\sec \beta = \frac{c}{a}.$$

Choosing the trigonometric ratio so that the unknown side is in the numerator,

$$\cos \beta = \frac{a}{c}$$

$$\cos 41°20' = \frac{a}{5.2}$$

$$a = 5.2 \cos 41°20'$$

$$a = 5.2(.7509)$$

$$a = 3.90.$$

Side b is opposite β, and the hypotenuse c is given. Since, for any angle θ,

$$\sin \theta = \frac{\text{opposite side}}{\text{hypotenuse}}$$

and

$$\csc \theta = \frac{\text{hypotenuse}}{\text{opposite side}},$$

we may choose

$$\sin \beta = \frac{b}{c}$$

or

$$\csc \beta = \frac{c}{b}.$$

Again choosing the unknown side in the numerator,

$$\sin \beta = \frac{b}{c}$$

$$\sin 41°20' = \frac{b}{5.2}$$

$$b = 5.2 \sin 41°20'$$

$$b = 5.2(.6604)$$

$$b = 3.43.$$

The solution is $\alpha = 48°40'$, $a = 3.90$, and $b = 3.43$. Angle $\beta = 41°20'$, and $c = 5.2$ were given. Again observe that the smallest side is opposite the smallest angle, and both sides a and b are smaller than the hypotenuse c.

To check using the Pythagorean theorem,

$$c^2 = a^2 + b^2$$

$$5.2^2 \overset{?}{\approx} 3.90^2 + 3.43^2$$

$$27.04 \overset{?}{\approx} 15.21 + 11.7649$$

$$27.0 \approx 27.0.$$

In the case that two sides are given, we could use the Pythagorean theorem to find the third side; however, we would still have to use the trigonometric ratios to find one of the acute angles. We will do the entire solution using the trigonometric ratios, and reserve the Pythagorean theorem to check.

EXAMPLE 2.3.   Solve the right triangle with $a = 10$ and $c = 12.5$.

Solution.   Always draw and correctly label a diagram of the triangle:

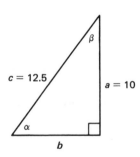

First, considering angle α, we see that the given sides a and c are the side opposite α and the hypotenuse. Since

$$\sin \theta = \frac{\text{opposite side}}{\text{hypotenuse}}$$

and

$$\csc \theta = \frac{\text{hypotenuse}}{\text{opposite side}},$$

we can choose

$$\sin \alpha = \frac{a}{c}$$

or

$$\csc \alpha = \frac{c}{a}$$

Dividing by 10 is easy, so we choose

$$\csc \alpha = \frac{c}{a}$$

$$\csc \alpha = \frac{12.5}{10}$$

$$\csc \alpha = 1.250$$

$$\alpha = 53°10'$$

to the nearest degrees and tens of minutes. Subtracting to find β,

$$\begin{array}{r} 89°60' \\ -53°10' \\ \hline 36°50' \end{array}$$

and so $\beta = 36°50'$.

   Again, if you are using a scientific calculator, you can find the angle α in one series of operations. Divide 12.5 by 10 to obtain 1.25, press "$1/x$," "inv," and "sin" to obtain 53.130102, which is α in decimal form. Then, transform α to degrees and minutes, subtracting 53 and multiplying by 60, to obtain 53° and 7.8061415, or 53°10' to the nearest degrees and tens of minutes.

   To find b using trigonometric ratios, we can use α with a or c, or β with a or c. We will use α in case of a subtraction error in finding β, and a because it is easy to compute with 10. (Sometimes there will be no obvious reason to make one choice over another.) Side a is opposite α and b is adjacent to α, and

$$\tan \theta = \frac{\text{opposite side}}{\text{adjacent side}}$$

$$\cot \theta = \frac{\text{adjacent side}}{\text{opposite side}}.$$

We will continue to choose the trigonometric ratio so that the unknown side is in the numerator:

$$\cot \alpha = \frac{b}{a}$$

$$\cot 53°10' = \frac{b}{10}$$

$$b = 10 \cot 53°10'$$

$$b = 10(.7490)$$

$$b = 7.49.$$

   The solution is $\alpha = 53°10'$, $\beta = 36°50'$, and $b = 7.49$. The sides $a = 10$ and $c = 12.5$ were given. As before, we observe that the smallest side is opposite the smaller acute angle, and the largest side is the hypotenuse.

Using the Pythagorean theorem to check,

$$c^2 = a^2 + b^2$$

$$12.5^2 \stackrel{?}{\approx} 10^2 + 7.49^2$$

$$156.25 \stackrel{?}{\approx} 100 + 56.1001$$

$$156 \approx 156.$$

Exercise 2.1

Solve the right triangle:

1. $\alpha = 36°30'$, $a = 10$ 2. $\alpha = 72°20'$, $b = 4.2$ 3. $a = 9.1$, $c = 15.3$

4. $b = 3.2$, $c = 5.1$ 5. $\beta = 12°10'$, $b = 2.4$ 6. $\beta = 42°50'$, $a = 18.2$

7. $\beta = 66°30'$, $c = 10$ 8. $\alpha = 59°50'$, $c = 22.5$ 9. $a = 10.2$, $b = 14.4$

10. $b = 16$, $c = 21$

Section 2.2   Applications of Solving Right Triangles

The techniques for solving right triangles can be applied to problems which cannot be solved using techniques of algebra. Such problems are usually described verbally. The first step is to read the description carefully, and draw a diagram of the triangle described. The diagram should involve a right triangle with one acute angle and one side, or two sides, given by the information in the problem. The question posed by the problem should require finding another side or an acute angle. You need find only the side or angle required by the problem; you do not need to solve the entire triangle.

EXAMPLE 2.4.   Suppose it is desired to find the distance across a river from a tree on one side to a tree on the other side, and that you are on the same side of the river as the first tree. You measure down the river bank to a point P, 100 meters from the first tree and perpendicular to the line of the two trees. At point P you use a surveying instrument to find that the angle between the lines from P to the first tree and P to the second tree is $54°30'$. What is the distance between the trees?

Solution.   The diagram described is

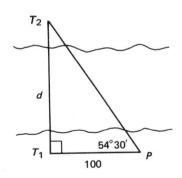

T_1 is the first tree, and T_2 is the second tree. We must find d, which is the side opposite the given angle. The adjacent side is also given, so we may use tangent or cotangent. Choosing so that the unknown side is in the numerator,

$$\tan 54°30' = \frac{d}{100}$$

$$d = 100 \tan 54°30'$$

$$d = 100(1.402)$$

$$d = 140 \text{ meters}$$

to three significant digits.

EXAMPLE 2.5.   A $5\frac{1}{2}$-foot flagpole extends horizontally from the side of a building. It is supported by an 8-foot wire attached to the far end of the pole and to the wall of the building above the pole. At what angle must the wire meet the building?

Solution.   We draw a diagram, where the vertical line represents the wall of the building, the horizontal line represents the flagpole, and the diagonal line represents the supporting wire:

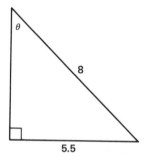

Since the wall of the building is vertical and the flagpole is horizontal, we have a right triangle. The angle θ represents the angle at which the wire meets the building. The flagpole is the side opposite θ and the wire is the hypotenuse. Therefore,

$$\sin \theta = \frac{5.5}{8}$$

$$\sin \theta = .6875$$

$$\theta = 43°30'$$

to the nearest degrees and tens of minutes.

Some applications of solving right triangles involve the terms **angle of elevation** and **angle of depression**. The angle of elevation starts from the horizontal and goes upward. The angle of depression starts from the horizontal and goes downward:

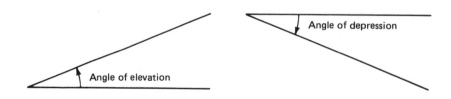

Angle of depression

Angle of elevation

The point to remember is that the angle of elevation and the angle of depression are *always measured from the horizontal*.

EXAMPLE 2.6. From a boat out on a bay, the angle of elevation to a point at the top of a cliff is 23°30′. The cliff is known to rise 58 feet above the water at that point. How far is the boat from the base of the cliff?

Solution.   The diagram is

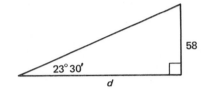

We are given the opposite side and must find the adjacent side. Choosing the trigonometric ratio so that the unknown side is in the numerator,

$$\cot 23°30′ = \frac{d}{58}$$

$$d = 58 \cot 23°30′$$

$$d = 58(2.300)$$

$$d = 133 \text{ feet.}$$

EXAMPLE 2.7. The angle of depression from an observation balloon to a marker on the ground is 78°20′. If the marker is 30 meters from the point directly below the balloon, how high is the balloon?

Solution.   We draw a diagram, remembering that the angle of depression is measured from the horizontal. Therefore, the angle of depression is outside the triangle:

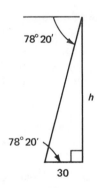

However, observe that the angle of depression is equal to the angle of elevation observed at the marker. Therefore,

$$\tan 78°20′ = \frac{h}{30}$$

$$h = 30 \tan 78°20′$$

$$h = 30(4.843)$$

$$h = 145 \text{ meters.}$$

Exercise 2.2 ODD

1. You are standing on one side of a ravine, and need a rope long enough to throw over a rock on the other side. You measure 25 feet along the edge of the ravine, on a line perpendicular to the line of the point where you were standing and the rock. The angle between the lines from your new position to the first point and to the rock is 18°50′. What is the distance between the point where you were standing and the rock?

2. You know that the distance between two trees is 11.2 meters. You measure a distance of 10 meters from the first tree to a point, perpendicular to the line of the trees. What is the angle between the line from the point to the first tree and the line from the point to the second tree?

3. An 8-foot flagpole is placed horizontally from the side of a building. It is to be supported by a wire attached to the far end of the pole, and meeting the side of the building above the pole at an angle of 54°. How long a wire is needed?

4. A post is placed vertically in the ground. It is supported by wires which are 7.5 feet long and attached to the post 7.3 feet above the ground. Assuming the ground is horizontal, at what angle do the wires meet the ground?

5. You want to reach a window 6.2 feet above the ground, and you have a ladder 7 feet long. At what angle to the ground must you place the ladder?

6. A ladder is to reach 13 feet up the side of a house. It should be placed at an angle of 60° to the ground. Assuming that the side of the house is vertical and the ground is horizontal, how long a ladder is needed?

7. A flagpole casts a shadow 21.3 meters long when the sun is at an angle of 25°10′ from the horizontal. How tall is the flagpole?

8. If an $8\frac{1}{2}$-foot statue casts a shadow 2.7 feet long, what is the angle of the sun from the horizontal?

9. From a point on the ground, the angle of elevation to the top of a wall is 37°50′. If the wall rises 3.6 meters above the ground, how far is the point on the ground from the top of the wall?

10. A distance is measured horizontally to a point 100 feet from the center of the base of the Bunker Hill monument. The angle of elevation from that point is found to be 65°40′. How high is the Bunker Hill monument?

11. From the top of a 30-meter observation tower, the angle of depression to a landmark on the ground is 18°20′. How far is the landmark from the center of the base of the tower?

12. The angle of depression from the top of the Washington monument to a marker on the ground is 57°10′. The Washington monument is 555 feet high. How far is the marker from an observer at the top of the monument?

Self-test

1. If in a right triangle $a = 20$ and $\alpha = 50°$, find b to three significant digits.

   1._____

2. If in a right triangle $b = 40$ and $c = 50$, find α to the nearest degrees and tens of minutes.

   2._____

3. If in a right triangle $\alpha = 22°30'$ and $b = 40.2$, find c to three significant digits.

   3._____

4. The angle of depression from a point at the top of a cliff to a rock below is $59°50'$. If the rock is known to be 65 feet from the base of the cliff, measured horizontally, how high is the cliff at that point to three significant digits?

   4._____

5. A 12-foot ladder is to reach 11.3 feet up the side of a house. At what angle to the ground must the ladder be placed?

   5._____

Angles in a
Coordinate System

INTRODUCTION

In this unit you will see how an angle may be defined as an amount of rotation. This definition makes it possible to represent the trigonometric ratios in terms of points in the Cartesian coordinate system. You will also learn a form of definition of the trigonometric ratios in terms of points on a circle called the unit circle. The new forms of the definitions will make it possible to solve a wider range of applied problems, and also provide the basis for the topics in trigonometry called "analytic trigonometry." In this unit you will use the unit circle to find values of the trigonometric ratios for a type of special angles called the quadrantal angles.

OBJECTIVES

When you have finished this unit you should be able to:

1. State whether or not an angle is in standard position.
2. State the trigonometric ratios in terms of a point (x, y) and a radius r, or in terms of (x, y) where (x, y) is a point on the unit circle.
3. Determine whether a trigonometric ratio of a given angle in standard position is positive or negative.
4. Use a unit point to find the value of a trigonometric ratio of a quadrantal angle, or state that the value is undefined.

Section
3.1
Angles in Standard Position

In the first two units, we used angles without attempting to define the term "angle." Of course, you know from previous courses the meaning of angles of a triangle. Without a different perspective on the meaning of "angle," however, we cannot proceed beyond right triangle trigonometry.

We begin with an ordinary line which extends indefinitely in both directions:

Any point on the line divides the line into two **half-lines**:

A **ray** consists of the dividing point and one of the half-lines:

The point is called the **endpoint** of the ray, and the direction of the half-line is the **direction** of the ray.

   Now, we imagine two rays with the same endpoint and the same direction. The two rays look as if they were one ray:

Holding one of the two rays in its initial position, we rotate the other ray about the common endpoint:

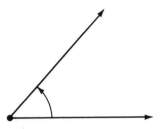

The amount of rotation can be any amount we wish.

Definition: An **angle** is the amount of rotation of a ray from an initial position to a terminal position.

The ray which is held in the initial position is called the **initial side** of the angle. The ray which is rotated is the **terminal side** of the angle. The common endpoint is the **vertex** of the angle:

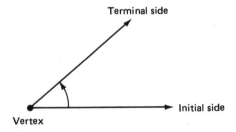

We have used right angles, which are 90°, and acute angles, which are between 0° and 90°. We will also use obtuse angles, which are between 90° and 180°:

Using the rotation definition, we also have an angle which is 180°:

and angles which are more than 180°:

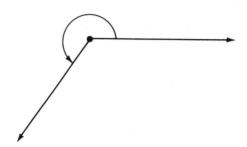

There is an angle which is 360°. The terminal side is rotated through a full circle until it coincides again with the initial side:

There are angles which are more than 360°:

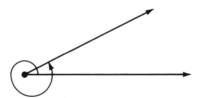

There is an angle which is 0°. The 0° angle looks very much like the 360° angle, but the amount of rotation is different. The 0° degree angle has no rotation at all:

We may even define negative angles. Except for the 0° angle, which has no rotation, the rotation of the angles we have shown so far is in the **counterclockwise** direction; that is, the rotation is opposite in direction to the rotation of the hands of a clock. Such angles are taken to be **positive**. If the rotation is **clockwise**, in the same direction as the hands of a clock, the angle is taken to be **negative**:

Throughout algebra, you have used a system for drawing graphs called the **Cartesian** (or **rectangular**) **coordinate system**. The Cartesian coordinate system consists of two number lines called the x- and y-axes, which are perpendicular to one another and cross at a point called the origin. We state a definition which relates our definition of an angle to the Cartesian coordinate system.

Definition: An angle in the Cartesian coordinate system is in **standard position** if its vertex is at the origin and its initial side coincides with the positive x-axis.

EXAMPLE 3.1. Which of these angles is in standard position:

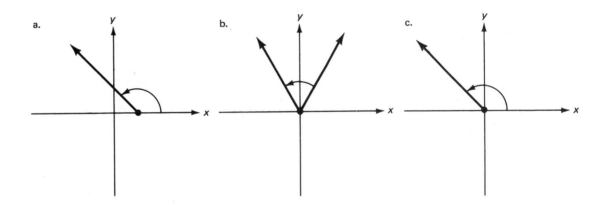

Solutions.

a. The vertex of the angle is not at the origin, so the angle is not in standard position.
b. The initial side of the angle does not coincide with the positive x-axis, so the angle is not in standard position.
c. The vertex of the angle is at the origin and the initial side coincides with the positive x-axis, so the angle is in standard position.

It is important to observe the curved arrow which indicates the rotation. Remember that the angle is the rotation.

EXAMPLE 3.2.   State whether or not this angle is in standard position:

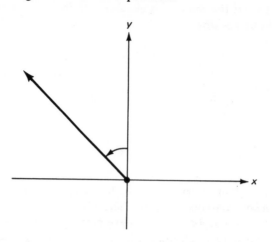

Solution.   The angle is not in standard position. Observe that the rotation starts from the *y*-axis, so the initial side of the angle coincides with the positive *y*-axis and not the positive *x*-axis.

Exercise 3.1

State whether or not the angle is in standard position:

1.

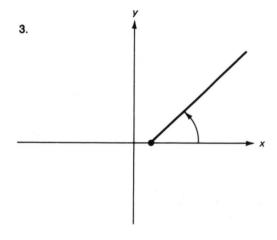

2.

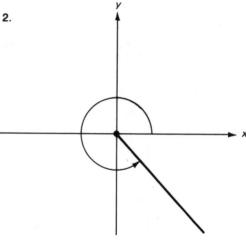

3.

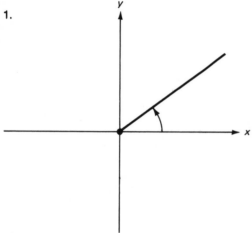

4.

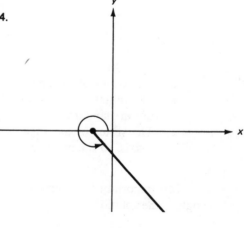

5.

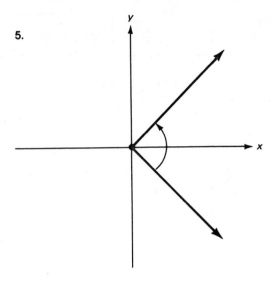

6.

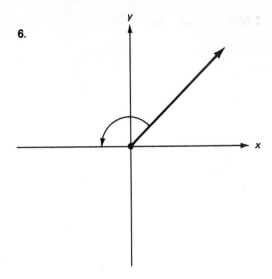

7.

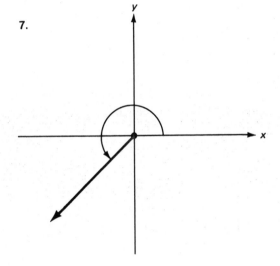

8.

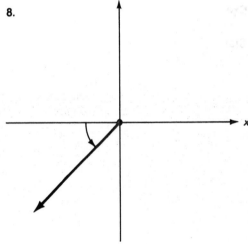

9.

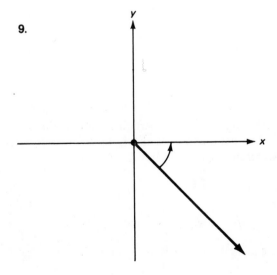

10.

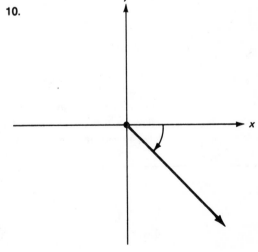

Section
3.2

The Ratios in the Coordinate System

Suppose θ is an acute angle in standard position. Then the terminal side of θ is in the first quadrant. Let (x, y) be any point on the terminal side of θ:

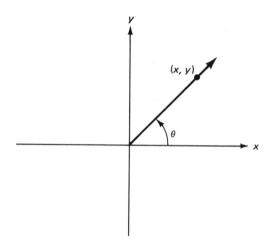

Observe that x and y are both positive. Since θ is an acute angle, if we draw a line from the point (x, y) perpendicular to the x-axis, we have an ordinary right triangle:

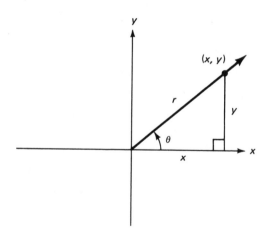

The side opposite θ is y and the side adjacent to θ is x. It is common to call the distance from the origin to the point (x, y) the **radius**, r. The hypotenuse of the right triangle is r. Using the Pythagorean theorem, we can find r if we know the coordinates (x, y):

$$r^2 = x^2 + y^2$$

$$r = \sqrt{x^2 + y^2} \, .$$

Observe that r is taken to be the positive square root, so r is always positive.

We use the definitions from Section 1.2 to state the trigonometric ratios in terms of (x, y) and r:

$$\sin \theta = \frac{\text{opposite side}}{\text{hypotenuse}} = \frac{y}{r}$$

$$\cos \theta = \frac{\text{adjacent side}}{\text{hypotenuse}} = \frac{x}{r}$$

$$\tan \theta = \frac{\text{opposite side}}{\text{adjacent side}} = \frac{y}{x}$$

$$\cot \theta = \frac{\text{adjacent side}}{\text{opposite side}} = \frac{x}{y}$$

$$\sec \theta = \frac{\text{hypotenuse}}{\text{adjacent side}} = \frac{r}{x}$$

$$\csc \theta = \frac{\text{hypotenuse}}{\text{opposite side}} = \frac{r}{y}.$$

You should be able to derive trigonometric ratios in terms of (x, y) and r, as we have here, using an acute angle θ in standard position. However, if you know the first three ratios, you can derive the last three using reciprocals. Recall that $\cot \theta$ is the reciprocal of $\tan \theta$. Therefore, since

$$\tan \theta = \frac{y}{x},$$

we have

$$\cot \theta = \frac{x}{y}.$$

Similarly, $\sec \theta$ is the reciprocal of $\cos \theta$. Therefore, since

$$\cos \theta = \frac{x}{r},$$

we have

$$\sec \theta = \frac{r}{x}.$$

Finally, $\csc \theta$ is the reciprocal of $\sin \theta$. Therefore, since

$$\sin \theta = \frac{y}{r},$$

we have

$$\csc \theta = \frac{r}{y}.$$

Recall from Section 1.2 that, in a right triangle, the value of a trigonometric ratio depends only on the size of the acute angle and not on the size of the triangle. Similarly, when an angle is defined as an amount of rotation, the value of a trigonometric ratio of the angle depends only on the amount of rotation and not on the length of the rays. If we suppose that the initial and terminal sides of an angle each are one unit, then they are radii of a circle with radius 1 and center at the origin. This circle is called the **unit circle**:

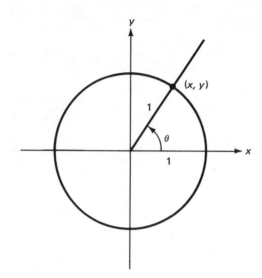

Since $r = 1$, the equation of the unit circle is

$$x^2 + y^2 = 1.$$

Also, since $r = 1$,

$$\sin \theta = \frac{y}{r} = \frac{y}{1} = y,$$

and

$$\cos \theta = \frac{x}{r} = \frac{x}{1} = x.$$

Therefore, if (x, y) is a point on the unit circle and on the terminal side of θ, $\sin \theta = y$ and $\cos \theta = x$. The reciprocals of these ratios are then $\csc \theta = \frac{1}{y}$ and $\sec \theta = \frac{1}{x}$. However, $\tan \theta = \frac{y}{x}$ and $\cot \theta = \frac{x}{y}$, as before.

Exercise 3.2

Use (x, y) and r to complete the statement:

1. $\cos \theta =$ 2. $\csc \theta =$ 3. $\tan \theta =$ 4. $\sin \theta =$

5. $\sec \theta =$ 6. $\cot \theta =$

If (x, y) is a point on the unit circle, use x or y to complete the statement:

7. $\sin \theta =$ 8. $\csc \theta =$ 9. $\sec \theta =$ 10. $\cos \theta =$

(Check the preceding pages to see how you did.)

Section
3.3

Signs of the Trigonometric Ratios

Observe that the axes of the Cartesian coordinate system divide the system into four parts:

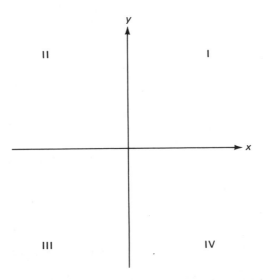

We call these four parts **quadrants**. The quadrants are numbered for reference, beginning with the upper right section, which is always the first quadrant, and proceeding counterclockwise. The use of Roman numerals is traditional.

Suppose θ is an angle in standard position in the Cartesian coordinate system. We say that θ is **in the quadrant** in which its terminal side falls. In the Cartesian coordinate system, x and y can be either positive or negative. Therefore, a trigonometric ratio of an angle θ in the Cartesian coordinate system can be positive or negative. Whether a trigonometric ratio of an angle θ in standard position is positive or negative depends on the quadrant θ is in.

We can show that the signs of the trigonometric ratios of an angle θ in standard position depend only on the quadrant the angle is in. Suppose θ is any angle in the first quadrant. Then, if θ is positive but less than 360°, θ is between 0° and 90°:

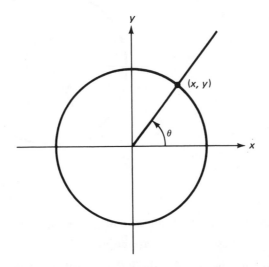

Since both x and y are positive in the first quadrant, all the trigonometric ratios of θ are positive.

If θ is in the second quadrant, and is positive but less than 360°, θ is between 90° and 180°:

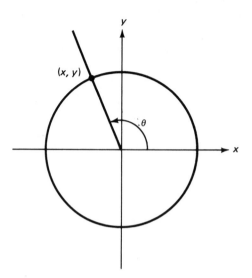

In the second quadrant, y is positive but x is negative. Therefore, since $\sin \theta = y$ and $\csc \theta = \dfrac{1}{y}$, $\sin \theta$ and $\csc \theta$ are positive. But, since $\cos \theta = x$ and $\sec \theta = \dfrac{1}{x}$, $\cos \theta$ and $\sec \theta$ are negative. Also, $\tan \theta = \dfrac{y}{x}$ and $\cot \theta = \dfrac{x}{y}$. Since x and y have opposite signs, $\tan \theta$ and $\cot \theta$ are negative. In the second quadrant, only $\sin \theta$ and $\csc \theta$ are positive.

If θ is in the third quadrant, and is positive but less than 360°, θ is between 180° and 270°:

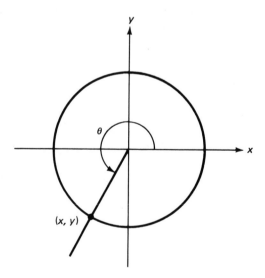

In the third quadrant, both x and y are negative. Therefore, since $\sin \theta = y$, $\cos \theta = x$, $\sec \theta = \dfrac{1}{x}$, and $\csc \theta = \dfrac{1}{y}$, all are negative. However, $\tan \theta = \dfrac{y}{x}$ and $\cot \theta = \dfrac{x}{y}$, and since x and y both are negative, $\tan \theta$ and $\cot \theta$ are positive. In the third quadrant, only $\tan \theta$ and $\cot \theta$ are positive.

If θ is in the fourth quadrant, and is positive but less than 360°, θ is between 270° and 360°:

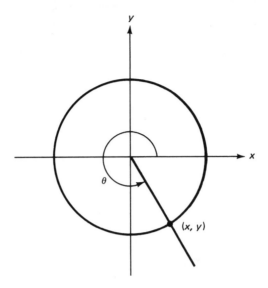

In the fourth quadrant, x is positive but y is negative. Therefore, since $\cos \theta = x$ and $\sec \theta = \dfrac{1}{x}$, $\cos \theta$ and $\sec \theta$ are positive. But, since $\sin \theta = y$ and $\csc \theta = \dfrac{1}{y}$, $\sin \theta$ and $\csc \theta$ are negative. Also, since x and y have opposite signs, $\tan \theta = \dfrac{y}{x}$ and $\cot \theta = \dfrac{x}{y}$ are negative. In the fourth quadrant, only $\cos \theta$ and $\sec \theta$ are positive.

There is a diagram which is helpful in remembering the quadrants where each ratio is positive. We use a "backwards Z":

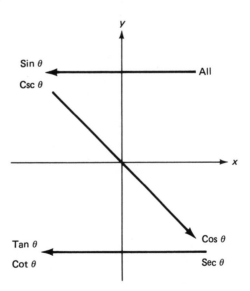

In the first quadrant, all the trigonometric ratios are positive. Then, $\sin \theta$ and its reciprocal $\csc \theta$ are positive in the second quadrant; $\cos \theta$ and its reciprocal $\sec \theta$ are positive in the fourth quadrant; $\tan \theta$ and its reciprocal $\cot \theta$ are positive in the third quadrant. In quadrants where a trigonometric ratio is not positive, it is of course negative.

EXAMPLE 3.3.   Determine whether sin 150° is positive or negative.

Solution.   We draw a diagram of the angle in standard position. Since 150° is more than 90° but less than 180°, 150° is in the second quadrant:

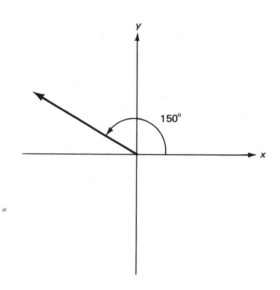

Then, since *y* is positive in the second quadrant, sin 150° is positive.

EXAMPLE 3.4.   Determine whether cos 120° is positive or negative.

Solution.   Since 120° is also between 90° and 180°, 120° is also in the second quadrant:

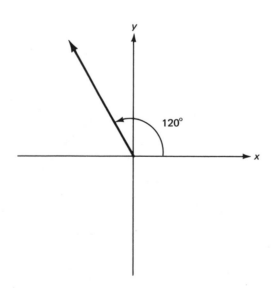

Then, since *x* is negative in the second quadrant, cos 120° is negative.

EXAMPLE 3.5.   Determine whether cot 135° is positive or negative.

Solution.   135° is in the second quadrant:

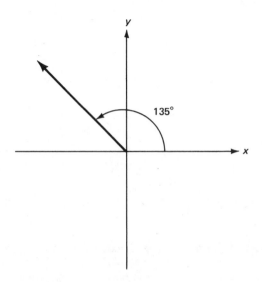

But, *x* is negative and *y* is positive in the second quadrant. Therefore, since *x* and *y* have opposite signs, cot 135° is negative.

EXAMPLE 3.6.   Determine whether tan 210° is positive or negative.

Solution.   Since 210° is more than 180° but less than 270°, 210° is in the third quadrant:

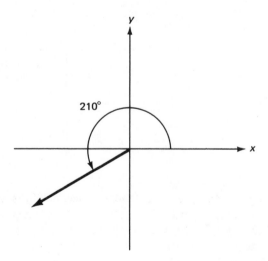

But *x* is negative in the third quadrant and *y* is also negative in the third quadrant. Therefore, tan 210° is positive.

EXAMPLE 3.7.   Determine whether sin(− 150°) is positive or negative.

Solution.   Recall that a negative angle is drawn clockwise starting from the positive *x*-axis. Therefore, −150° is in the third quadrant:

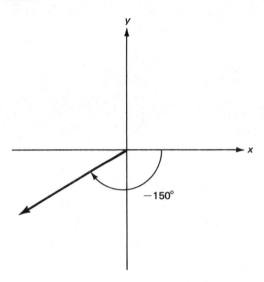

Since *y* is negative in the third quadrant, sin(−150°) is negative.

EXAMPLE 3.8.   Determine whether sec(−30°) is positive or negative.

Solution.   − 30° is in the fourth quadrant:

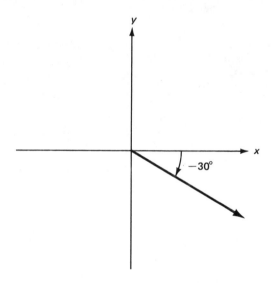

Since *x* is positive in the fourth quadrant, sec(−30°) is positive. It is a common error to think that a trigonometric ratio of a negative angle must be negative. You should see that a trigonometric ratio of a negative angle may be positive or negative, depending on the quadrant the angle is in.

Exercise 3.3

Determine whether the trigonometric ratio is positive or negative:

1.   cos 30°   2.   tan 45°   3.   sin 120°

4. cos 150°

5. tan 150°

6. tan 210°

7. sin 225°

8. cos 240°

9. cos 330°

10. sin 300°

11. cot 120°

12. sec 135°

13. csc 150°

14. cot 240°

15. sec 300°

16. csc 315°

17. sin(−30°)

18. cos(−45°)

19. tan(−150°)

20. cot(−120°)

21. csc(−135°)

22. sec(−210°)

23. cot(−240°)

24. sin(−300°)

Section	
3.4	# The Quadrantal Angles

A type of special angle has its terminal side on one of the coordinate axes. The angles which are positive or 0° but less than 360° which are of this type are 0°, 90°, 180°, and 270°. These angles are called the **quadrantal angles**. Observe that the quadrantal angles are not in any quadrant.

　　We find values of the trigonometric ratios of the quadrantal angles using points on the unit circle. These special points are called **unit points**. The unit points are on the x- and y-axes and are one unit from the origin:

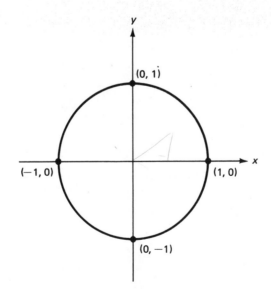

EXAMPLE 3.9.   Use a unit point to find sin 0°.

Solution.　　The terminal side of 0° is on the positive x-axis. The point one unit from the origin on the positive *x*-axis is (1, 0). Therefore, the unit point is (1, 0):

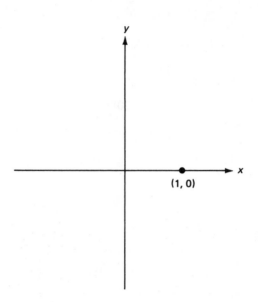

Since $y = 0$, sin 0° = 0.

EXAMPLE 3.10.   Use a unit point to find sec 0°.

Solution.   Again we use the unit point (1, 0):

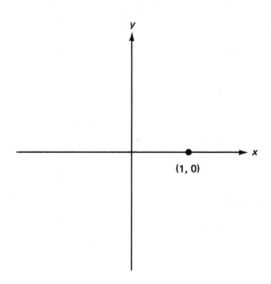

Since $x = 1$, sec 0° $= \dfrac{1}{1} = 1$.

EXAMPLE 3.11.   Use a unit point to find cos 90°.

Solution.   The terminal side of 90° is on the positive *y*-axis. Therefore, the unit point is (0, 1):

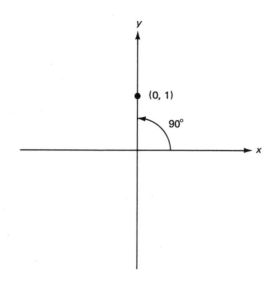

Since $x = 0$, cos 90° = 0.

EXAMPLE 3.12.   Use a unit point to find tan 90°.

Solution.   Again, the unit point is (0, 1):

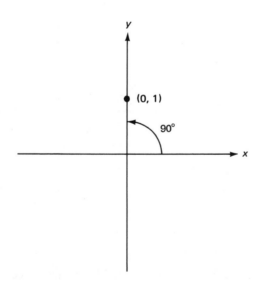

Since $x = 0$ and $y = 1$, tan 90° $= \dfrac{y}{x} = \dfrac{1}{0}$. But, recall from arithmetic that we cannot divide by zero. Therefore, tan 90° does not exist. We say that tan 90° is **undefined**.

EXAMPLE 3.13.   Use a unit point to find cos 180°.

Solution.   The terminal side of 180° is on the negative *x*-axis. Therefore, the unit point is $(-1, 0)$:

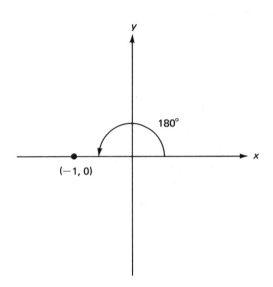

Since $x = -1$, cos 180° $= -1$.

EXAMPLE 3.14.   Use a unit point to find sec 270°.

Solution.   The terminal side of 270° is on the negative *y*-axis. Therefore, the unit point is $(0, -1)$:

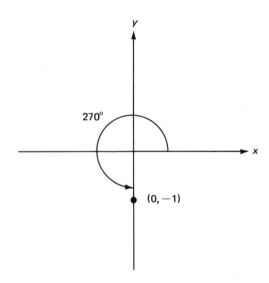

Since $x = 0$, sec 270° $= \dfrac{1}{0}$, and so sec 270° is undefined.

Exercise 3.4

Use a unit point to find:

1. cos 0°   2. tan 0°   3. sin 90°   4. cot 90°

5. sec 90°

6. cot 0°

7. csc 0°

8. tan 180°

9. sin 180°

10. csc 180°

11. sec 180°

12. csc 270°

13. cos 270°

14. sin 270°

15. cot 180°

16. cot 270°

Self-test

1. Use (x, y) and r to complete the statement:
   $\sec \theta =$

   1._____

2. If (x, y) is a point on the unit circle, use x and y to complete the statement:
   $\csc \theta =$

   2._____

3. Use a unit point to find:
   a. $\csc 90°$
   b. $\tan 270°$

   3a._____

   3b._____

4. Determine whether the trigonometric ratio is positive or negative:
   a. $\sin 315°$
   b. $\tan(-135°)$

   4a._____

   4b._____

5. Which of these angles is in standard position:

   a. 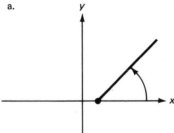   b. 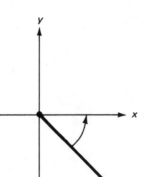   c.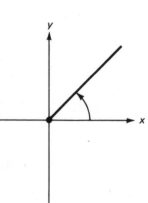

   5._____

Unit
4 Ratio Values for Any Angle

INTRODUCTION

In this unit you will learn how to find values of the trigonometric ratios for any angle in standard position. The angle can be positive or negative, and the angle can be greater than 360° or less than −360°. Also, you can find angles in standard position when a positive or negative value of a trigonometric ratio is given.

OBJECTIVES

When you have finished this unit you should be able to:

1. Find the least positive coterminal angle and the reference angle for any angle in standard position.
2. Use a special triangle to find the value of a trigonometric ratio of an angle whose reference angle is a special angle, and a unit point to find the value of a trigonometric ratio of an angle whose least positive coterminal angle is a quadrantal angle.
3. Use Table II to find the value of a trigonometric ratio of any angle given in degrees and tens of minutes.
4. Given a value of a trigonometric ratio, find all angles between 0° and 360° which have that value.

Section 4.1   Coterminal and Reference Angles

In Section 3.1 we observed that the 0° and the 360° angle look the same except for the amount of rotation:

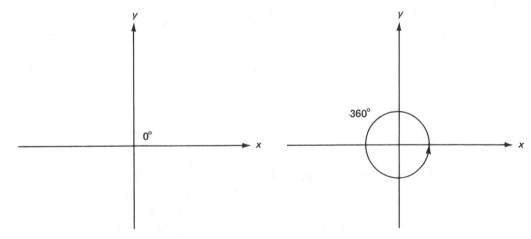

55

It is a common error to think that these are the same angle. They are not the same because 0 is not the same as 360. The angles clearly are related, however.

Definition: Two angles in standard position are **coterminal angles** if they have the same terminal side.

All angles in standard position have the same initial side. Since coterminal angles also have the same terminal side, coterminal angles differ only in their amount of rotation. The 0° angle has no rotation and the 360° angle is one full rotation. The 0° and 360° angles are coterminal since they have the same terminal side, the positive x-axis.

Any angle which is a multiple of 360° is also coterminal with 0°. For example, 720° is twice 360°, and is two full rotations. Similarly, 1080° is three full rotations, and so on.

For any angle greater than 360° which is not a multiple of 360°, there is exactly one coterminal angle between 0° and 360°. This coterminal angle is called the **least positive coterminal angle**.

EXAMPLE 4.1. Find the least positive coterminal angle for:

a. 450° b. 833°30′

Solutions.

a. If we subtract one full rotation from 450°, we have an angle with the same terminal side:

$$\begin{array}{r} 450° \\ -360° \\ \hline 90°. \end{array}$$

Therefore, the least positive coterminal angle is 90°. We draw the 450° angle by drawing one full rotation and then 90°:

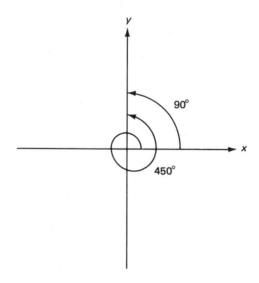

b. For 833°30′, we subtract two full rotations, or 720°:

$$\begin{array}{r} 833°30' \\ -720° \\ \hline 113°30'. \end{array}$$

The least positive coterminal angle is 113°30′. The diagram is

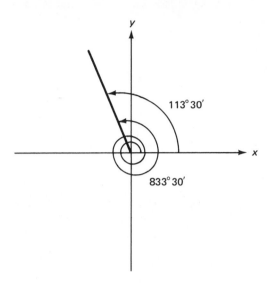

For any negative angle between 0° and −360°, there is exactly one positive coterminal angle between 0° and 360°. This angle is the least positive coterminal angle.

EXAMPLE 4.2.   Find the least positive coterminal angle for −138°50′.

Solution.   We know that −138°50′ is between −90° and −180°, so it is in the third quadrant:

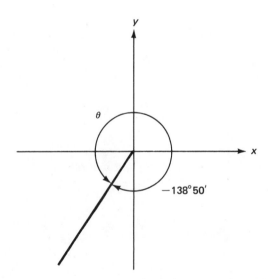

The least positive coterminal angle θ completes a full rotation of 360°. Therefore, we subtract 138°50′ from 360°, writing 360° as 359°60′:

$$\begin{array}{r} 359°60′ \\ -\,138°50′ \\ \hline 221°10′ \end{array}.$$

The least positive coterminal angle is 221°10′.

Negative multiples of 360° also are coterminal with 0°. To find the least positive coterminal angle for a negative angle less than −360° and not a multiple of −360°, we use the methods of the preceding examples.

EXAMPLE 4.3.   Find the least positive coterminal angle for −442°40′.

Solution.   We have a full rotation in the negative direction. If we add 360° to −442°40′, we have a negative angle between 0° and −360°:

$$
\begin{array}{r}
-\ 442°40' \\
360° \\
\hline
-\ \ \ 82°40'\ , \\
\end{array}
$$

so we know the angle is in the fourth quadrant:

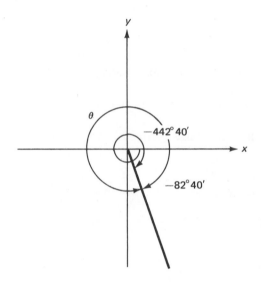

The least positive coterminal angle θ completes a full rotation of 360°, so we subtract 82°40′ from 360°:

$$
\begin{array}{r}
359°60' \\
-82°40' \\
\hline
277°20'\ . \\
\end{array}
$$

The least positive coterminal angle is 277°20′.

There is a second important type of angle related to most angles.

Definition: For any angle in standard position which is not coterminal with a quadrantal angle, the **reference angle** is the smallest positive angle (not necessarily in standard position) between the terminal side and the x-axis.

We can interpret this definition by a diagram which shows an angle in each quadrant and its reference angle:

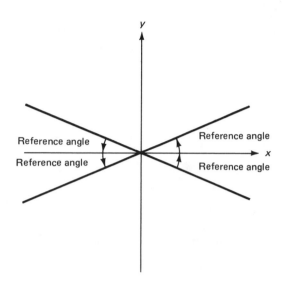

Observe that the reference angle is always an acute angle.

If we are given an acute angle, the reference angle is the same as the given angle.

EXAMPLE 4.4.   Find the reference angle for 36°.

Solution.

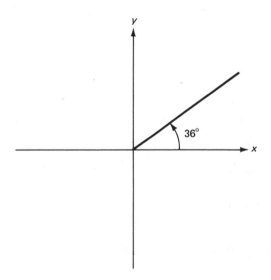

Since 36° is an acute angle, the reference angle is 36°.

For any other first quadrant angle, the reference angle is the least positive coterminal angle.

EXAMPLE 4.5.   Find the least positive coterminal angle and the reference angle for:

      a.   391°20′   b.   −303°40′

Solutions.   a.

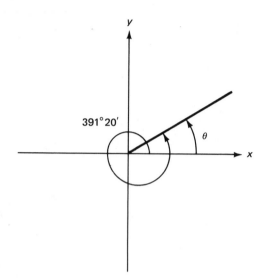

The least positive coterminal angle θ is:

$$391°20'$$
$$\underline{-360°}$$
$$31°20' .$$

Therefore, the reference angle also is 31°20′.

b.

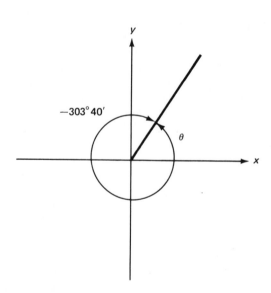

The least positive coterminal angle θ is:

$$359°60'$$
$$\underline{-303°40'}$$
$$56°20' .$$

Therefore, the reference angle also is 56°20′.

For second quadrant angles, we find the reference angle by subtracting the least positive coterminal angle from 180°.

EXAMPLE 4.6.   Find the reference angle for 135°50′.

Solution.

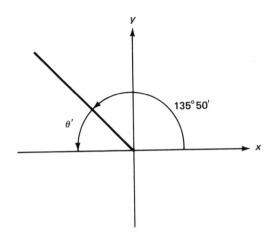

To find the reference angle θ' (read "theta prime") we subtract 135°50′ from 180°:

$$
\begin{array}{r}
179°60' \\
-135°50' \\
\hline
44°10' \ .
\end{array}
$$

The reference angle is 44°10′.

For third quadrant angles, we find the reference angle by subtracting 180° from the least positive coterminal angle.

EXAMPLE 4.7.   Find the reference angle for 235°50′.

Solution.

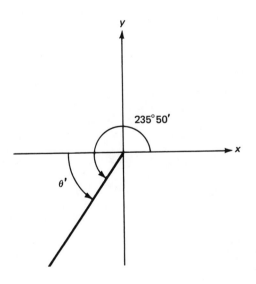

To find the reference angle θ' we subtract $180°$ from $235°50'$:

$$\begin{array}{r} 235°50' \\ -\ 180° \\ \hline 55°50' \, . \end{array}$$

The reference angle is $55°50'$.

It is a common error to write $180°$ automatically as $179°60'$. You do not need to use this form when you are subtracting $180°$ from another angle, or any other time that you do not have to "borrow" minutes in order to subtract.

To find the reference angle for a fourth quadrant angle, we subtract the least positive coterminal angle from $360°$.

EXAMPLE 4.8.   Find the reference angle for $287°40'$.

Solution.

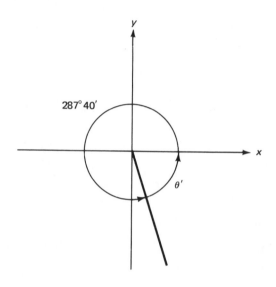

To find the reference angle θ', we subtract $287°40'$ from $360°$, where we must write $360°$ as $359°60'$:

$$\begin{array}{r} 359°60' \\ -\ 287°40' \\ \hline 72°20' \, . \end{array}$$

The reference angle is $72°20'$.

For angles which are greater than $360°$, or which are negative, we will usually find the least positive coterminal angle and then the reference angle. Often a diagram will indicate how to do the calculations.

EXAMPLE 4.9.   Find the least positive coterminal angle and the reference angle for $507°50'$.

Solution.   Since subtracting $360°$ from $507°50'$ gives

$$\begin{array}{r} 507°50' \\ -\ 360° \\ \hline 147°50' \, , \end{array}$$

the least positive coterminal angle is 147°50′, and the angle is in the second quadrant:

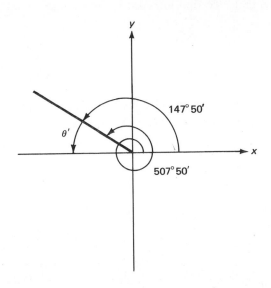

To find the reference angle θ' we subtract 147°50′ from 180°:

$$\begin{array}{r} 179°60' \\ -\,147°50' \\ \hline -\,32°10'. \end{array}$$

The reference angle is 32°10′.

EXAMPLE 4.10.   Find the least positive coterminal angle and the reference angle for −415°50′.

Solution.   Since adding 360° to −415°50′ gives

$$\begin{array}{r} -\,415°50' \\ 360° \\ \hline -\,55°50', \end{array}$$

the angle is in the fourth quadrant:

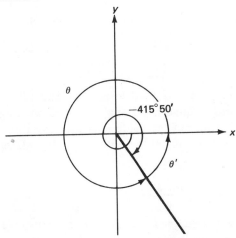

Observe that, since $-55°50'$ is negative, it is neither the least positive coterminal angle nor the reference angle. To find the least positive coterminal angle θ we subtract $55°50'$ from $360°$:

$$\begin{array}{r} 359°60' \\ -\ 55°50' \\ \hline 304°10' \end{array}$$

The least positive coterminal angle is $304°10'$. Clearly, the reference angle θ' is the positive angle $55°50'$.

Exercise 4.1

Find the reference angle:

1. $150°$ 2. $101°$ 3. $221°$ 4. $249°$

5. $322°$ 6. $347°$ 7. $144°20'$ 8. $193°50'$

9. $224°20'$ 10. $324°20'$ 11. $358°50'$ 12. $91°50'$

Find the least positive coterminal angle and the reference angle:

13. $420°$ 14. $585°$ 15. $494°20'$ 16. $659°50'$

17. $-148°20'$ 18. $-219°20'$ 19. $-348°30'$ 20. $-81°30'$

21. $-393°50'$ 22. $-644°10'$ 23. $-466°20'$ 24. $-575°40'$

25. $759°$ 26. $1330°$ 27. $-965°$ 28. $-1140°$

Section 4.2

Values Related to Special Angles

In Section 4.1, we saw that coterminal and reference angles relate any angle, except those coterminal with a quadrantal angle, to an acute angle. If we know how to find values for the trigonometric ratios of the acute angle, we can find values of the trigonometric ratios of angles related to the acute angle. We will begin with angles related to the special angles $45°$, $30°$, and $60°$.

EXAMPLE 4.11.   Find $\sin 135°$ and $\cos 135°$.

Solution. $135°$ is in the second quadrant. The reference angle is $45°$. A $45°$ angle which is the mirror image of the reference angle may be drawn in the first quadrant. Recalling the 45–45–90 triangle from Section 1.3, we see that we have such a triangle in the first quadrant. Therefore, the point

$(-1, 1)$ is on the terminal side of 135°, and $r = \sqrt{2}$:

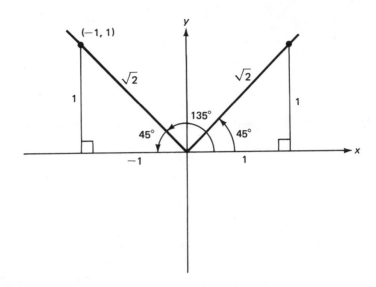

Now, we may use the (x, y) and r form of the trigonometric ratios, from Section 3.2, to find any trigonometric ratio of 135°. Since $x = -1$, $y = 1$, and $r = \sqrt{2}$,

$$\sin 135° = \frac{y}{r} = \frac{1}{\sqrt{2}}$$

$$\cos 135° = \frac{x}{r} = \frac{-1}{\sqrt{2}} = -\frac{1}{\sqrt{2}}.$$

Observe that, since 135° is in the second quadrant, the sine is positive but the cosine negative, as we found in Section 3.3.

EXAMPLE 4.12.   Find tan 150°.

Solution.   150° is in the second quadrant and its reference angle is 30°. A 30° angle which is the mirror image of the reference angle may be drawn in the first quadrant. Then, we have a 30–60–90 triangle in the first quadrant. Therefore, the point $(-\sqrt{3}, 1)$ is on the terminal side of 150°, and $r = 2$:

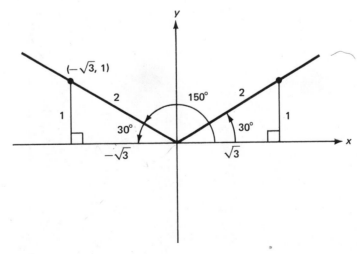

Therefore, $x = -\sqrt{3}$ and $y = 1$. Then,

$$\tan 150° = \frac{y}{x} = \frac{1}{-\sqrt{3}} = -\frac{1}{\sqrt{3}}.$$

Observe that the tangent is negative in the second quadrant.

EXAMPLE 4.13.   Find tan 225°.

Solution.   225° is in the third quadrant and its reference angle is again 45°. We may draw a 45° angle in the first quadrant and a 45–45–90 triangle:

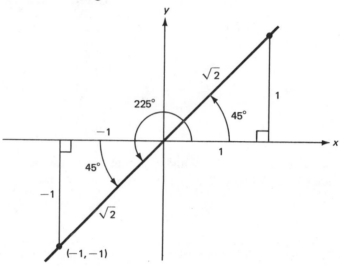

Therefore,

$$\tan 225° = \frac{y}{x} = \frac{-1}{-1} = 1.$$

In the third quadrant, the tangent is positive.

EXAMPLE 4.14.   Find sin 300° and cos 300°.

Solution.   300° is in the fourth quadrant and its reference angle is 60°. We may draw a 60° angle in the first quadrant and a 30–60–90 triangle:

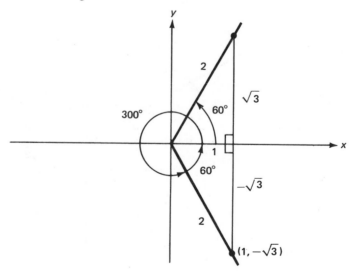

Therefore,

$$\sin 300° = \frac{y}{r} = \frac{-\sqrt{3}}{2} = -\frac{\sqrt{3}}{2}.$$

$$\cos 300° = \frac{x}{r} = \frac{1}{2}.$$

In the fourth quadrant, the sine is negative but the cosine is positive.

It is very important that we always take the reference angle to the *x*-axis. You must *never* take a reference angle to the *y*-axis. This is because the (x, y) and r form of the trigonometric ratios derived in Section 3.2 depends on *y* being related to the opposite side and *x* to the adjacent side of a right triangle. If a reference angle were taken to the *y*-axis, these sides would be reversed. We would have *y* related to the adjacent side and *x* related to the opposite side. For example, suppose our angle is in the second quadrant:

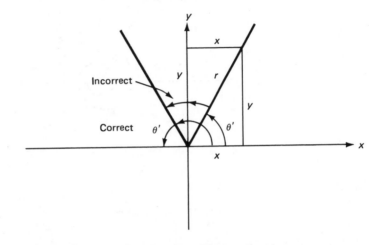

If the reference angle is incorrectly taken to the *y*-axis, then the values of the sine and the cosine would be reversed, the values of the tangent and cotangent would be reversed, and the values of the secant and cosecant would be reversed.

If an angle is greater than 360°, or is negative, we may find the least positive coterminal angle and then the reference angle to find values of the trigonometric ratios.

EXAMPLE 4.15.   Find cot 675°.

Solution.   Subtracting 360°, we have the least positive coterminal angle 315°. Then the reference angle is 45°:

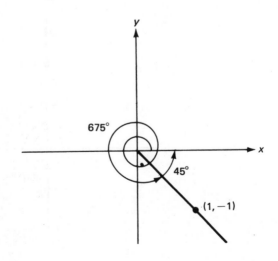

Therefore,

$$\cot 675° = \frac{x}{y} = \frac{1}{-1} = -1.$$

EXAMPLE 4.16.   Find $\csc(-600°)$.

Solution.   Adding 360° to −600°, we have the negative angle −240°. Then, adding another 360°, the least positive coterminal angle is 120°. The reference angle is then 60°. Remember always to take the reference angle to the *x*-axis:

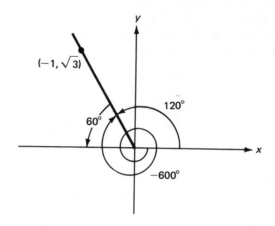

Therefore,

$$\csc(-600°) = \frac{r}{y} = \frac{2}{\sqrt{3}}.$$

The sine, and its reciprocal the cosecant, are positive in the second quadrant. Recall from Section 3.3 that a negative angle can have a positive trigonometric ratio. The sign of the trigonometric ratio depends only on the quadrant the angle is in, and not on whether the angle is positive or negative.

For angles which are coterminal with quadrantal angles, we have no reference angle. However, we can find values of the trigonometric ratios using the coterminal quadrantal angle.

EXAMPLE 4.17.   Find sin 450°.

Solution.   The coterminal angle is the 90° quadrantal angle:

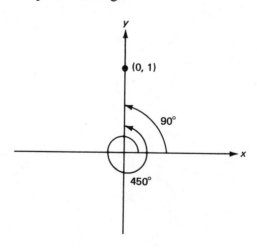

Recall the unit point (0, 1) from Section 3.4. We have $y = 1$, thus $\sin 450° = 1$.

EXAMPLE 4.18.   Find $\sec(-450°)$.

Solution.   We add 360° to $-450°$ to find the negative angle $-90°$. Then the least positive coterminal angle is 270°:

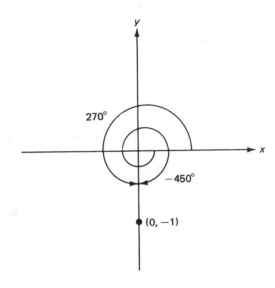

The unit point is $(0, -1)$, and $x = 0$. Therefore,

$$\sec(-450°) = \frac{1}{0},$$

so $\sec(-450°)$ is undefined.

Exercise 4.2

Use a special triangle to find:

1. $\sin 120°$ 2. $\cos 120°$ 3. $\tan 210°$ 4. $\cot 210°$

5. $\tan 315°$ 6. $\cos 315°$ 7. $\sec 390°$ 8. $\csc 690°$

9. $\cot 510°$ 10. $\sec 600°$ 11. $\sin(-30°)$ 12. $\cos(-45°)$

13. $\tan(-135°)$ 14. $\cot(-240°)$ 15. $\sec(-510°)$ 16. $\csc(-585°)$

Use a unit point to find:

17. $\sin 360°$ 18. $\cos 630°$ 19. $\tan 450°$ 20. $\cot 540°$

21. $\sec(-540°)$ 22. $\csc(-630°)$

Values Related to Other Angles

Suppose θ is any second quadrant angle, and its reference angle is θ'. Then θ' is an acute angle which we can draw in standard position in the first quadrant:

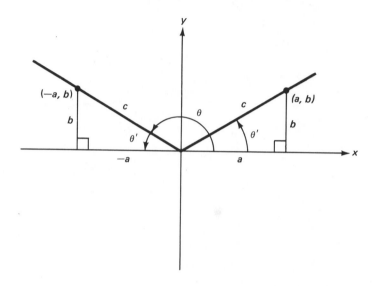

We choose a point (a, b) on the terminal side of θ', and we have a triangle in the first quadrant giving the trigonometric ratios of θ'. All of the trigonometric ratios of θ' are positive.

There is a point $(-a, b)$ on the terminal side of θ which is the mirror image of (a, b). Using the (x, y) and r form of the trigonometric ratios, where $r = c$ and is positive, we can find the trigonometric ratios for θ:

$$\sin \theta = \frac{y}{r} = \frac{b}{c} = \sin \theta'$$

$$\cos \theta = \frac{x}{r} = \frac{-a}{c} = -\cos \theta'$$

$$\tan \theta = \frac{y}{x} = \frac{b}{-a} = -\tan \theta'$$

$$\cot \theta = \frac{x}{y} = \frac{-a}{b} = -\cot \theta'$$

$$\sec \theta = \frac{r}{x} = \frac{c}{-a} = -\sec \theta'$$

$$\csc \theta = \frac{r}{y} = \frac{c}{b} = \csc \theta'.$$

We see that the absolute value of each trigonometric ratio of θ is the same as the trigonometric ratio of θ'. The sign of the trigonometric ratio may be different. In particular, in the second quadrant, $\cos \theta$, $\tan \theta$, $\cot \theta$, and $\sec \theta$ are negative. However, we may use the fact that the absolute value is the same to find values of trigonometric ratios for second quadrant angles.

EXAMPLE 4.19.   Find:

    a.   sin 165°   b.   tan (−213°20′)

Solutions.

    a.   The reference angle for 165° is 15°:

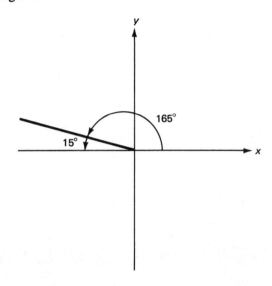

    Therefore, we use Table II to find that

$$\sin 15° = .2588.$$

Since sin θ is positive in the second quadrant, we also have

$$\sin 165° = .2588.$$

    b.   The least positive coterminal angle for −213°20′ is 146°40′, and the reference angle is 33°20′:

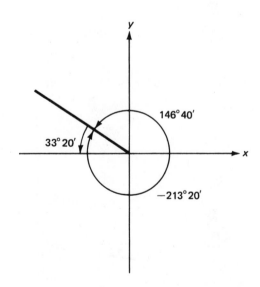

    We use Table II to find that

$$\tan 33°20′ = .6577.$$

Since tan θ is negative in the second quadrant,

$$\tan(-213°20') = -.6577.$$

Remember that the value is not negative because the angle is negative, but because the angle is in the second quadrant.

Using similar methods, we can show that the absolute values of the trigonometric ratios of a third quadrant angle θ are the same as the trigonometric ratios of its reference angle θ'. However, in the third quadrant, sin θ, cos θ, sec θ, and csc θ are negative.

EXAMPLE 4.20.   Find:

    a.   sin 245°   b.   cot 189°30′

Solutions.

    a.   The reference angle for 245° is 65°:

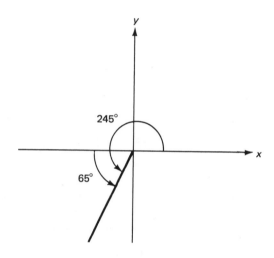

Since

$$\sin 65° = .9063,$$

and sin θ is negative in the third quadrant,

$$\sin 245° = -.9063.$$

    b.   The reference angle for 189°30′ is 9°30′:

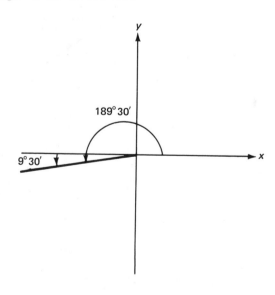

We have

$$\cot 9°30' = 5.976,$$

and cot θ is also positive in the third quadrant, so

$$\cot 189°30' = 5.976.$$

We have a similar situation in the fourth quadrant. However, in the fourth quadrant, sin θ, tan θ, cot θ, and csc θ are negative.

EXAMPLE 4.21. Find:

a. cos 289°20'   b. cot(−23°50')

Solutions.

a. The reference angle for 289°20' is 70°40':

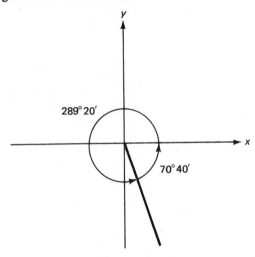

We have

$$\cos 70°40' = .3311.$$

But cos θ is also positive in the fourth quadrant, so

$$\cos 289°20' = .3311.$$

b. The reference angle for −23°50' is 23°50':

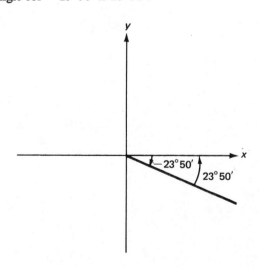

Since

$$\cot 23°50' = 2.264,$$

and cot θ is negative in the fourth quadrant,

$$\cot(-23°50') = -2.264.$$

Using the least positive coterminal angle and the reference angle, we can find values of trigonometric ratios for angles greater than 360° and for any negative angle.

EXAMPLE 4.22. Find csc 569°20′.

Solution. Subtracting 360°, the least positive coterminal angle is 209°20′. Therefore, the angle is in the third quadrant, and the reference angle is 29°20′:

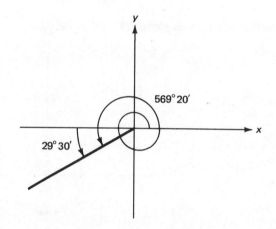

From Table II, we have

$$\csc 29°20' = 2.041.$$

But csc θ is negative in the third quadrant; therefore,

$$\csc 569°20' = -2.041.$$

EXAMPLE 4.23. Find sin(−663°30′).

Solution. Adding 360°, we have −303°30′ and the angle is in the first quadrant. Adding 360° again, the least positive coterminal angle and reference angle both are 56°30′:

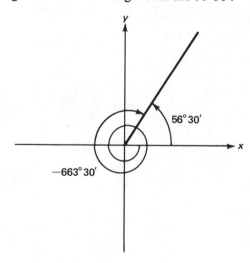

From Table II, we have

$$\sin 56°30' = .8339.$$

Since all trigonometric ratios of angles in the first quadrant are positive, we also have

$$\sin(-663°30') = .8339.$$

Exercise 4.3

Use Table II to find:

1. sin 143°

2. tan 229°

3. cos 128°40'

4. sin 153°10'

5. tan 235°20'

6. sec 197°50'

7. sin 332°10'

8. cot 283°30'

9. sin(−149°50')

10. cos(−206°50')

11. sin(−253°30')

12. tan(−101°40')

13. cot(−16°50')

14. csc(−297°10')

15. cos 528°20'

16. tan 697°30'

17. sec(−415°30')

18. cot(−559°20')

19. tan 1041°

20. sin(−1177°)

Section 4.4   Angles Related to Values

In the preceding sections of this unit, we have found values of trigonometric ratios given an angle θ. In Sections 1.4 and 1.5, we also found an acute angle θ given a value of a trigonometric ratio. Using the techniques of Unit 3 and the preceding sections of this unit, we can find angles which are not necessarily acute angles. Given a value, positive or negative, of a trigonometric ratio, we can find many angles θ in standard position which have that value.

EXAMPLE 4.24.   Find all angles θ which are positive but less than 360° such that $\sin \theta = \frac{1}{2}$.

Solution.   We know from a special triangle that $\sin 30° = \frac{1}{2}$. However, from Section 3.3, we know that $\sin \theta$ is also positive in the second quadrant. Thus there is another angle θ such that $\sin \theta = \frac{1}{2}$ in the second quadrant:

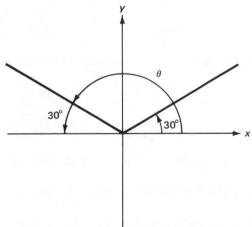

The reference angle for this second quadrant angle is 30°. Therefore, the other angle θ is

$$
\begin{array}{r}
180° \\
-30° \\
\hline
150° \, .
\end{array}
$$

Now we have two angles, $\theta = 30°$ and $\theta = 150°$, such that $\sin \theta = \frac{1}{2}$. Observe that all positive and negative angles θ coterminal with these angles also have $\sin \theta = \frac{1}{2}$. There are infinitely many such angles. We will usually find all those angles which are positive or 0° but less than 360°.

EXAMPLE 4.25. Find all angles θ which are positive but less than 360° such that $\cos \theta = -\frac{1}{2}$.

Solution. First, we use special triangles to find that the reference angle is 60°. Now, we recall from Section 3.3 that $\cos \theta$ is positive in the first and fourth quadrants. Therefore, $\cos \theta$ is negative in the second and third quadrants. Thus we must find angles in the second and third quadrants which have a reference angle of 60°:

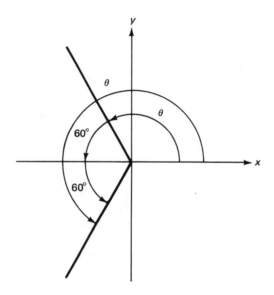

To find the second quadrant angle we subtract 60° from 180°:

$$
\begin{array}{r}
180° \\
-60° \\
\hline
120° \, .
\end{array}
$$

To find the third quadrant angle we add 60° to 180°:

$$
\begin{array}{r}
180° \\
60° \\
\hline
240° \, .
\end{array}
$$

Thus the angles θ between 0° and 360° such that $\cos \theta = -\frac{1}{2}$ are 120° and 240°.

EXAMPLE 4.26. Find all angles θ which are positive or 0° but less than 360° such that $\sin \theta = 1$.

Solution. From Section 3.4, we know that if $\sin \theta = 1$, θ is a quadrantal angle associated with a unit point:

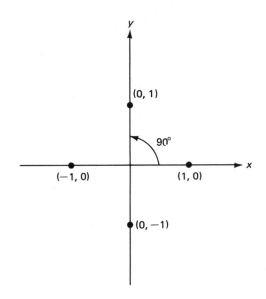

Since

$$\sin \theta = y = 1,$$

we must have $y = 1$. The only unit point for which $y = 1$ is $(0, 1)$. Thus $90°$ is the only angle between $0°$ and $360°$ such that $\sin \theta = 1$.

For $\sin \theta = \pm 1$ and $\cos \theta = \pm 1$, θ is a quadrantal angle. It is a common error to assume that $\tan \theta = \pm 1$ also is associated with a quadrantal angle. However, if we draw the 45–45–90 special triangle,

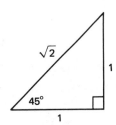

we see that $\tan 45° = 1$ and $\cot 45° = 1$. Thus, if $\tan \theta = 1$ or $\cot \theta = 1$, θ is $45°$ or $225°$. If $\tan \theta = -1$ or $\cot \theta = -1$, θ is $135°$ or $315°$. You should check that these are the angles θ which are positive but less than $360°$ by finding angles which have a reference angle of $45°$ in the appropriate quadrants.

For the remaining examples, we will use Table II.

EXAMPLE 4.27.   Use Table II to find all angles θ which are positive but less than $360°$ such that $\cos \theta = .6065$.

Solution.   Using Table II, we find that

$$\cos 52°40' = .6065.$$

Since $\cos \theta$ is positive, one angle is in the first quadrant, so $\theta = 52°40'$ is one solution. Also, $\cos \theta$

is positive in the fourth quadrant:

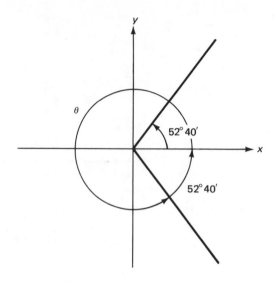

We subtract 52°40′ from 360°:

$$
\begin{array}{r}
359°60′ \\
-\ 52°40′ \\
\hline
307°20′\ .
\end{array}
$$

The angles are 52°40′ and 307°20′.

EXAMPLE 4.28. Use Table II to find all angles θ which are positive but less than 360° such that $\tan \theta = 3.962$.

Solution. From Table II,

$$\tan 75°50′ = 3.962.$$

Since $\tan \theta$ is positive, one angle is in the first quadrant, so $\theta = 75°50′$ is one solution. Also, $\tan \theta$ is positive in the third quadrant:

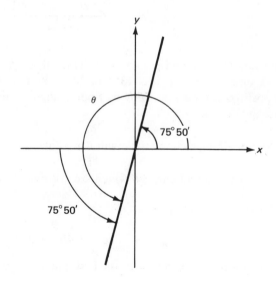

We add 75°50′ to 180°:

$$180°$$
$$75°50′$$
$$\overline{255°50′}.$$

The angles are 75°50′ and 255°50′.

   If the given value is negative, the method is essentially the same. However, since all the trigonometric ratios are positive in the first quadrant, the angle from Table II is the reference angle but is not one of the angles.

EXAMPLE 4.29. Use Table II to find all angles θ which are positive but less than 360° such that $\cot \theta = -2.539$.

Solution.   From Table II,

$$\cot 21°30′ = 2.539.$$

Since $\cot \theta$ is negative in the second and fourth quadrants, we use 21°30′ as a reference angle:

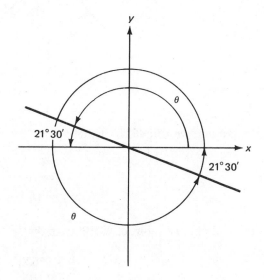

In the second quadrant, we subtract 21°30′ from 180°:

$$179°60′$$
$$-21°30′$$
$$\overline{158°30′}.$$

In the fourth quadrant, we subtract 21°30′ from 360°:

$$359°60′$$
$$-21°30′$$
$$\overline{338°30′}.$$

The angles are 158°30′ and 338°30′.

EXAMPLE 4.30. Use Table II to find all angles θ which are positive but less than 360° such that csc $\theta = -1.342$.

Solution.   From Table II,

$$\text{csc } 48°10' = 1.342.$$

Since csc θ is negative in the third and fourth quadrants, we use 48°10′ as a reference angle:

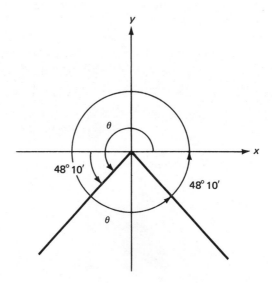

In the third quadrant, we add 48°10′ to 180°:

$$
\begin{array}{r}
180° \\
48°10' \\
\hline
228°10' .
\end{array}
$$

In the fourth quadrant, we subtract 48°10′ from 360°:

$$
\begin{array}{r}
359°60' \\
-48°10' \\
\hline
311°50' .
\end{array}
$$

The angles are 228°10′ and 311°50′.

Exercise 4.4

Use a special triangle to find all angles θ which are positive but less than 360°:

1. $\sin \theta = \dfrac{\sqrt{3}}{2}$ 2. $\tan \theta = \dfrac{1}{\sqrt{3}}$ 3. $\sec \theta = -\dfrac{2}{\sqrt{3}}$

4. $\csc \theta = -2$ 5. $\cos \theta = \dfrac{1}{\sqrt{2}}$ 6. $\cot \theta = -1$

Use a unit point to find all angles θ which are positive or $0°$ but less than $360°$:

7. $\cos \theta = 1$

8. $\sin \theta = -1$

9. $\sin \theta = 0$

10. $\cos \theta = 0$

11. $\cos \theta = -1$

12. $\tan \theta = 0$

Use Table II to find all angles θ which are positive but less than $360°$:

13. $\sin \theta = .7660$

14. $\cos \theta = .8192$

15. $\cos \theta = -.9511$

16. $\tan \theta = -1.072$

17. $\tan \theta = 2.394$

18. $\cot \theta = -1.530$

19. $\sin \theta = -.8587$

20. $\csc \theta = 1.239$

21. $\cos \theta = -.0843$

22. $\sec \theta = 4.745$

23. $\cot \theta = -3.647$

24. $\csc \theta = -6.636$

Another Note on Calculators

You have probably already discovered that you can use your scientific calculator to find the value of a trigonometric ratio of any angle without using the coterminal and reference angles. For example, to find $\tan(-213°20')$, first convert the angle to the decimal form 213.33333 and use the "$+/-$" key to change the sign to -213.33333. Then, press "tan" to obtain $-.65771034$.

But beware! The process does not reverse. Enter .65771034 and press "$+/-$" to change the sign to $-.65771034$. Then, press "inv" and "tan." Your display will show -33.333333. This angle is the negative of the reference angle, $33°20'$.

Similarly, entering a negative number for the sine of an angle results in a negative angle. For example, enter 1.342 (which is $\csc \theta$ in Example 4.30) and press "$+/-$" and "$1/x$" to obtain $-.74515648$. Then, press "inv" and "sin" to obtain the angle -48.172541, which is the negative of the reference angle $48°10'$.

However, if you enter a negative number for the cosine of an angle, the result is a second quadrant angle. (Recall that the cosine is positive in the fourth quadrant, and so the negative of the reference angle is impossible.) For example, enter .5, press "$+/-$," "inv," and "cos." The result is $120°$.

The negative of the reference angle for sine and tangent, and the second quadrant angle for cosine, are called the principal values of the inverse sine, inverse tangent, and inverse cosine. You will meet them in Unit 15. From them, you can find the reference angle, but it is your job to locate the appropriate quadrants for positive angles less than $360°$, and to calculate these angles from the reference angle. Your calculator will not do this for you.

Self-test

1. Find the least positive coterminal angle and the reference angle for $-483°40'$.

   1._____

2. Use a special triangle to find $\sin(-315°)$.

   2._____

3. Use Table II to find $\csc 554°50'$.

   3._____

4. Use Table II to find $\cos(-560°30')$.

   4._____

5. Use Table II to find all angles θ which are positive but less than $360°$ such that $\tan \theta = -.3185$.

   5._____

Unit 5

Cumulative Review

INTRODUCTION

In this unit you should make certain you remember all the material in all of the preceding units.

OBJECTIVE

When you have finished this unit you should be able to demonstrate that you can fulfill every objective of each preceding unit.

To prepare for this unit you should review the Self-Tests for Units 1 through 4. Do each problem of each Self-Test over again. If you cannot do a problem, or even if you have the slightest difficulty, you should:

1. Find out from the answer section the Objective for the unit to which the problem relates.
2. Review all the material in the section which has the same number as the objective, and redo all the Exercises for the section.
3. Try the Self-Test for the unit again.

Repeat these steps until you can do each problem in each Self-Test for each of Units 1 through 4 easily and accurately.

Self-test

1. Use opposite side, adjacent side, and hypotenuse to complete the statement:
   $\cos \theta =$

   1._____

2. If (x, y) is a point on the unit circle, use x and y to complete the statement:

   $\sec \theta =$

   2._____

3. Use a special triangle to find:

   a. $\cot 45°$

   b. $\sec 60°$

   3a._____

   3b._____

4. Use a unit point to find:

   a. $\sin 0°$

   b. $\sec 270°$

   4a._____

   4b._____

5. Use a unit point to find all angles θ which are positive or $0°$ but less than $360°$ such that $\cos \theta = 1$.

   5._____

6. Find the least positive coterminal angle and the reference angle for 502°20′.

6._____

For the remaining problems use Table II.

7. Find sec 502°20′.

7._____

8. Find all angles θ which are positive but less than 360° such that sin θ = $-$ .9272.

8._____

9. If in a right triangle a = 20 and b = 40, find α to the nearest degrees and tens of minutes.

9._____

10. A flagpole casts a shadow 15 feet long. The angle of elevation from the tip of the shadow to the top of the pole is 61°. How high is the pole, to two significant digits?

10._____

Unit 6

Laws of Sines and Cosines

INTRODUCTION

In Unit 2, you used the trigonometric ratios to find angles and sides of right triangles. However, the trigonometric ratios can be applied directly only to right triangles. In this unit you will learn two formulas which apply to triangles which are not right triangles, and how to use the formulas to solve such triangles.

OBJECTIVES

When you have finished this unit you should be able to:

1. Use the law of sines to find the other two sides of a triangle, given two angles and one side.
2. Use the law of cosines to find the third side of a triangle, given two sides and the included angle.
3. Use the law of cosines to find any angle of a triangle, given three sides.
4. Use the law of sines to determine if any triangles exist, how many exist, and to find any angles of those that do exist, given two sides and a nonincluded angle.

Section 6.1

The Law of Sines

Triangles which are not right triangles are called **oblique** triangles. Examples of oblique triangles are:

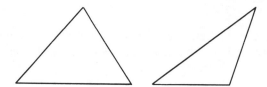

To solve an oblique triangle, three parts of the triangle, where at least one is a side, must be given.

Whenever two angles of an oblique triangle are given, the third angle is easy to find because the sum of the three angles is 180°. Remember that the third angle of an oblique triangle is not a right angle. To find the third angle, we subtract the sum of the two given angles from 180°. For example, suppose two angles of an oblique triangle are 25°30′ and 59°50′. We add 25°30′, and 59°50′:

$$
\begin{array}{r}
25°30' \\
59°50' \\
\hline
84°80'.
\end{array}
$$

But, 60′ = 1°, so 80′ = 1°20′. Therefore,

$$84°80' = 85°20'.$$

Then, subtracting 85°20′ from 180°,

$$
\begin{array}{r}
179°60' \\
- \ 85°20' \\
\hline
94°40'.
\end{array}
$$

The third angle is 94°40′. The third angle may be an acute angle or an obtuse angle. Recall that an acute angle is between 0° and 90°, and an obtuse angle is between 90° and 180°. The 94°40′ angle is an obtuse angle. An oblique triangle can have no more than one obtuse angle because the sum of two obtuse angles is more than 180°.

If two angles of an oblique triangle are called α and β, the sides opposite them are a and b. The third angle is called γ (gamma), the third letter of the Greek alphabet. The side opposite γ is c. It is a common error to associate γ with a right angle and continue to think of c as an hypotenuse. In an oblique triangle, there is no right angle and no hypotenuse. Each angle has an opposite side, but each angle has two adjacent sides where neither is an hypotenuse. Therefore, we cannot use the trigonometric ratios to solve an oblique triangle.

If we are given two angles and one side of an oblique triangle, we use a rule called the **law of sines** to find the other two sides.

Law of Sines:	$\dfrac{a}{\sin \alpha} = \dfrac{b}{\sin \beta} = \dfrac{c}{\sin \gamma}$
or	$\dfrac{\sin \alpha}{a} = \dfrac{\sin \beta}{b} = \dfrac{\sin \gamma}{c}.$

Any two parts of the same form of the law of sines can be used together. For example,

$$\frac{a}{\sin \alpha} = \frac{b}{\sin \beta}$$

can be used together as an equality. Similarly,

$$\frac{a}{\sin \alpha} = \frac{c}{\sin \gamma}$$

can be used together, and

$$\frac{b}{\sin \beta} = \frac{c}{\sin \gamma}$$

can be used together. The second form of the law of sines is used in a similar way. The proof of the first equality of the first form is given at the end of this unit.

EXAMPLE 6.1.   Solve the triangle with $\alpha = 48°50'$, $\beta = 74°50'$, and $c = 112$.

Solution.   First, we always draw a diagram:

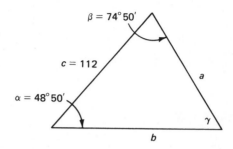

Observe that side a is opposite angle α, side b is opposite angle β, and side c is opposite angle γ. However, there is no special property which distinguishes any angle or side from any other.
     We find γ using the given angles α and β. Adding $48°50'$ and $74°50'$,

$$\begin{array}{r} 48°50' \\ 74°50' \\ \hline 122°100'. \end{array}$$

Remember not to "carry" the 1 of $100'$ into the degrees. Since $100' = 1°40'$,

$$122°100' = 123°40'.$$

Now, subtracting $123°40'$ from $180°$,

$$\begin{array}{r} 179°60' \\ -123°40' \\ \hline 56°20'. \end{array}$$

Therefore, $\gamma = 56°20'$. In this triangle, β is the largest angle.
     To find a and b, we use the law of sines. Using the equality

$$\frac{a}{\sin \alpha} = \frac{c}{\sin \gamma},$$

we know α, γ, and c, so we can find a:

$$\frac{a}{\sin 48°50'} = \frac{112}{\sin 56°20'}$$

$$\frac{a}{.7528} = \frac{112}{.8323}.$$

Solving for a,

$$a = \frac{(.7528)(112)}{.8323}$$

$$= 101,$$

to three significant digits.

Similarly, to find b, we use

$$\frac{b}{\sin \beta} = \frac{c}{\sin \gamma},$$

where we know β, γ, and c:

$$\frac{b}{\sin 74°50'} = \frac{112}{\sin 56°20'}$$

$$\frac{b}{.9652} = \frac{112}{.8323}.$$

Solving for b,

$$b = \frac{(.9652)(112)}{.8323}$$

$$= 130,$$

to three significant digits.

The solution of the triangle is $\gamma = 56°20'$, $a = 101$, and $b = 130$. Angles $\alpha = 48°50'$, $\beta = 74°50'$, and side $c = 122$ were given. Observe that the smallest side a is opposite the smallest angle α, and the largest side b is opposite the largest angle β. Since the triangle is not a right triangle, we cannot use the Pythagorean theorem to check. However, observe that

$$\frac{a}{\sin \alpha} = \frac{101}{\sin 48°50'} = \frac{101}{.7528} \approx 134,$$

$$\frac{b}{\sin \beta} = \frac{130}{\sin 74°50'} = \frac{130}{.9652} \approx 135,$$

$$\frac{c}{\sin \gamma} = \frac{112}{\sin 56°20'} = \frac{112}{.8323} \approx 135.$$

Therefore,

$$\frac{a}{\sin \alpha} \approx \frac{b}{\sin \beta} \approx \frac{c}{\sin \gamma}$$

as in the law of sines.

EXAMPLE 6.2.   Solve the triangle with $\alpha = 25°30'$, $\beta = 59°50'$, and $a = 12.2$.

Solution.   First, always draw a diagram:

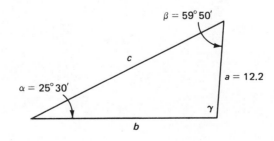

Subtracting the sum of α and β from 180°, we find $\gamma = 94°40'$. The computation was done at the beginning of this unit. Recall that γ is an obtuse angle.

To find side b, we use the form of the law of sines

$$\frac{b}{\sin \beta} = \frac{a}{\sin \alpha}.$$

It is usually convenient to put the unknown part in the upper left position. Observe that all the other parts are known parts. Then,

$$\frac{b}{\sin 59°50'} = \frac{12.2}{\sin 25°30'}$$

$$\frac{b}{.8646} = \frac{12.2}{.4305}$$

$$b = \frac{(.8646)(12.2)}{.4305}$$

$$= 24.5.$$

To find side c, we use the value we calculated for γ and the form of the law of sines

$$\frac{c}{\sin \gamma} = \frac{a}{\sin \alpha}$$

$$\frac{c}{\sin 94°40'} = \frac{12.2}{\sin 25°30'}.$$

Since 94°40′ is an obtuse angle, to find sin 94°40′ we must use the reference angle:

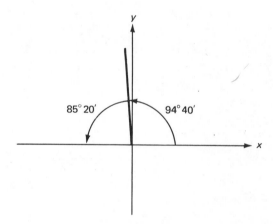

Since sin 85°20′ = .9967, and the sine is positive in the second quadrant, sin 94°40′ = .9967. An obtuse angle is always between 90° and 180°, so the sine of an obtuse angle is always positive. Then,

$$\frac{c}{.9967} = \frac{12.2}{.4305}$$

$$c = \frac{(.9967)(12.2)}{.4305}$$

$$= 28.2.$$

The solution of the triangle is $\gamma = 94°40'$, $b = 24.5$, and $c = 28.2$. Angles $\alpha = 25°30'$, $\beta = 59°50'$, and side $a = 12.2$ were given. The smallest side is opposite the smallest angle and the largest side is opposite the largest angle. You can check to see that

$$\frac{a}{\sin \alpha} \approx \frac{b}{\sin \beta} \approx \frac{c}{\sin \gamma}.$$

If you are using a scientific calculator and have set up the law of sines so that the unknown side is in a numerator, you can solve for the unknown side in one series of operations. For example, to find c, we have

$$\frac{c}{\sin 94°40'} = \frac{12.2}{\sin 25°30'}.$$

On your calculator, enter 12.2, press the division key, then enter 25.5 and press "sin" (since $25°30' = 25.5°$). Then press the multiplication key. The display so far is 28.33841. Now, to multiply by sin 94°40′, press the left parenthesis, divide 40 by 60 and add 94, and press the right parenthesis. Be sure not to press the equals key during any part of this series as doing so would end the series. Press "sin" and finally the equals key to obtain 28.244465, or 28.2 to three significant digits.

If the angle in the denominator on the right-hand side is not readily converted to a decimal, you can use the parentheses in the same way to convert it. You may try this procedure for Example 6.1. Alternatively, you can avoid the parentheses altogether by calculating the sine of the angle on the right-hand side and pressing "sto" to store the result. Then, find the sine of the angle on the left-hand side, multiply by the given side, and press the division key, "rcl," and finally the equals key.

Exercise 6.1

Solve the triangle:

1. $\alpha = 55°10'$, $\beta = 65°20'$, $c = 20.2$ 2. $\alpha = 42°40'$, $\beta = 66°30'$, $c = 110$

3. $\alpha = 50°50'$, $\beta = 85°30'$, $a = 9.5$ 4. $\alpha = 73°50'$, $\beta = 22°50'$, $b = 42$

5. $\alpha = 110°$, $\beta = 43°20'$, $c = 30$ 6. $\alpha = 62°50'$, $\beta = 98°50'$, $b = 252$

7. $\alpha = 34°30'$, $\gamma = 40°30'$, $b = 15.5$ 8. $\alpha = 82°20'$, $\gamma = 28°20'$, $c = 105$

9. $\beta = 35°50'$, $\gamma = 45°40'$, $a = 2.56$ 10. $\beta = 47°50'$, $\gamma = 46°50'$, $b = 50$

Section
6.2

The Law of Cosines

There are two cases of oblique triangles in which two sides and one angle are given. The easier case is the case where the given angle is the **included angle**. An angle is the included angle if it is formed by the two given sides. For example, if sides b and c are given, angle α is the included angle:

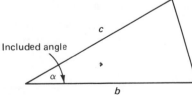

To find the side of the triangle opposite the included angle, we use the **law of cosines**.

Law of Cosines:	$a^2 = b^2 + c^2 - 2bc \cos \alpha,$
	$b^2 = a^2 + c^2 - 2ac \cos \beta,$
	$c^2 = a^2 + b^2 - 2ab \cos \gamma.$

The proof of the first form of the law of cosines is given at the end of this unit.

EXAMPLE 6.3.   Solve the triangle with $a = 22$, $b = 30$, and $\gamma = 55°$.

Solution.   A diagram is

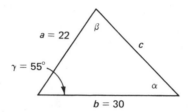

We have two sides a and b, and the included angle γ. To find c, we use the third form of the law of cosines:

$$c^2 = a^2 + b^2 - 2ab \cos \gamma$$

$$c^2 = (22)^2 + (30)^2 - 2(22)(30)(\cos 55°)$$

$$c^2 = 484 + 900 - 2(22)(30)(.5736)$$

$$c^2 = 484 + 900 - 757.15$$

$$c^2 = 627.$$

We have found c^2. To find c,

$$c = \sqrt{627}$$

$$c = 25.0.$$

To calculate the law of cosines on your scientific calculator, enter 22 and press "x^2," press plus, then 30 and "x^2," then minus and 2 times 22 times 30 and the multiplication key again. Enter 55 and press "cos." Now, press the equals key. You have 626.8791, which is c^2. Press "$\sqrt{x}$" to obtain 25.037554, or 25.0 to three significant digits. (On the TI-35 there is no "$\sqrt{x}$" key. Press "inv" and then "x^2" to find a square root. This is the "second function," given in blue.) If the angle involves minutes, either use parentheses to convert to a decimal or calculate and store the cosine of the angle before you begin.

Once we have a pair consisting of an angle and its opposite side, we may use the law of sines. We now have γ and c, which are such a pair. To find α, we use the second form of the law of sines:

$$\frac{\sin \alpha}{a} = \frac{\sin \gamma}{c}.$$

Again, it is convenient to put the unknown in the upper left. All the other parts are known:

$$\frac{\sin \alpha}{22} = \frac{\sin 55°}{25}$$

$$\frac{\sin \alpha}{22} = \frac{.8192}{25}$$

$$\sin \alpha = \frac{(22)(.8192)}{25}$$

$$\sin \alpha = .7209.$$

We have found $\sin \alpha$. Using Table II,

$$\alpha = 46°10'$$

to the nearest degrees and tens of minutes.

We could find β by subtracting the sum of γ and α from 180°. However, we may also use the law of sines again:

$$\frac{\sin \beta}{b} = \frac{\sin \gamma}{c}$$

$$\frac{\sin \beta}{30} = \frac{\sin 55°}{25}$$

$$\sin \beta = \frac{30(.8192)}{25}$$

$$\sin \beta = .9830$$

$$\beta = 79°30'.$$

The solution is $c = 25$, $\alpha = 46°10'$, and $\beta = 79°30'$. Angle $\gamma = 55°$, and sides $a = 22$ and $b = 30$ were given. The smallest side is opposite the smallest angle and the largest side is opposite the largest angle. To check, we observe that

$$\alpha + \beta + \gamma = 46°10' + 79°30' + 55°$$

$$= 180°40'$$

$$\approx 180°.$$

Because of the approximations involved, the sum usually will not be exactly 180°. You should also check that

$$\frac{a}{\sin \alpha} \approx \frac{b}{\sin \beta} \approx \frac{c}{\sin \gamma}.$$

EXAMPLE 6.4.   Solve the triangle with $\alpha = 32°20'$, $b = 64.3$, and $c = 129$.

Solution.   A diagram is

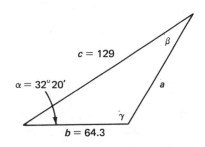

We find side a by using the first form of the law of cosines:

$$a^2 = b^2 + c^2 - 2bc \cos \alpha$$

$$a^2 = (64.3)^2 + (129)^2 - 2(64.3)(129)(\cos 32°20')$$

$$a^2 = (64.3)^2 + (129)^2 - 2(64.3)(129)(.8450)$$

$$a^2 = 6758.$$

Remembering to find the square root,

$$a = 82.2.$$

We may now turn to the law of sines to find β and γ, using the pair α and a:

$$\frac{\sin \beta}{b} = \frac{\sin \alpha}{a}$$

$$\frac{\sin \beta}{64.3} = \frac{\sin 32°20'}{82.2}$$

$$\sin \beta = \frac{(64.3)(.5348)}{82.2}$$

$$\sin \beta = .4183$$

$$\beta = 24°40'.$$

Similarly,

$$\frac{\sin \gamma}{c} = \frac{\sin \alpha}{a}$$

$$\frac{\sin \gamma}{129} = \frac{\sin 32°20'}{82.2}$$

$$\sin \gamma = \frac{(129)(.5348)}{82.2}$$

$$\sin \gamma = .8393.$$

It appears that γ is 57°. But observe that

$$\alpha + \beta + \gamma = 32°20' + 24°40' + 57°$$

$$= 114°$$

$$\ne 180°.$$

Recall that the sine is positive in the second quadrant as well as the first. One of the angles must be an obtuse angle. Since the largest side is opposite the largest angle, and $c = 129$ is the largest side, γ must be the obtuse angle. Thus 57° is the reference angle:

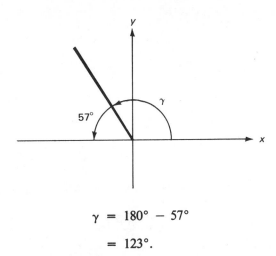

$$\gamma = 180° - 57°$$

$$= 123°.$$

The solution is $a = 82.2$, $\beta = 24°40'$, and $\gamma = 123°$.

We may also show that γ is an obtuse angle using the law of cosines. Recall that the cosine is negative in the second quadrant. Thus the cosine of any obtuse angle is negative. Using the third form of the law of cosines,

$$c^2 = a^2 + b^2 - 2ab \cos \gamma,$$

we solve for $\cos \gamma$:

$$2ab \cos \gamma = a^2 + b^2 - c^2$$

$$\cos \gamma = \frac{a^2 + b^2 - c^2}{2ab}.$$

Now, we substitute for a, b, and c:

$$\cos \gamma = \frac{(82.2)^2 + (64.3)^2 - (129)^2}{2(82.2)(64.3)}$$

$$= \frac{-5749.67}{10570.92}$$

$$= -.5439.$$

The reference angle is 57°. But, since $\cos \gamma$ is negative, γ is an obtuse angle. Thus $\gamma = 123°$.

If the given angle in the law of cosines is obtuse, we must be sure to write the cosine as a negative.

EXAMPLE 6.5.   Solve the triangle with $\alpha = 98°$, $b = 5.2$, and $c = 6.3$.

Solution.   A diagram is

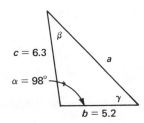

Observe that α is an obtuse angle. Using the first form of the law of cosines:

$$a^2 = b^2 + c^2 - 2bc \cos \alpha$$

$$a^2 = (5.2)^2 + (6.3)^2 - 2(5.2)(6.3)(\cos 98°).$$

The reference angle for 98° is 82° and cos 82° = .1392. Therefore,

$$\cos 98° = -.1392.$$

Thus

$$a^2 = (5.2)^2 + (6.3)^2 - 2(5.2)(6.3)(-.1392)$$

$$a^2 = (5.2)^2 + (6.3)^2 + 2(5.2)(6.3)(.1392)$$

$$a^2 = 75.85,$$

and the square root is

$$a = 8.71.$$

Since there can be no more than one obtuse angle in a triangle, angles β and γ are acute angles. You can use the law of sines to find that $\beta = 36°10'$ and $\gamma = 45°40'$, to the nearest degrees and tens of minutes.

It is a very common error to forget to use a negative cosine for an obtuse angle in the law of cosines. You should keep in mind this diagram:

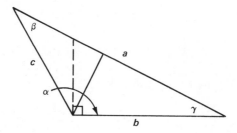

The law of cosines is

$$a^2 = b^2 + c^2 - 2bc \cos \alpha.$$

If α were a right angle, we would have

$$a^2 = b^2 + c^2.$$

If α is an acute angle, we want a shorter side a, so we subtract. However, if α is an obtuse angle, we need a longer side a, so we must add. Using a negative cosine in the law of cosines adds $2bc \cos \alpha$ rather than subtracting it.

Exercise 6.2

Solve the triangle:

1. $a = 5.5$, $b = 7.5$, $\gamma = 78°20'$

2. $a = 17$, $b = 19$, $\gamma = 38°30'$

3. $\alpha = 54°30'$, $b = 3.1$, $c = 4.2$

4. $\beta = 62°50'$, $a = 8.2$, $c = 8.4$

5. $\alpha = 45°20'$, $b = 22$, $c = 64$ 6. $a = 12.5$, $b = 9.2$, $\gamma = 35°30'$

7. $\alpha = 110°$, $b = 11$, $c = 9$ 8. $a = 10.1$, $b = 10.3$, $\gamma = 92°$

9. $\beta = 41°40'$, $a = 8.3$, $c = 2.3$ 10. $\beta = 96°30'$, $a = 24$, $c = 16$

Section 6.3   Another Use of the Law of Cosines

In Example 6.4, we saw that the law of cosines may be used to find an unknown angle when all three sides of a triangle are known. Although the calculations are more complicated than the calculations using the law of sines, using the law of cosines has the advantage that we can immediately identify an obtuse angle. Remember that the cosine of an obtuse angle is always negative.

EXAMPLE 6.6.   Solve the triangle with sides $a = 9.2$, $b = 15.5$, and $c = 11.8$.

Solution.   A diagram is

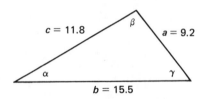

For our first calculation we cannot use the law of sines because we do not know any pair consisting of an angle and its opposite side. However, since we know all three sides, we can use the law of cosines. We will find angle α using

$$a^2 = b^2 + c^2 - 2bc \cos \alpha.$$

First, we solve for $\cos \alpha$:

$$2bc \cos \alpha = b^2 + c^2 - a^2$$

$$\cos \alpha = \frac{b^2 + c^2 - a^2}{2bc}.$$

Then, substituting for a, b, and c,

$$\cos \alpha = \frac{(15.5)^2 + (11.8)^2 - (9.2)^2}{2(15.5)(11.8)}.$$

Doing out the calculations, we find

$$\cos \alpha = .8060.$$

Since $\cos \alpha$ is positive, α is an acute angle. From Table II,

$$\alpha = 36°20'$$

to the nearest degrees and tens of minutes.

If you are using a calculator, be careful in handling the denominator. You have three options. You can calculate the numerator and press the equals key, divide by 2, divide by 15.5, divide by 11.8, and press the equals key again. (Observe that you must not use the multiplication key between 2 and 15.5 and 11.8.) Alternatively, you can calculate the numerator, and press the equals key, the division key, and the left parenthesis. Now, you can multiply 2 times 15.5 times 11.8, press the right parenthesis and the equals key. Finally, you can calculate the denominator first, store the result using "sto," then calculate the numerator, press the equals key, divide by the denominator using "rcl," and press the equals key. On a scientific calculator, after using any of these methods, your display shows .80604155. Now, if you are using a scientific calculator, press "inv" and "cos" to obtain the angle 36.289036, or 36°20′ to the nearest degrees and tens of minutes.

Since we now have a pair consisting of α and a, we could use the law of sines to find β. However, observe that b is the largest side, and so β is the largest angle and may be an obtuse angle. We use the law of cosines to see whether cos β is negative:

$$b^2 = a^2 + c^2 - 2ac \cos \beta$$

$$2ac \cos \beta = a^2 + c^2 - b^2$$

$$\cos \beta = \frac{a^2 + c^2 - b^2}{2ac}.$$

Substituting for a, b, and c, and doing the calculations, we find

$$\cos \beta = \frac{(9.2)^2 + (11.8)^2 - (15.5)^2}{2(9.2)(11.8)}$$

$$\cos \beta = -.0754.$$

Since cos β is negative, β is an obtuse angle. We find

$$\cos 85°40′ = .0754$$

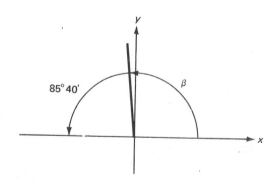

The angle 85°40′ is the reference angle, and

$$\beta = 94°20′$$

to the nearest degrees and tens of minutes.

There are several ways to find angle γ. We could subtract the sum of α and β from 180°, but we will reserve the sum of the angles for a check. Since there can be no more than one obtuse angle in a triangle, we know γ is an acute angle, so we could use the law of sines to find γ. Also, we could use the law of cosines to find γ, as we did to find α and β. You should use the law of sines and the law of cosines to find

$$\gamma = 49°30'$$

using the law of sines with the pair α and a, or

$$\gamma = 49°20'$$

using the law of cosines, to the nearest degrees and tens of minutes. The different answers are due to approximations in rounding and in Table II, and either answer is correct. However, we will use $\gamma = 49°20'$, since we have used the law of cosines to find the other angles. The solution is $\alpha = 36°20'$, $\beta = 94°20'$, and $\gamma = 49°20'$. Observe that

$$\alpha + \beta + \gamma = 36°20' + 94°20' + 49°20'$$

$$= 180°.$$

You should also check that

$$\frac{a}{\sin \alpha} \approx \frac{b}{\sin \beta} \approx \frac{c}{\sin \gamma}.$$

We emphasized following Example 6.5 that it is a common error to forget that the cosine of an obtuse angle is negative in using the law of cosines to find a side opposite an obtuse angle. It is also a common error to forget to take a negative cosine into account in using the law of cosines to find an angle.

EXAMPLE 6.7.   Find the largest angle of the triangle with sides 14, 18, and 25.

Solution.   The largest angle is opposite the largest side. If we call this angle α, then $a = 25$. The other sides are $b = 14$ and $c = 18$. A diagram is

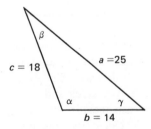

Using the law of cosines to find α,

$$a^2 = b^2 + c^2 - 2bc \cos \alpha$$

$$2bc \cos \alpha = b^2 + c^2 - a^2$$

$$\cos \alpha = \frac{b^2 + c^2 - a^2}{2bc}$$

$$= \frac{(14)^2 + (18)^2 - (25)^2}{2(14)(18)}$$

$$= -.2083.$$

Since

$$\cos 78° = .2083,$$

78° is the reference angle and

$$\cos 102° = -.2083.$$

The largest angle is the obtuse angle 102°.

Exercise 6.3

Solve the triangle:

1. $a = 10, b = 15, c = 12$ 2. $a = 1.7, b = 1.9, c = 2.3$

3. $a = 16, b = 22, c = 13$ 4. $a = 2.8, b = 4.8, c = 7$

5. $a = 19.2, b = 11.5, c = 8.6$ 6. $a = 14.5, b = 10.1, c = 10.2$

Find the largest angle of the triangle with sides:

7. 25, 30, 35   8. 5.9, 6.9, 9

9. 10.2, 12.3, 16.4   10. 1.6, 2.4, 3.2

Section 6.4   The Ambiguous Case

In Section 6.2, we used the law of cosines to solve triangles where two sides and the included angle were given. We now turn to the other case in which two sides and one angle of an oblique triangle are given. In this case, the given angle is not the included angle. A triangle is not uniquely determined by two sides and an angle other than the included angle, and so this case is called **the ambiguous case**, meaning there is more than one possibility. In the ambiguous case, the given sides and angle can determine two, one, or no triangles. We use the law of sines to determine which of these possibilities we have when the case is the ambiguous case.

EXAMPLE 6.8.   If $\alpha = 52°50'$, $a = 11.3$, and $b = 20.2$, find and solve all possible triangles.

Solution.   A diagram is

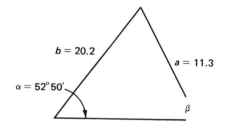

Observe that we have not drawn an angle β. We must determine whether any angle β exists. Using the law of sines,

$$\frac{\sin \beta}{b} = \frac{\sin \alpha}{a}$$

$$\frac{\sin \beta}{20.2} = \frac{\sin 52°50'}{11.3}$$

$$\frac{\sin \beta}{20.2} = \frac{.7069}{11.3}$$

$$\sin \beta = \frac{(20.2)(.7069)}{11.3}$$

$$\sin \beta = 1.4245.$$

Since there is no angle β for which $\sin \beta$ is more than 1, no angle β exists. Therefore, there is no triangle.

You may think of side a in the diagram as if it were attached to side b by a hinge, and free to swing back and forth:

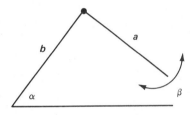

In the preceding example, a is too short and does not meet the third side at all as it swings. Thus there is no triangle.

EXAMPLE 6.9.   If $\alpha = 52°50'$, $a = 17.3$, and $b = 20.2$, find and solve all possible triangles.

Solution.   A diagram is

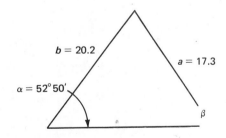

We use the law of sines to determine whether any angle β exists:

$$\frac{\sin \beta}{b} = \frac{\sin \alpha}{a}$$

$$\frac{\sin \beta}{20.2} = \frac{\sin 52°50'}{17.3}$$

$$\frac{\sin \beta}{20.2} = \frac{.7969}{17.3}$$

$$\sin \beta = \frac{(20.2)(.7969)}{17.3}$$

$$\sin \beta = .9305.$$

One possible angle is

$$\beta = 68°30'.$$

Since the sine is positive in the second quadrant, there also is a possible obtuse angle

$$\beta' = 111°30'.$$

Recall the diagram of a swinging on a hinge attached to b. Since a is long enough to meet the third side, it may be possible to swing a to meet the third side again forming an obtuse angle:

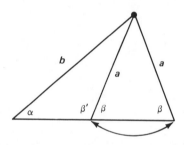

Clearly, if an obtuse angle β' is formed in this way, we must have $\beta' = 180° - \beta$.
First, we solve the triangle with $\beta = 68°30'$:

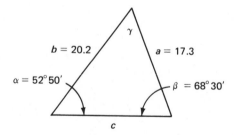

We cannot find either γ or c using the laws of sines and cosines. Therefore, we must find γ using the sum of the angles of the triangle. Adding α and β, and subtracting from 180°, we find

$$\gamma = 58°40'.$$

Then, to find c we use the law of sines:

$$\frac{c}{\sin \gamma} = \frac{a}{\sin \alpha}$$

$$\frac{c}{\sin 58°40'} = \frac{17.3}{\sin 52°50'}$$

$$\frac{c}{.8542} = \frac{17.3}{.7969}$$

$$c = \frac{(.8542)(17.3)}{.7969}$$

$$c = 18.5.$$

One solution is $\beta = 68°30'$, $\gamma = 58°40'$, and $c = 18.5$.

Now, we look for a triangle with $\beta' = 111°30'$:

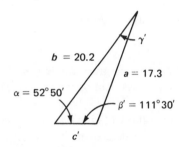

Adding α and β', and subtracting from 180°,

$$\gamma' = 15°40'.$$

Then,

$$\frac{c'}{\sin \gamma'} = \frac{a}{\sin \alpha}$$

$$\frac{c'}{\sin 15°40'} = \frac{17.3}{\sin 52°50'}$$

$$\frac{c'}{.2700} = \frac{17.3}{.7969}$$

$$c' = \frac{(.2700)(17.3)}{.7969}$$

$$c' = 5.86.$$

There is a second triangle, and the second solution is $\beta' = 111°30'$, $\gamma' = 15°40'$, and $c' = 5.86$.

EXAMPLE 6.10. If $\gamma = 79°20'$, $a = 7.4$, and $c = 12.8$, find and solve all possible triangles.

Solution. A diagram is

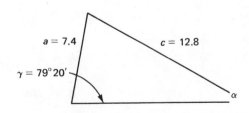

We use the law of sines to determine whether any angle α exists:

$$\frac{\sin \alpha}{a} = \frac{\sin \gamma}{c}$$

$$\frac{\sin \alpha}{7.4} = \frac{\sin 79°20'}{12.8}$$

$$\frac{\sin \alpha}{7.4} = \frac{.9827}{12.8}$$

$$\sin \alpha = \frac{(7.4)(.9827)}{12.8}$$

$$\sin \alpha = .5681.$$

One possible angle is

$$\alpha = 34°40'.$$

The corresponding obtuse angle is

$$\alpha' = 145°20'.$$

Observe that the sum of the given angle $\gamma = 79°20'$ and the obtuse angle $\alpha' = 145°20'$ is more than 180°. Two such angles cannot exist in the same triangle, so α' is extraneous. Therefore, there is just one triangle.

This time we may think of side c as being on a hinge attached to side a. Since side c is longer than side a, if we swing side c over, it will not meet the third side a second time:

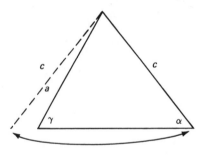

Thus there is one triangle.

Adding γ and α, and subtracting from 180°,

$$\beta = 66°.$$

Then,

$$\frac{b}{\sin \beta} = \frac{c}{\sin \gamma}$$

$$\frac{b}{\sin 66°} = \frac{12.8}{\sin 79°20'}$$

$$\frac{b}{.9135} = \frac{12.8}{.9827}$$

$$b = \frac{(.9135)(12.8)}{.9827}$$

$$b = 11.9.$$

The one solution is $\alpha = 34°40'$, $\beta = 66°$, and $b = 11.9$.

If the given angle is an obtuse angle, we can have no triangle or one triangle. We cannot have two triangles because the second triangle would have two obtuse angles which is not possible.

EXAMPLE 6.11.   If $\beta = 98°$, $b = 20$, and $c = 14$, find and solve all possible triangles.

Solution.   A diagram is

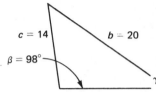

We use the law of sines to determine whether any angle γ exists:

$$\frac{\sin \gamma}{c} = \frac{\sin \beta}{b}$$

$$\frac{\sin \gamma}{14} = \frac{\sin 98°}{20}$$

$$\frac{\sin \gamma}{14} = \frac{.9903}{20}$$

$$\sin \gamma = \frac{(14)(.9903)}{20}$$

$$\sin \gamma = .6932$$

$$\gamma = 43°50'.$$

The corresponding obtuse angle is

$$\gamma' = 136°10'.$$

Clearly this angle cannot exist in the same triangle as the given angle $\beta = 98°$, so γ' is extraneous.

Adding β and γ, and subtracting from $180°$,

$$\alpha = 38°10'.$$

Then,

$$\frac{a}{\sin \alpha} = \frac{b}{\sin \beta}$$

$$\frac{a}{\sin 38°10'} = \frac{20}{\sin 98°}$$

$$\frac{a}{.6180} = \frac{20}{.9903}$$

$$a = \frac{(.6180)(20)}{.9903}$$

$$a = 12.5.$$

The one solution is $\gamma = 43°50'$, $\alpha = 38°10'$, and $a = 12.5$.

Exercise 6.4

Find and solve all possible triangles:

1. $\alpha = 58°$, $a = 10$, $b = 12$

2. $\alpha = 40°$, $a = 9$, $c = 16$

3. $\alpha = 58°$, $a = 10.2$, $b = 12$

4. $\alpha = 45°$, $a = 36$, $c = 50$

5. $\alpha = 58°$, $a = 12$, $b = 10$

6. $\gamma = 70°$, $a = 11.2$, $c = 15$

7. $\beta = 112°$, $b = 25$, $c = 16.2$

8. $\alpha = 108°$, $a = 5.7$, $b = 6.3$

9. $\gamma = 37°50'$, $a = 18$, $c = 10.2$

10. $\gamma = 43°20'$, $a = 17.5$, $c = 15.2$

11. $\beta = 62°40'$, $b = 14.3$, $c = 16.7$

12. $\beta = 23°30'$, $b = 9.4$, $c = 7.2$

13. $\beta = 53°20'$, $a = 4.3$, $b = 3.5$

14. $\gamma = 66°50'$, $b = 5.5$, $c = 5.6$

15. $\gamma = 74°40'$, $b = 8.5$, $c = 17$

16. $\beta = 28°30'$, $a = 19$, $b = 9.1$

Proof of the Law of Sines

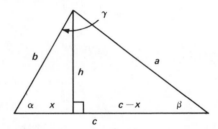

The height h separates the oblique triangle into two right triangles. In the left-hand triangle, h is opposite α and b is the hypotenuse. Thus,

$$\sin \alpha = \frac{h}{b}$$

$$b \sin \alpha = h.$$

In the right-hand triangle, h is opposite β and a is the hypotenuse. Thus,

$$\sin \beta = \frac{h}{a}$$

$$a \sin \beta = h.$$

Therefore,

$$a \sin \beta = b \sin \alpha,$$

and, dividing by $\sin \alpha \sin \beta$,

$$\frac{a}{\sin \alpha} = \frac{b}{\sin \beta}.$$

Proof of the Law of Cosines

Using the diagram above and the Pythagorean theorem, in the left-hand triangle,

$$x^2 + h^2 = b^2$$
$$h^2 = b^2 - x^2.$$

In the right-hand triangle,

$$(c - x)^2 + h^2 = a^2$$
$$h^2 = a^2 - (c - x)^2$$
$$h^2 = a^2 - (c^2 - 2cx + x^2)$$
$$h^2 = a^2 - c^2 + 2cx - x^2.$$

Therefore,

$$a^2 - c^2 + 2cx - x^2 = b^2 - x^2$$
$$a^2 - c^2 + 2cx = b^2$$
$$a^2 = b^2 + c^2 - 2cx$$

But in the left-hand triangle, x is the side adjacent to α and b is the hypotenuse, and so

$$\cos \alpha = \frac{x}{b}$$
$$b \cos \alpha = x.$$

Substituting $b \cos \alpha$ for x,

$$a^2 = b^2 + c^2 - 2c(b \cos \alpha)$$
$$a^2 = b^2 + c^2 - 2bc \cos \alpha.$$

Self-test

1. If $a = 7$, $b = 8$, and $c = 10$, find γ to the nearest degrees and tens of minutes.

1._____

2. If $\alpha = 94°30'$, $b = 10$, and $c = 12$, find a to three significant digits.

2._____

3. If $\alpha = 24°50'$, $\beta = 65°30'$, and $a = 84$, find b to three significant digits.

3._____

4. If $\alpha = 24°50'$, $a = 20$, and $b = 40$, find all possible angles β to the nearest degrees and tens of minutes.

4._____

5. If $\beta = 36°$, $b = 50$, and $c = 90$, find all possible angles γ to the nearest degrees and tens of minutes.

5._____

Unit
7

Applications

INTRODUCTION

In Unit 2 you learned how to solve verbal problems where the problem described a right triangle. Using the laws of sines and cosines, you can solve a much wider variety of problems. In this unit you will learn how to solve verbal problems where the triangle described may be an oblique or a right triangle. Some specific types of problems concern measurement, vectors, and bearings.

OBJECTIVES

When you have finished this unit you should be able to:

1. Solve verbal problems involving oblique or right triangles.
2. Solve problems involving indirect measurement.
3. Solve problems involving resultant and component vectors.
4. Solve problems involving navigational bearings.

Section 7.1   Applications of Solving Triangles

The solutions of many types of applied problems involve triangles. The triangle may be an oblique triangle or a right triangle. We must decide whether to use the law of sines, the law of cosines, or a trigonometric ratio. For each problem, we first draw a diagram, showing as accurately as possible the triangle involved, and labeling all the known parts. Also, we label with a letter the part or parts to be found.

Recall from Unit 2 that a right triangle can be solved using a trigonometric ratio. Two parts of the triangle besides the right angle, where at least one part is a side, must be known. Whenever the triangle described is a right triangle, we use a trigonometric ratio involving the two known parts and the part to be found.

If the triangle described is an oblique triangle, three parts, where at least one part is a side, must be known. We use the law of sines when a pair consisting of an angle and its opposite side is given. Sometimes the angle needed for the pair is not given but can be found if the other two angles are given. We use the law of cosines when two sides and the included angle or three sides are given.

EXAMPLE 7.1.   A house has a peaked roof. The back side of the roof is 54 feet long and the front is 64 feet. If the back is to make an angle of 40° with the horizontal, what angle must the front make with the horizontal?

Solution.   First, we draw a diagram of the triangle described:

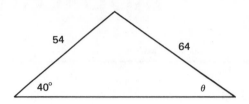

Observe that we have a pair consisting of the 40° angle and the side opposite, so we use the law of sines to find the angle labeled θ:

$$\frac{\sin \theta}{54} = \frac{\sin 40°}{64}$$

$$\sin \theta = \frac{54 \sin 40°}{64}$$

$$\sin \theta = \frac{(54)(.6428)}{64}$$

$$\sin \theta = .5424$$

$$\theta = 32°50'$$

to the nearest degrees and tens of minutes. Therefore, the roof makes an angle of 32°50′ with the horizontal at the front of the house. In this unit we will generally find angles to the nearest degrees and tens of minutes, and sides to three significant digits.

EXAMPLE 7.2.   Pleasant Street and Main Street meet at an angle of 60°. A pedestrian island is to be built between the streets, with the side along the Pleasant Street 12.5 feet long and the side along Main Street 14 feet long. How long must the third side of the island be?

Solution.   The triangle described is

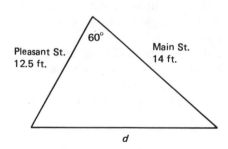

Since we do not have a pair consisting of an angle and its opposite side, we use the law of cosines to find the side labeled d:

$$d^2 = (12.5)^2 + (14)^2 - 2(12.5)(14)(\cos 60°)$$

$$d^2 = 156.25 + 196 - 2(12.5)(14)(.5)$$

$$d^2 = 156.25 + 196 - 175$$

$$d^2 = 177.25$$

$$d = 13.3$$

to three significant digits. The third side must be 13.3 feet long.

EXAMPLE 7.3.   Suppose 5th Avenue runs directly north–south, and 52nd Street runs directly east–west. You are standing on the roof of a building on 52nd Street, 80 feet west of the intersection of 5th Avenue and 52nd Street. You are looking at a point on the roof of a second building on 5th Avenue, 120 feet south of the intersection. What angle does the line of the two buildings make with 52nd Street?

Solution.   The diagram is

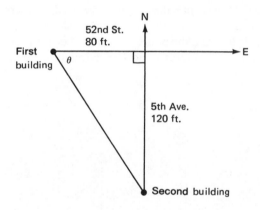

Since north–south is perpendicular to east–west, we have a right triangle and can use a trigonometric ratio. The angle between the line of the buildings and 52nd Street is labeled θ. 5th Avenue is the side opposite θ, and 52nd Street is the side adjacent to θ. Therefore,

$$\tan \theta = \frac{120}{80}$$

$$\tan \theta = 1.5$$

$$\theta = 56°20'.$$

The angle between the line of the buildings and 52nd Street is 56°20'.

Exercise 7.1

1. An isosceles triangle has a height of 18 inches and a base of 11 inches. What are its base angles? (The height of an isosceles triangle divides the triangle into two identical right triangles.)

2. The sides of a triangular trough are each $1\frac{1}{2}$ feet, and meet at an angle of 75°. How deep is the trough?

3. A triangular shelter is rigged by stretching a rope between two trees and spreading a piece of canvas over the rope so that the canvas makes a top angle over the rope and meets the ground at an angle on each side. Suppose 6.5 feet of canvas are on the left side of the rope and 8.5 feet are on the right, and the angle at the top is 55°. How wide will the opening along the ground at the bottom of the shelter be?

4. In problem 3, what is the angle between the left side of the shelter and the ground?

5. In problem 3, how high should the rope be placed to build the shelter described?

6. Suppose another shelter like the one in problem 3 is to be built. This time the left side is to be 9 feet, the right side is to be 7 feet, and the angle between the left side and the ground is to be 52°. What should the angle between the right side and the ground be?

Section 7.2   Indirect Measurement

In Unit 2, we used triangles to measure distances which cannot be measured directly.

EXAMPLE 7.4.   It is required to measure the distance between two trees, one on each side of a river. You measure 20 meters down the side of the river along a line perpendicular to the line of the trees, and place a rock to mark the point. The angle between the line from the rock to the first tree and the line from the rock to the second tree is found to be 48°40′. What is the distance between the trees?

Solution.   We call the first tree T_1, the second tree T_2, and the rock R:

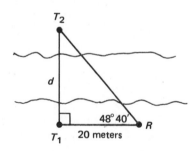

Then,

$$\tan 48°40′ = \frac{d}{20}$$

$$d = 20 \tan 48°40′$$

$$d = 20(1.137)$$

$$d = 22.7.$$

The distance between the trees is 22.7 meters.

It may not always be possible to measure on a line perpendicular to the required distance. In this case, we may measure at any angle to the required distance and use the law of sines. The method used is called the **method of triangulation.**

EXAMPLE 7.5.   The method of triangulation is used to measure the distance between two trees, one on each side of a river. Because of the direction of the river, you find you must measure at an angle of 64°50′ from the first tree. You measure 20 meters to a point which you mark with a rock. The angle between the lines from the rock to the first tree and the rock to the second tree is found to be 83°40′. What is the distance between the trees?

Solution.   The diagram is

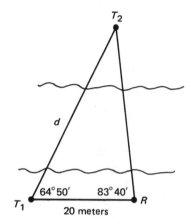

We do not have a pair consisting of an angle and the side opposite. However, we can find the third angle of the triangle:

$$64°50′ + 83°40′ = 147°90′ = 148°30′,$$

and

$$180° - 148°30′ = 179°60′ - 148°30′ = 31°30′.$$

Then, using the law of sines,

$$\frac{d}{\sin 83°40′} = \frac{20}{\sin 31°30′}$$

$$d = \frac{(\sin 83°40′)(20)}{\sin 31°30′}$$

$$d = \frac{(.9939)(20)}{.5225}$$

$$d = 38.0.$$

The distance is 38.0 meters.

Exercise 7.2

1. To measure the distance across a river from a tree on one side to a tree on the other side, a distance of 10 meters is measured from the first tree perpendicular to the line of the trees, and marked by a stake. The angle between the line from the stake to the first tree and the line from the stake to the second tree is found to be 54°30′. What is the distance between the trees?

2. In problem 1, a distance from the first tree is not measured perpendicular to the line of the trees because of the direction of the river. Using the method of triangulation, a distance of 8 meters is measured at an angle of 75° to the line of the trees and marked by a stake. The angle between the line from the stake to the first tree and the line from the stake to the second tree is found to be 72°10′. What is the distance between the trees?

3. Suppose it is attempted to find the distance between two posts in the following way: A distance of 10 feet is measured from the first post at an angle of 65° to the line of the posts and marked. It is then found that the marker is 8.4 feet from the second post. What is the angle between the line from the second post to the marker and the line of the posts?

4. The distance between the posts in problem 3 is found by triangulation. A distance of 10 feet is measured from the first post at an angle of 65° to the line of the posts and marked. The angle between the line from the marker to the first post and the line from the marker to the second post is found to be 45°50′. What is the distance between the posts?

5. Forest fires may be located by the method of triangulation. Suppose two watch towers are located 22.5 miles apart. A fire is spotted, and the ranger in the first tower measures the angle between his line to the fire and the line of the towers to be 81°. The ranger in the second tower measures the angle between his line to the fire and the line of the towers to be 61°20′. How far is the fire from the first tower?

6. Two observation towers are 16.2 kilometers apart. From the first tower, a monument marker is located on a line at an angle of 63°20′ from the line of the towers. The angle between the line from the second tower to the marker and the line of the towers is found to be 48°50′. How far is the monument marker from the first tower?

Section 7.3   Vectors

The part of a line bounded by two points on the line is called a **line segment**:

A line segment with a direction, where the direction may be described by an angle, is a **vector**:

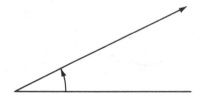

One important example of a vector is a **force**. The length of the line segment of a vector which represents a force is the amount of the force, often given in pounds. For example, weight is a force which can be represented by a vector directed downward.

We will consider problems involving two forces and their **resultant**. The two forces are represented by two adjacent sides of a parallelogram, and the resultant is represented by the longer diagonal of the parallelogram:

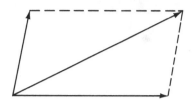

Recall from geometry that a parallelogram has opposite pairs of sides parallel. The opposite pairs of sides are also equal, as are opposite pairs of angles. The sum of two adjacent angles is 180°.

EXAMPLE 7.6.   Two forces are 20 pounds and 24 pounds, and are at an angle of 40° to one another. Find their resultant.

Solution.   We draw the forces as two adjacent sides of a parallelogram meeting at an angle of 40°, and the resultant as the longer diagonal:

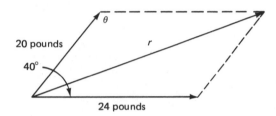

It is a common error to assume that the resultant splits the 40° angle into two 20° angles. This is not true unless the two forces are equal. In this example, the angle opposite the 20-pound force is smaller than the angle opposite the 24-pound force. We know that the sum of these angles is 40°, but we do not know either angle and we cannot find either angle without knowing r. However, we do know that the 40° angle and the angle labeled θ together make 180°. Therefore,

$$\theta = 180° - 40°$$

$$\theta = 140°.$$

Now, the top side of the parallelogram is equal to the bottom side, and so using the top triangle and the law of cosines,

$$r^2 = 20^2 + 24^2 - 2(20)(24)(\cos 140°).$$

Recall from Unit 6 that it is a very common error to forget to write the cosine of an obtuse angle as a negative number. Since 140° is an obtuse angle, cos 140° is negative:

$$r^2 = 20^2 + 24^2 - 2(20)(24)(-.7660)$$

$$r^2 = 20^2 + 24^2 + 2(20)(24)(.7660)$$

$$r^2 = 1711.36$$

$$r = 41.4.$$

The resultant is 41.4 pounds.

EXAMPLE 7.7. Two forces are 32 pounds and 45 pounds, and their resultant is 56 pounds. Find the angle between the resultant and the smaller force.

Solution.   The diagram is

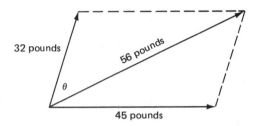

The angle between the resultant and the smaller force is labeled θ. Since the top side of the parallelogram is equal to the bottom side, the three sides of the top triangle are known. Therefore, we use the law of cosines:

$$45^2 = 32^2 + 56^2 - 2(32)(56)(\cos \theta)$$

$$\cos \theta = \frac{32^2 + 56^2 - 45^2}{2(32)(56)}$$

$$\cos \theta = .5957$$

$$\theta = 53°30'.$$

Thus, the angle between the resultant and the smaller force is 53°30'.

It is possible for any type of verbal problem which describes an oblique triangle to describe a triangle which is an example of the ambiguous case. Then the problem may have two, one, or no solutions. Some such problems have been included in Exercises for this unit.

EXAMPLE 7.8. You wish to combine a force of 12 pounds with a force of 18 pounds to give a resultant force, and you want the angle between the 18-pound force and the resultant to be 43°. What should the angle between the 12-pound force and the resultant be?

Solution.   The diagram is

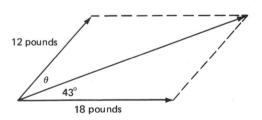

The top of the parallelogram also represents 18 pounds, and the second acute angle of the top triangle is also 43°. Therefore,

$$\frac{\sin \theta}{18} = \frac{\sin 43°}{12}$$

$$\sin \theta = \frac{(18)(\sin 43°)}{12}$$

$$\sin \theta = \frac{(18)(.6820)}{12}$$

$$\sin \theta = 1.023.$$

Since $\sin \theta$ is more than 1, there is no angle θ. Thus the problem has no solution.

In many applications we must find perpendicular **components** of a vector. The vector is then the resultant of the components.

EXAMPLE 7.9. Find the horizontal and vertical components of a vector with length 5 and at an angle of 20° to the x-axis.

Solution. The horizontal and vertical components are taken on the positive x- and y-axes, and are labeled v_x and v_y in the diagram:

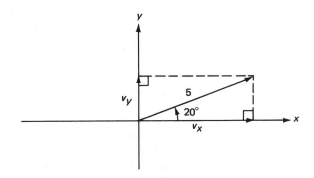

Since the triangles formed are right triangles,

$$\cos 20° = \frac{v_x}{5}$$

$$v_x = 5 \cos 20°$$

$$v_x = 5(.9397)$$

$$v_x = 4.70,$$

and

$$\sin 20° = \frac{v_y}{5}$$

$$v_y = 5 \sin 20°$$

$$v_y = 5(.3420)$$

$$v_y = 1.71.$$

The horizontal and vertical components of the vector are 4.70 and 1.71. Since the triangles are right triangles, you can use the Pythagorean theorem to check that

$$v_x^2 + v_y^2 \approx 5^2.$$

We mentioned at the beginning of this section that weight can be represented by a vector directed downward. An application of the perpendicular components of a vector involves an object resting on an inclined ramp. The weight of the object is represented by a vector directed downward. The component of this vector parallel to the ramp represents the force needed to keep the object from sliding down the ramp.

EXAMPLE 7.10.   What force is needed to keep a 120-pound object from sliding down a ramp inclined at an angle of 42°30′?

Solution.   We draw a diagram of the ramp, with the object represented by a box, and its weight represented by a vector directed downward from the center of the box:

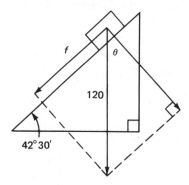

Using theorems of geometry, it can be shown that the angle θ between the vector representing the weight and the component perpendicular to the ramp is equal to the angle of inclination of the ramp. Thus, $\theta = 42°30′$. Then, using the right-hand right triangle, we find the component f parallel to the ramp:

$$\sin \theta = \frac{f}{120}$$

$$\sin 42°30′ = \frac{f}{120}$$

$$f = 120 \sin 42°30′$$

$$= 120(.6756)$$

$$= 81.1.$$

The force needed to keep the object from sliding down the ramp is 81.1 pounds.

Exercise 7.3

1. If two adjacent sides of a parallelogram are 8 inches and 12 inches, and its obtuse angle is 150°, what is its longer diagonal?

2. Two forces are 30 pounds and 40 pounds. If the angle between the forces is 60°, what is their resultant?

3.   Two forces are 15 pounds and 18 pounds. Their resultant is 24 pounds. Find the angle between the resultant and the 15-pound force.

4.   In problem 3, find the angle at which the forces meet.

5.   The resultant of two forces is 80 pounds. The angle between the resultant and the larger force is 33°10′, and the angle between the resultant and the smaller force is 41°50′. What are the two forces?

6.   Two forces have a resultant of 200 pounds. The larger force is 140 pounds. If the angle between the resultant and the smaller force is 45°, what is the smaller force?

7.   Find the horizontal and vertical components of a vector with length 32 and at an angle of 35° to the x-axis.

8.   Find the horizontal and vertical components of a vector with length 16.5 and at an angle of 82°50′ to the x-axis.

9.   What force is needed to keep a 10-pound object from sliding down a ramp inclined at an angle of 30°?

10.   What does an object weigh if an 8.5-pound force is needed to keep it from sliding down a ramp inclined at an angle of 34°30′?

Section 7.4	Navigational Bearings

A distance or speed, along with a direction, provide a common application of vectors. The distance or speed is represented by the length of the vector, and the direction by the direction of the vector. In this section we will use **navigational bearings** to indicate the direction.

We have used the four compass points, north, south, east, and west, in Example 7.3. As is common practice on maps, we will draw north toward the top of the page, south downward, east to the right, and west to the left. Now, we need a way to indicate directions to the northeast, southeast, northwest, and southwest. Northeast, for example, may mean a direction exactly between north and east, 45° from each:

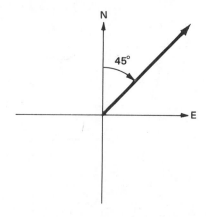

However, we also need to indicate directions more toward the north or more toward the east.

Consider the phrases "from the north 30° to the east" and "from the north 60° to the east." We abbreviate these phrases N 30°E and N 60°E. The symbol N 30°E describes the diagram:

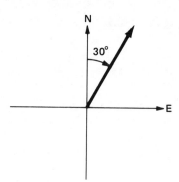

The symbol N 60°E describes the diagram:

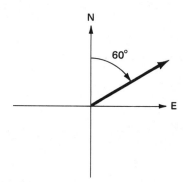

Similarly, S 30°E describes the diagram:

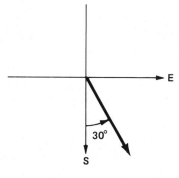

and S 60°E describes the diagram:

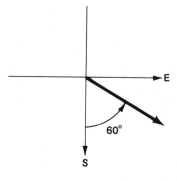

We use similar symbols for angles west of north and west of south. Observe that bearings are always taken from the vertical. The primary navigational direction, north and south, is traditionally taken to be vertical. Recall that mathematical reference angles, by contrast, are always taken from the horizontal since the primary axis, the *x*-axis, is generally taken to be horizontal.

We can often use the fact that two main compass points may be perpendicular in solving problems.

EXAMPLE 7.11. A ship leaves a harbor on a bearing of S 75°W. There is a lighthouse due west of the harbor. After the ship has sailed 4.2 nautical miles, it is due south of the lighthouse. How far is the ship from the lighthouse?

Solution. We draw a careful diagram, indicating the relevant main compass points, south and west, and the locations of the harbor, lighthouse, and ship:

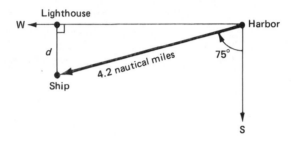

Since south and west are perpendicular, we have a right triangle. Also, the acute angle of the triangle at the harbor is

$$90° - 75° = 15°.$$

Therefore,

$$\sin 15° = \frac{d}{4.2}$$

$$d = 4.2 \sin 15°$$

$$d = 4.2(.2588)$$

$$d = 1.09.$$

The ship is 1.09 nautical miles from the lighthouse.

EXAMPLE 7.12. An island is 2.35 nautical miles due south of a buoy. A boat is 1 nautical mile due west of the buoy. What bearing should the boat take to reach the island?

Solution. The diagram is

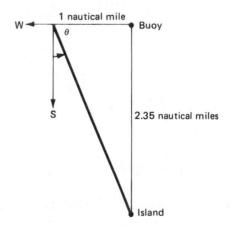

To find the bearing, first we find the angle θ of the triangle:

$$\tan \theta = \frac{2.35}{1}$$

$$\tan \theta = 2.35$$

$$\theta = 67°.$$

Now, the bearing is the angle starting from the south and going toward the east. From the diagram, the bearing is clearly S 23°E.

When we do not have two perpendicular compass points, we have an oblique triangle and we need one further piece of information. You will need to read the problem very carefully to draw an accurate diagram.

EXAMPLE 7.13.   Two harbors are known to be 112 nautical miles apart. The northern harbor is on a bearing of N 12°50′W from the southern harbor. A ship has sailed 100 nautical miles due north from the southern harbor. How far is the ship from the northern harbor?

Solution.   Reading the problem carefully, we draw the diagram:

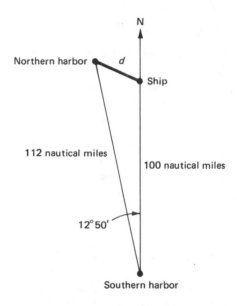

Now, we do not know we have a right triangle, and we do not have a pair of an angle and its opposite side, or two angles, given. Therefore, we use the law of cosines:

$$d^2 = 112^2 + 100^2 - 2(112)(100)(\cos 12°50′)$$

$$d^2 = 112^2 + 100^2 - 2(112)(100)(.9750)$$

$$d^2 = 704$$

$$d = 26.5.$$

The distance is 26.5 nautical miles.

EXAMPLE 7.14.   In Example 7.13, what bearing should the ship take to reach the northern harbor?

Solution.   From Example 7.13, we know the distance from the ship to the northern harbor, and the diagram is

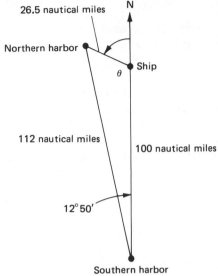

Since we have a pair consisting of an angle and its opposite side, we may use the law of sines:

$$\frac{\sin \theta}{112} = \frac{\sin 12°50'}{26.5}$$

$$\sin \theta = \frac{112(\sin 12°50')}{26.5}$$

$$\sin \theta = \frac{112(.2221)}{26.5}$$

$$\sin \theta = .9387.$$

It is possible that θ is 69°50′. However, it also appears possible that θ is an obtuse angle. If this is the case, then 69°50′ is the reference angle, and

$$\theta = 180° - 69°50'$$

$$\theta = 179°60' - 69°50'$$

$$\theta = 110°10'.$$

To determine whether θ is an obtuse angle, we may find the third angle of the triangle and see which sum of all the angles makes approximately 180°. Alternatively, we may check θ using the three sides and the law of cosines. We will use the law of cosines:

$$112^2 = 100^2 + (26.5)^2 - 2(100)(26.5)\cos \theta$$

$$2(100)(26.5)\cos \theta = 100^2 + (26.5)^2 - 112^2$$

$$\cos \theta = \frac{100^2 + (26.5)^2 - 112^2}{2(100)(26.5)}$$

$$\cos \theta = -.3475.$$

Recall from Unit 6 that, when we use the law of cosines, if $\cos \theta$ is negative then θ is an obtuse angle. Since $\cos 69°40' = .3475$, the reference angle is $69°40$, and so

$$\theta = 110°20'.$$

The difference between the two answers is due to approximations and rounding. Either answer is correct. The ship should take a bearing of between N 69°50'W and N 69°40'W. It will then come close enough to the harbor to find buoys and other markers.

EXAMPLE 7.15.   Two ships leave a harbor at the same time. One sails on a bearing of N 14°20'E at an average rate of 9.6 knots (nautical miles per hour), and the second sails on a bearing of S 43°E at an average rate of 10.3 knots. How far apart are the ships at the end of one hour?

Solution.   In one hour, the first ship goes 9.6 nautical miles and the second ship goes 10.3 nautical miles. The diagram is

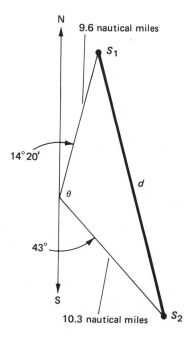

The angle θ between the bearings of the ships is

$$\theta = 180° - (14°20' + 43°)$$

$$= 180° - 57°20'$$

$$= 122°40'.$$

Using the law of cosines, and remembering that the cosine of an obtuse angle is negative,

$$d^2 = (9.6)^2 + (10.3)^2 - 2(9.6)(10.3)(\cos 122°40')$$

$$d^2 = (9.6)^2 + (10.3)^2 - 2(9.6)(10.3)(-.5398)$$

$$d^2 = (9.6)^2 + (10.3)^2 + 2(9.6)(10.3)(.5398)$$

$$d^2 = 305.0$$

$$d = 17.5.$$

The distance is 17.5 nautical miles. Observe that, if you do forget to write cos 122°40′ as a negative number, your answer will be 9.57 nautical miles. An accurate diagram tells you that this answer is clearly too small.

EXAMPLE 7.16. A ship sails from a harbor on a bearing of N 70°E to one buoy. It then turns and sails due south 1.6 nautical miles to a second buoy. The second buoy is 1.4 nautical miles from the harbor. What is the new bearing of the ship from the harbor?

Solution. The diagram is

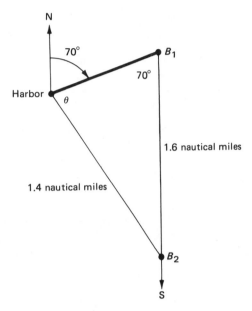

We must find θ in order to find the new bearing. Because due south is parallel to due north, we know that the angle at the first buoy is 70°. Therefore, we use the law of sines:

$$\frac{\sin \theta}{1.6} = \frac{\sin 70°}{1.4}$$

$$\sin \theta = \frac{(1.6)(\sin 70°)}{1.4}$$

$$\sin \theta = \frac{(1.6)(.9397)}{1.4}$$

$$\sin \theta = 1.074.$$

Since $\sin \theta$ is greater than one, there is no angle θ. Therefore, the problem has no solution. We must assume the captain of the ship has read the charts or the instruments incorrectly.

Exercise 7.4

1. A lighthouse is 4.0 nautical miles due east of a harbor. A ship is due south of the lighthouse. If the bearing of the ship from the harbor is S 72°E, how far is the ship from the lighthouse?

2. A boat starts from 1.25 nautical miles due south of a buoy and sails on a bearing of N 35°30′W until it is due west of the buoy. How far does the boat sail?

3. A boat is due south of a lighthouse, and a harbor is due west of the lighthouse. The lighthouse is 2.2 nautical miles from the harbor, and the boat is .25 nautical miles from the lighthouse. What bearing should the boat take to reach the harbor?

4. A boat sails from a buoy in a southwesterly direction for 355 meters. It is then due south of another buoy. The second buoy is due west of the first buoy and the buoys are 100 meters apart. On what bearing did the boat sail?

5. A ship is due north of a harbor. It is to sail 5.2 nautical miles on a bearing of S 43°50′E to an island, and then sail 3.4 nautical miles to the harbor. What is the bearing from the harbor to the island?

6. A buoy is 4.5 nautical miles due north of a harbor. A ship is .5 nautical miles from the buoy, and the bearing from the buoy to the ship is S 35°E. How far is the ship from the harbor?

7. A ship is located 3 nautical miles form a lighthouse, and the bearing from the lighthouse to the ship is S 12°40′W. A harbor is 15 nautical miles from the lighthouse, on a bearing of S 82°50′E from the lighthouse. How far is the ship from the harbor?

8. Since north and south are parallel, the bearing from the ship to the lighthouse in problem 7 is N 12°40′E. What is the bearing from the ship to the harbor?

9. Two ships leave a harbor at the same time. One sails due east at 9.2 knots and the other sails on a bearing of S 45°E at 8.8 knots. How far apart are the ships at the end of an hour?

10. A ship leaves a harbor on a bearing of N 27°10′E at 14 knots. Half an hour later, another ship leaves the same harbor on a bearing of S 52°20′E at 18 knots. How far apart are the ships an hour after the first ship leaves?

Self-test

1. A ship is 1.4 nautical miles from a lighthouse. The bearing from the lighthouse to the ship is S 18°10′W. There is a harbor 12 nautical miles due east of the lighthouse. How far is the ship from the harbor?

    1.＿＿＿＿＿＿＿＿＿＿

2. To find the distance between two trees, one on each side of a river, a distance of 100 meters is measured from the first tree at an angle of 87°30′ to the line of the trees and marked. The angle between the line from the marker to the first tree and the line from the marker to the second tree is found to be 64°20′. What is the distance between the trees?

    2.＿＿＿＿＿＿＿＿＿＿

3. A hike is planned 4 miles north from a town to a picnic ground, and then 3 miles northeast to a campsite. The angle between the line from the town to the picnic ground and the line from the town to the campsite is 52°. What is the angle between the line from the campsite to the picnic ground and the line from the campsite to the town?

    3.＿＿＿＿＿＿＿＿＿＿

4. An isosceles triangle has 8-inch sides and base angles 64°10′. What is its height?

    4.＿＿＿＿＿＿＿＿＿＿

5. Two forces of 6.4 pounds and 10.2 pounds have a resultant of 14.5 pounds. What is the angle between the two forces?

    5.＿＿＿＿＿＿＿＿＿＿

Unit 8

Radian Measure

INTRODUCTION

In the preceding units you have learned about trigonometric ratios of angles given in degree measure. For many applications of trigonometry in mathematics and science, however, the trigonometric ratios are interpreted as functions of a variable which may not be an angle. For example, the swing of a pendulum and the voltage of alternating current both are described by the sine interpreted as a function where the variable is time. But you cannot measure a quantity such as time in degrees. In this unit you will learn about a measure called a radian, which relates angular measurement to linear measurement. You will learn how to measure angles using radians, how to convert angular measurement between degrees and radians, and use radians in applications.

OBJECTIVES

When you have finished this unit you should be able to:

1. Draw an approximate diagram of an angle in standard position given in radians.
2. Convert the measure of an angle from degrees to radians and from radians to degrees.
3. Solve problems involving arclength or angular velocity.
4. Find the value of a trigonometric ratio of an angle given in radians.

Section 8.1

Definition of the Radian

Recall from geometry that the circumference of a circle is given by the formula

$$C = 2\pi r.$$

That is, if we were to cut the circle and straighten it into a line segment, the length of the line segment would be $2\pi r$:

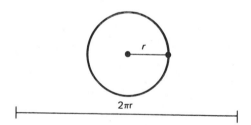

128

The letter r stands for the radius of the circle and π (pi) is a constant.

You have probably seen π approximated by the fraction $\frac{22}{7}$ or the decimal 3.14. You should remember, however, that π represents a nonterminating decimal and is not exactly equal to either of these numbers. It is a common error to think that π can be approximated by dividing out the fraction $\frac{22}{7}$. A calculator with a π key tells us that, to seven decimal places,

$$\frac{22}{7} \approx 3.1428571$$

and

$$\pi \approx 3.1415927,$$

so actually only the first few places are the same. Using sophisticated methods, and with the aid of computers, π has been approximated to thousands of places. Throughout this unit, we will use the approximation

$$\pi \approx 3.1416.$$

(Since this approximation is well beyond three significant digits, you should get the same answers if you use the π key on a calculator.)

Now, we return to the circle we have "unwrapped" into a line segment. Since

$$\pi \approx 3.1416,$$

we have

$$2\pi \approx 6.2832.$$

Therefore, the length of the line segment is

$$2\pi r \approx 6.2832r,$$

or a little more than six times the radius:

We "wrap" the line segment back around the circle:

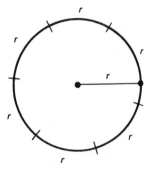

Now, we may state the following definition.

Definition: An angle with measure one **radian** is an angle which is rotated around the circumference an amount equal to the radius.

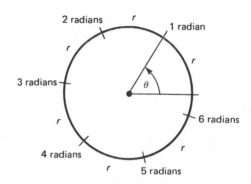

Angle θ has a measure of one radian. Observe that one radian is considerably larger than one degree. There are 360° in a circle, but there are only a little more than six radians in a circle.

To get an idea of the approximate sizes of angles measured in radians, we will estimate the diagrams of some angles in standard position.

EXAMPLE 8.1. Draw an approximate diagram of an angle in standard position with measure 1 radian.

Solution. From our "wrapped" circle, we see that 1 radian is approximately two-thirds of the way through the first quadrant. Therefore, an approximate diagram is

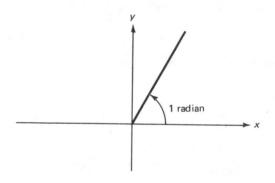

EXAMPLE 8.2. Draw an approximate diagram of an angle in standard position with measure 3 radians.

Solution. We see that 3 radians is not quite one-half of the way around the circle. Therefore, an approximate diagram is

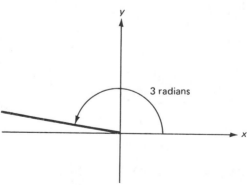

EXAMPLE 8.3. Draw an approximate diagram of an angle in standard position with measure 4.5 radians.

Solution. From the locations of 4 radians and 5 radians, we see that 4.5 radians is not quite three-quarters of the way around the circle. An approximate diagram is

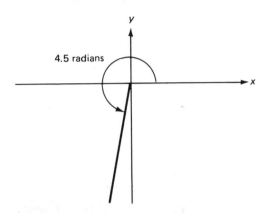

If we go beyond one full rotation, we must be very careful. It is a common error to start counting again at the x-axis, and so to draw an angle of 7 radians as if it were coterminal with an angle of 1 radian. Actually, we must start at 6 radians and measure one more length of the radius, so the terminal side of an angle of 7 radians is slightly before the terminal side of an angle of 1 radian.

EXAMPLE 8.4. Draw an approximate diagram of an angle in standard position with measure 7 radians.

Solution. We start from an approximation of 6 radians and measure one more length of the radius:

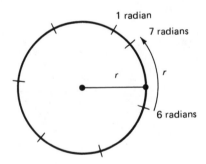

An approximate diagram is

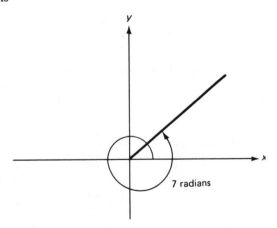

Exercise 8.1

Draw an approximate diagram of an angle in standard position with measure:

1. 2 radians
2. 4 radians
3. 5 radians
4. 6 radians

5. .5 radian
6. 1.5 radians
7. 3.5 radians
8. 5.5 radians

9. 8 radians
10. 9 radians
11. 12 radians
12. 13 radians

1 radian ≈ 57°

Section 8.2	Radian and Degree Measure

Now that we have an idea of the size of a radian, we need to know how radians and degrees are related, and how to convert from degree measure to radian measure and from radian measure to degree measure. We recall the unit circle from Section 3.2:

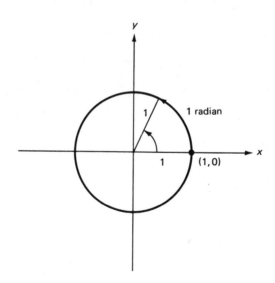

Since the circumference of a circle with radius r is

$$C = 2\pi r,$$

the circumference of the unit circle is

$$C = 2\pi(1) = 2\pi.$$

One rotation about the unit circle is a rotation of 2π and also a rotation of $360°$:

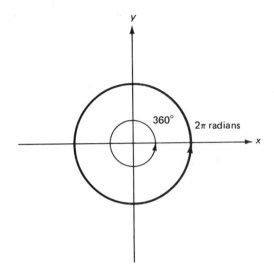

Therefore,

$$2\pi \text{ radians} = 360°$$

or

$$1 \text{ radian} = \frac{360°}{2\pi} \approx 57.3°.$$

Because of the first equation above, we often write radian measure in terms of π. Since $\pi \approx 3.1416$, π radians are just a little more than 3 radians. Also, dividing the first equation by 2,

$$\pi \text{ radians} = 180°.$$

Dividing by 2 again,

$$\frac{\pi}{2} \text{ radians} = 90°.$$

If π radians is a little more than 3 radians, clearly $\frac{\pi}{2}$ radians is a little more than 1.5 radians. Finally, $\frac{\pi}{2}$ radians added to π radians gives

$$\frac{3\pi}{2} \text{ radians} = \left(\pi + \frac{\pi}{2}\right) \text{ radians} = 180° + 90° = 270°,$$

and $\frac{3\pi}{2}$ radians is a little more than $3 + 1.5 = 4.5$ radians. Clearly,

$$0 \text{ radians} = 0°,$$

Thus we have all the quadrantal angles in radian measure:

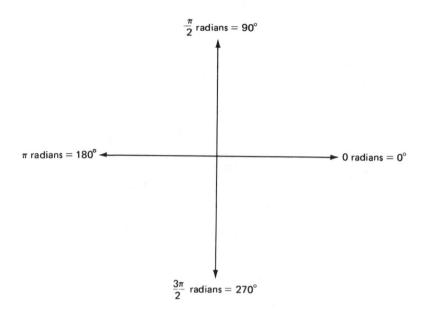

Now, we take one-half of $\frac{\pi}{2}$ radians:

$$\frac{1}{2}\left(\frac{\pi}{2}\right) = \frac{\pi}{4},$$

and so

$$\frac{\pi}{4} \text{ radians } = 45°.$$

Counting around one full rotation in fourths of π, we have the diagram:

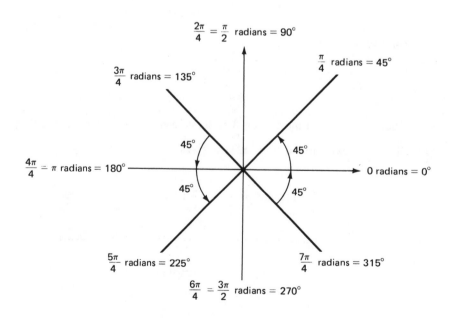

We see that an angle whose radian measure in terms of π has a denominator of 4, has a reference angle of 45°.

If we take one-third of $\frac{\pi}{2}$ radians,

$$\frac{1}{3}\left(\frac{\pi}{2}\right) = \frac{\pi}{6},$$

and so

$$\frac{\pi}{6} \text{ radians } = 30°.$$

Counting around one full rotation in sixths of π we have denominators of 6 and 3:

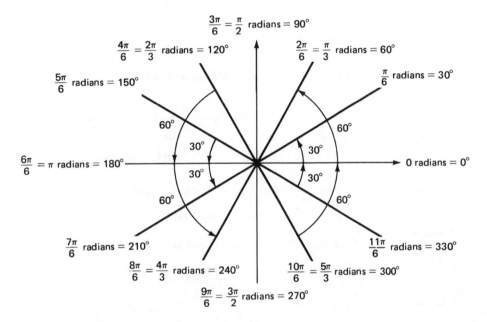

An angle whose radian measure in terms of π has a denominator of 6, has a reference angle of 30°. An angle whose radian measure in terms of π has a denominator of 3, has a reference angle of 60°.

In particular, you should remember the quadrantal and special angles:

$$0 \text{ radians } = 0°$$

$$\pi \text{ radians } = 180°$$

$$\frac{\pi}{2} \text{ radians } = 90°$$

$$\frac{\pi}{4} \text{ radians } = 45°$$

$$\frac{\pi}{6} \text{ radians } = 30°$$

$$\frac{\pi}{3} \text{ radians } = 60°.$$

From these, we can convert the measure of any angle related to a special or quadrantal angle.

EXAMPLE 8.5. Write 120° in radian measure in terms of π.

Solution. We may observe that 120° has a reference angle of 60°, and therefore the denominator is 3. Using the reference angle, we write

$$120° = 180° - 60°$$

$$= \pi \text{ radians} - \frac{\pi}{3} \text{ radians}$$

$$= \frac{3\pi}{3} \text{ radians} - \frac{\pi}{3} \text{ radians}$$

$$= \frac{2\pi}{3} \text{ radians}.$$

EXAMPLE 8.6. Write $\frac{5\pi}{4}$ radians in degree measure.

Solution. Since the denominator is 4, the reference angle is 45°. We may write

$$\frac{5\pi}{4} \text{ radians} = \frac{4\pi}{4} \text{ radians} + \frac{\pi}{4} \text{ radians}$$

$$= \pi \text{ radians} + \frac{\pi}{4} \text{ radians}$$

$$= 180° + 45°$$

$$= 225°.$$

To convert the measures of angles which are not related to special or quadrantal angles, we begin with

$$\pi \text{ radians} = 180°.$$

Then,

$$\frac{\pi \text{ radians}}{180°} = 1$$

and

$$\frac{180°}{\pi \text{ radians}} = 1.$$

Therefore, we can multiply by either of these fractions without changing the value of the measure. To convert from degree measure to radian measure, we must introduce the radians and divide out the degrees, so we multiply by

$$\frac{\pi \text{ radians}}{180°}.$$

EXAMPLE 8.7. Write 70° in radian measure in terms of π.

Solution. We multiply by

$$\frac{\pi \text{ radians}}{180°},$$

so we calculate

$$70° = 70°\left(\frac{\pi \text{ radians}}{180°}\right)$$

$$= \frac{70\pi}{180} \text{ radians}$$

$$= \frac{7\pi}{18} \text{ radians.}$$

EXAMPLE 8.8.   Write 213° in radian measure to two decimal places.

Solution.

$$213° = 213°\left(\frac{\pi \text{ radians}}{180°}\right)$$

$$= \frac{213\pi}{180} \text{ radians.}$$

To write this expression as a decimal, we use $\pi \approx 3.1416$:

$$\frac{213\pi}{180} \text{ radians} = \frac{213(3.1416)}{180} \text{ radians}$$

$$= 3.72 \text{ radians}$$

to two decimal places.

EXAMPLE 8.9.   Write 36°50′ in radian measure to two decimal places.

Solution.   We must write 36°50′ as a decimal. The method is exactly the same as the method for converting from minutes to a decimal on your scientific calculator. Since there are 60′ in a degree,

$$50' = \left(\frac{50}{60}\right)°$$

$$= .833°.$$

Therefore,

$$36°50' = 36° + .833°$$

$$= 36.833°.$$

Now, we write:

$$36°50' = 36.833°$$

$$= 36.833°\left(\frac{\pi \text{ radians}}{180°}\right)$$

$$= \frac{36.833\pi}{180} \text{ radians}$$

$$= \frac{(36.833)(3.1416)}{180} \text{ radians}$$

$$= .64 \text{ radians.}$$

To convert from radian measure to degree measure, we must introduce degrees and divide out radians. Therefore, we multiply by

$$\frac{180°}{\pi \text{ radians}}.$$

EXAMPLE 8.10.   Write $\frac{2\pi}{9}$ radians in degree measure.

Solution.   We multiply by

$$\frac{180°}{\pi \text{ radians}},$$

so we have

$$\frac{2\pi}{9} \text{ radians} = \frac{2\pi}{9} \text{ radians}\left(\frac{180°}{\pi \text{ radians}}\right)$$

$$= \frac{2\pi(180)°}{9\pi}$$

$$= 40°.$$

EXAMPLE 8.11.   Write 3.6 radians in degree measure to the nearest degree.

Solution.

$$3.6 \text{ radians} = 3.6 \text{ radians}\left(\frac{180°}{\pi \text{ radians}}\right)$$

$$= \frac{(3.6)(180)°}{\pi}$$

$$= \frac{(3.6)(180)°}{3.1416}$$

$$= 206°.$$

EXAMPLE 8.12.   Write .459 radians in degree measure to the nearest degrees and tens of minutes.

Solution.

$$.459 \text{ radians} = .459 \text{ radians}\left(\frac{180°}{\pi \text{ radians}}\right)$$

$$= \frac{(.459)(180)°}{\pi}$$

$$= \frac{(.459)(180)°}{3.1416}$$

$$= 26.299°.$$

Our answer is written as a decimal. To convert the decimal part to minutes, the method is again the same as the method you use on your scientific calculator. If m is the number of minutes, then

$$\frac{m}{60} = .299$$

$$m = 60(.299)$$

$$m = 17.94$$

$$m = 20'$$

to the nearest tens of minutes. Therefore, the solution is 26°20′ to the nearest degrees and tens of minutes.

The instruction book for your scientific calculator may give an interesting method for converting angle measurement from degrees to radians. You enter the angle in degrees, converting minutes to decimal form. Then, in degree mode, press "sin" (or any other ratio name). Now, put the calculator into radian mode by pressing the "drg" key. Then press "inv" and "sin" (or whatever ratio name you used before). The angle comes back, but now it is given in radians. Similarly, to convert from radians to degrees, start with the angle in radian mode, press "sin," then put the calculator into degree mode and press "inv" and "sin." The angle comes back in degrees in decimal form, which you can convert to minutes. Although these methods are clever, and very fast when you have gotten used to them, you will probably find it easier simply to multiply by π and divide by 180 or vice versa.

Exercise 8.2

Write in radian measure in terms of π:

1. 150°
2. 240°
3. 315°
4. 330°

5. 50°
6. 130°
7. 220°
8. 340°

Write in radian measure to two decimal places:

9. 54°
10. 122°
11. 259°
12. 311°

13. 27°10′
14. 88°50′
15. 63°20′
16. 42°40′

17. 119°50′
18. 246°40′
19. 312°30′
20. 188°30′

Write in degree measure to the nearest degree:

21. $\frac{3\pi}{4}$ radians
22. $\frac{7\pi}{6}$ radians
23. $\frac{5\pi}{3}$ radians
24. $\frac{3\pi}{2}$ radians

25. $\frac{4\pi}{9}$ radians
26. $\frac{10\pi}{9}$ radians
27. $\frac{3\pi}{10}$ radians
28. $\frac{17\pi}{18}$ radians

29. 1.3 radians
30. 3.4 radians
31. 4.9 radians
32. 2.2 radians

Write in degree measure to the nearest degrees and tens of minutes:

33. .669 radians
34. .905 radians
35. 1.22 radians
36. 1.41 radians

37. 1.98 radians
38. 1.56 radians
39. 2.13 radians
40. 1.78 radians

Section
8.3

Arclength and Angular Velocity

Suppose we have a circle with radius r, and an angle θ with its vertex at the center of the circle. Then we say that θ **subtends** an arc of length s:

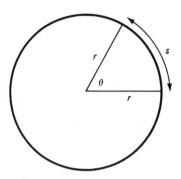

The ratio of the angle θ to one full rotation is equal to the ratio of the arclength s to the circumference of the circle. That is,

$$\frac{\theta}{2\pi} = \frac{s}{2\pi r}$$

$$\theta = \frac{s}{r}$$

or

$$s = r\theta.$$

Observe that, since we have written one full rotation as 2π, we must measure θ in radians.

EXAMPLE 8.13.   Find the arclength subtended by an angle of 20° on a circle of radius 12 inches.

Solution.   To use the formula

$$s = r\theta,$$

we must write θ in radian measure:

$$20°\left(\frac{\pi}{180}\right) = \frac{\pi}{9} \text{ radians.}$$

Then,

$$s = r\theta$$

$$= 12\left(\frac{\pi}{9}\right)$$

$$= \frac{4\pi}{3}$$

$$= 4.19.$$

The arclength is 4.19 inches to three significant digits.

EXAMPLE 8.14.   A pendulum 55 centimeters long swings through an angle of 12°40′. How far does the tip of the pendulum travel?

Solution.   The tip of the pendulum describes an arclength subtended by a 12°40′ angle in a circle with radius 55 centimeters:

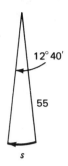

Converting to radians,

$$12°40' \left(\frac{\pi}{180} \right) = \frac{(12.667)(3.1416)}{180}$$

$$= .2211,$$

and

$$s = r\theta$$

$$= 55(.2211)$$

$$= 12.2.$$

The tip of the pendulum travels 12.2 centimeters.

The rate at which an angle rotates is called **angular velocity**. We use the Greek letter ω (omega) to denote angular velocity. The angular velocity ω is defined by the formula

$$\omega = \frac{\theta}{t}$$

where θ is measured in radians and t is a unit of time. Thus we may speak of radians per second, radians per minute, and so on. However, the motion of an object on the circumference of a circle is a **linear velocity**, denoted v, and defined by the formula

$$v = \frac{s}{t}.$$

Recall that an arc of a circle can be straightened out and thought of as a line segment, so this is merely the familiar formula "rate equals distance divided by time." Linear velocity is measured in feet per second, miles per hour, and so on.

Starting with the arclength formula

$$s = r\theta,$$

we divide by t:

$$\frac{s}{t} = \frac{r\theta}{t}$$

$$\frac{s}{t} = r\frac{\theta}{t}$$

$$v = r\omega.$$

Remember that ω must be in radians per unit time, not degrees, since θ was in radians and not degrees.

EXAMPLE 8.15.   An object rotates about a point at a distance of 2 feet and an angular velocity of 540° per second. What is its linear velocity?

Solution.   Since

$$540° = 3\pi,$$

$\omega = 3\pi$ radians per second. Therefore,

$$v = r\omega$$
$$v = 2(3\pi)$$
$$= 6\pi$$
$$= 18.8.$$

The linear velocity is 18.8 feet per second.

EXAMPLE 8.16.   A car is traveling at 90 kilometers per hour. If its tires have a radius of 37.5 centimeters, how many revolutions per second do the tires make? (90 km/hr = 2500 cm/sec.)

Solution.   In this example, we know v and r, and must find ω. Since

$$v = r\omega,$$

we have

$$\omega = \frac{v}{r}$$

$$= \frac{2500}{37.5}$$

$$= 66.67.$$

The angular velocity is 66.67 radians per second. But each revolution is 2π radians. Therefore, we divide by 2π:

$$\frac{66.67}{2\pi} = 10.6,$$

and so the tires make 10.6 revolutions per second.

Exercise 8.3

1. Find the arclength subtended by an angle of 54°40′ in a circle of radius 3 feet.

2. Find the arclength subtended by an angle of 100° in a circle of radius 25 centimeters.

3. A pendulum 30 inches long swings through an angle of 6°20′. How far does the tip of the pendulum travel?

4. The tip of a pendulum 48 centimeters long travels through an arclength of 8 centimeters. What angle does it travel through in degrees, to the nearest degrees and tens of minutes?

5. The longitudinal lines on the earth are 10° apart. If the radius of the earth is about 4000 miles, how many miles does the arclength between two longitudes represent?

6. It is about 4400 miles across the continental United States. How many degrees of longitude does the country span?

7. An object rotates about a point at a distance of 10 meters and an angular velocity of 10° per second. What is its linear velocity?

8. An object rotates about a point at a distance of 64 inches and an angular velocity of 90° per second. What is its linear velocity?

9. A 24-inch pendulum swings through an angle of 3°30′ in one second. How fast does its tip move?

10. A pendulum 60 centimeters long swings through an angle of 10° in 2 seconds. How fast does its tip move?

11. The second hand moves around a clock at a rate of one revolution per minute. If the second hand is 4 inches long, how fast does its tip move?

12. If the minute hand on a clock is 6 inches long, what is its linear velocity in inches per minute?

13. A car is traveling at 30 miles per hour. If the radius of the tires is 15 inches, how many revolutions per second do they make? (30 m/h = 44 ft/sec = 528 in/sec.)

14. A bicycle is traveling at 15 miles per hour. If the radius of the tires is 20 inches, how many revolutions per second do they make?

15. The radius of the earth is approximately 4000 miles. Since the earth rotates at a rate of one revolution in 24 hours, how fast are you moving if you are on a point on the equator?

16. In Boston, you are about 2950 miles from the axis of rotation of the earth. How fast are you going in miles per hour?

Section
8.4

Ratios of Angles in Radians

To find values of the trigonometric ratios for angles given in radians, we use coterminal and reference angles as in Unit 4. It is usual to write an angle given in radians without indicating the unit of measure. Thus $\frac{2\pi}{3}$ means an angle with measure $\frac{2\pi}{3}$ radians.

EXAMPLE 8.17.   Find the value of $\cos \dfrac{2\pi}{3}$.

Solution.   Recall from Section 8.2 that $\dfrac{2\pi}{3}$ is in the second quadrant:

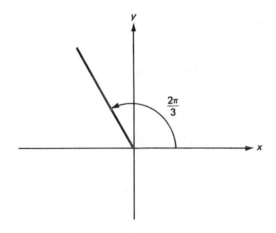

Since the denominator is 3, the reference angle is 60°. Recalling the 30–60–90 special triangle from Section 1.3,

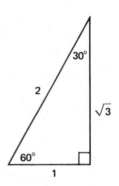

$$\cos 60° = \frac{1}{2}.$$

Since the cosine is negative in the second quadrant,

$$\cos \frac{2\pi}{3} = -\frac{1}{2}.$$

EXAMPLE 8.18.   Find the value of $\tan \dfrac{3\pi}{2}$.

Solution. $\dfrac{3\pi}{2}$ is a quadrantal angle, so we use a unit point from Section 3.4:

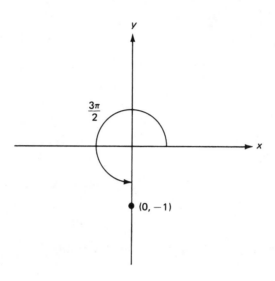

$$\tan \frac{3\pi}{2} = \frac{-1}{0}.$$

Therefore, $\tan \dfrac{3\pi}{2}$ is undefined.

EXAMPLE 8.19.   Find the value of $\sin \dfrac{11\pi}{4}$.

Solution.   We subtract 2π to find the least positive coterminal angle:

$$\frac{11\pi}{4} - 2\pi = \frac{11\pi}{4} - \frac{8\pi}{4}$$

$$= \frac{3\pi}{4}.$$

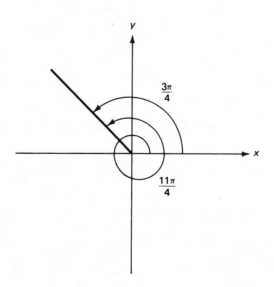

The angle is in the second quadrant and the reference angle is 45°. Using the 45–45–90 special triangle,

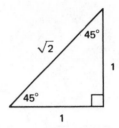

$$\sin 45° = \frac{1}{\sqrt{2}}.$$

Since the sine is positive in the second quadrant,

$$\sin \frac{11\pi}{4} = \frac{1}{\sqrt{2}}.$$

EXAMPLE 8.20.   Find the value of $\tan\left(-\frac{5\pi}{6}\right)$.

Solution.   Recall that a negative angle has a clockwise rotation. Therefore, $-\frac{5\pi}{6}$ is in the third quadrant:

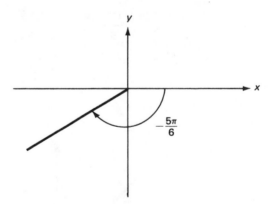

The reference angle is 30°, and using the 30–60–90 special triangle,

$$\tan 30° = \frac{1}{\sqrt{3}}.$$

The tangent is positive in the third quadrant, so

$$\tan\left(-\frac{5\pi}{6}\right) = \frac{1}{\sqrt{3}}.$$

EXAMPLE 8.21.   Find the value of cos 10.

Solution.   We must be careful not to confuse cos 10 with cos 10°. Since 10 has no unit of measure indicated, we have 10 radians, which is more than one full rotation. Indeed, 10 radians is somewhat more than 3π, so the angle is in the third quadrant:

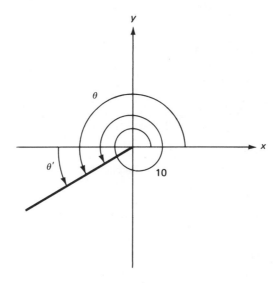

The least positive coterminal angle θ is

$$10 - 2\pi = 10 - 6.2832$$

$$= 3.7168,$$

and the reference angle θ' is

$$3.7168 - \pi = 3.7168 - 3.1416$$

$$= .5752.$$

The nearest value to .5752 in the *radians* column of Table II is

$$\cos .5760 = .8387.$$

Therefore, since the cosine is negative in the third quadrant,

$$\cos 10 = -.8387.$$

Using a scientific calculator, you can get more accurate values of the trigonometric ratios of angles given in radians because you do not have the approximation involved in using the nearest table value. Put the calculator into radian mode by pressing the "drg" key. Then, enter the angle in radians in decimal form, and press "sin," "cos," or "tan," as before. You can use "1/x" as before to find values of the reciprocals.

Exercise 8.4

Use a special triangle to find the value:

1. $\sin \dfrac{\pi}{6}$

2. $\tan \dfrac{\pi}{4}$

3. $\cos \dfrac{5\pi}{3}$

4. $\sin \dfrac{7\pi}{6}$

5. $\sec \dfrac{5\pi}{4}$

6. $\cos \dfrac{7\pi}{4}$

7. $\cot \dfrac{4\pi}{3}$

8. $\csc \dfrac{11\pi}{6}$

9. $\cos \dfrac{8\pi}{3}$

10. $\tan \dfrac{17\pi}{6}$

11. $\sin\left(-\dfrac{4\pi}{3}\right)$

12. $\cot\left(-\dfrac{5\pi}{4}\right)$

Use a unit point to find the value:

13. $\cos \pi$

14. $\sin\left(-\dfrac{3\pi}{2}\right)$

15. $\cot \dfrac{5\pi}{2}$

16. $\csc 5\pi$

Use Table II to find the value:

17. $\cos 12$

18. $\sin 30$

Self-test

1. Draw an approximate diagram of an angle in standard position with measure 2.5 radians.

2. Write 1.69 radians in degree measure to the nearest degrees and tens of minutes.

    2._____

3. Write 154° in radian measure to two decimal places.

    3._____

4. Use a special triangle to find the value of $\tan \dfrac{5\pi}{6}$.

    4._____

5. There is a fly on the end of a blade of a windmill. If the blade is 5 feet long, and the windmill turns at a rate of 30° a second, how fast is the fly moving?

    5._____

The Trigonometric Functions and Their Graphs

INTRODUCTION

In Unit 8 you learned how angles can be measured in terms of radian measure, and how you can apply radian measure to a quantity such as time. In this unit you will apply radian measure to the trigonometric ratios and learn to interpret the trigonometric ratios as functions. In algebra, you saw that it is useful to be able to draw the graph of a function. In this unit you will learn how to draw graphs of the trigonometric functions. With the introduction of the trigonometric functions, you will begin the study of the area of trigonometry called "analytic trigonometry."

OBJECTIVES

When you have finished this unit you should be able to:

1. Draw the graph and state the domain and range of $y = \sin x$ and $y = \cos x$.
2. Draw the graph and state the domain and range of $y = \tan x$ and $y = \cot x$.
3. On one set of axes, draw the graphs of $y = \cos x$ and $y = \sec x$, or $y = \sin x$ and $y = \csc x$, and state the domain and range of $y = \sec x$ and $y = \csc x$.

Section 9.1

The Functions $y = \sin x$ and $y = \cos x$

Let us consider the equation $y = \sin x$ as a relation between x and y. You should observe carefully the differences between this relation in x and y and our previous use of x and y. In Units 3 and 4 particularly, we used x and y as coordinates of a point (x, y) in the trigonometric ratios for an angle θ. However, when we write $y = \sin x$, x is the **angle** and y is the **value of the sine of the angle**. The numbers x and y are still coordinates of a point, but x is an angle and y is a value of a trigonometric function. In order to have the coordinates x and y both measured in a real number measure, we will always measure the angle x in radians when we discuss the relation $y = \sin x$.

To help in interpreting the relation $y = \sin x$, we use the unit circle from Section 3.2. Recall that, on the unit circle,

$$\sin \theta = y.$$

Therefore, $\sin \theta$ is simply the value of the y-coordinate:

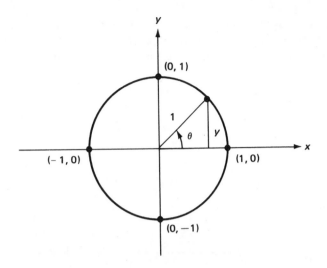

We observe that, as θ goes from 0 to $\dfrac{\pi}{2}$, y increases from 0 to 1. Then, as θ goes from $\dfrac{\pi}{2}$ to π, y decreases from 1 to 0; as θ goes from π to $\dfrac{3\pi}{2}$, y decreases through negative values from 0 to -1; and as θ goes from $\dfrac{3\pi}{2}$ to 2π, y increases from -1 back to 0.

To interpret these values as a relation $y = \sin x$, we use x in place of θ and y as the value of the sine. We may find as many other values as we choose using the special triangles and Table II. In the following chart, we have listed values for all the special and quadrantal angles from 0 to 2π:

x	0	$\dfrac{\pi}{6}$	$\dfrac{\pi}{4}$	$\dfrac{\pi}{3}$	$\dfrac{\pi}{2}$	$\dfrac{2\pi}{3}$	$\dfrac{3\pi}{4}$	$\dfrac{5\pi}{6}$	π	$\dfrac{7\pi}{6}$	$\dfrac{5\pi}{4}$	$\dfrac{4\pi}{3}$	$\dfrac{3\pi}{2}$	$\dfrac{5\pi}{3}$	$\dfrac{7\pi}{4}$	$\dfrac{11\pi}{6}$	2π
y	0	.5	.7	.9	1	.9	.7	.5	0	$-.5$	$-.7$	$-.9$	-1	$-.9$	$-.7$	$-.5$	0

We have approximated the values to just one decimal place because we cannot plot a point in the Cartesian coordinate system to greater accuracy than one-tenth of a unit. Thus,

$$\sin \frac{\pi}{6} = \frac{1}{2} = .5,$$

$$\sin \frac{\pi}{4} = \frac{1}{\sqrt{2}} \approx .7,$$

$$\sin \frac{\pi}{3} = \frac{\sqrt{3}}{2} \approx .9,$$

and so on.

A Note on Scale

You should draw graphs using an accurate scale. Recall that $\pi \approx 3.1416$ or, for graphing purposes, π is very approximately equal to 3. To draw a graph, first place π on the x-axis, and then place 1 on the y-axis, where 1 is about one-third of π. Then π is approximately 3:

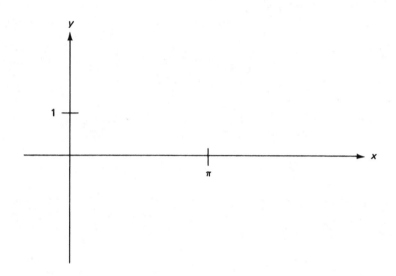

Now you can place $\dfrac{\pi}{2}$, $\dfrac{3\pi}{2}$, and 2π on the x-axis, and -1 on the y-axis:

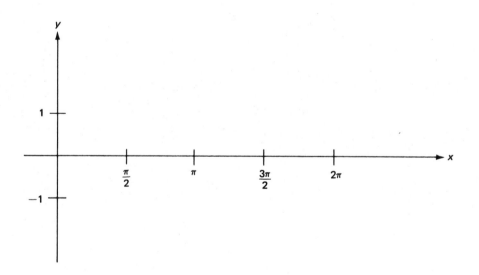

Then, $\dfrac{\pi}{4}$ is $\dfrac{1}{2}\left(\dfrac{\pi}{2}\right)$, so $\dfrac{\pi}{4}$ is halfway between 0 and $\dfrac{\pi}{2}$. Now, you can place $\dfrac{\pi}{4}$, $\dfrac{3\pi}{4}$, $\dfrac{5\pi}{4}$, and $\dfrac{7\pi}{4}$ on the x-axis. Finally, divide the space between 0 and $\dfrac{\pi}{2}$ into thirds. The first third is $\dfrac{\pi}{6}$, which is $\frac{1}{3}\left(\frac{\pi}{2}\right)$, and the second third is $\frac{\pi}{3}$, which is $\frac{2}{3}\left(\frac{\pi}{2}\right)$:

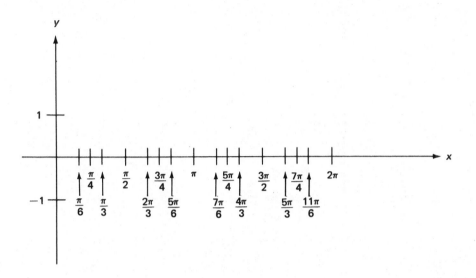

Be careful not to put $\frac{\pi}{6}$ halfway between 0 and $\frac{\pi}{4}$. Observe that $\frac{1}{2}\left(\frac{\pi}{4}\right)$ is $\frac{\pi}{8}$, not $\frac{\pi}{6}$. Similarly, $\frac{\pi}{3}$ is not halfway between $\frac{\pi}{4}$ and $\frac{\pi}{2}$. Be sure to divide the interval between 0 and $\frac{\pi}{2}$ into thirds to place $\frac{\pi}{6}$ and $\frac{\pi}{3}$.

We return to our chart of values for $y = \sin x$, and plot the y-values for the corresponding x-values:

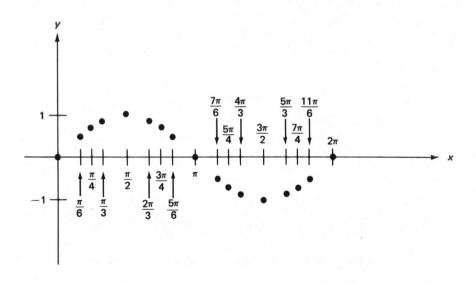

Connecting the points in a smooth curve, we have the graph of $y = \sin x$ from $x = 0$ to $x = 2\pi$:

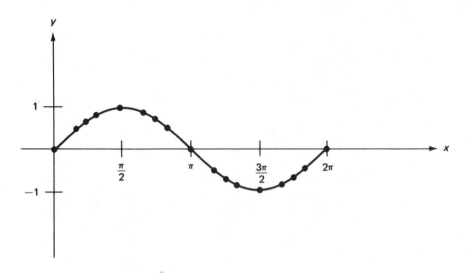

This graph represents one **cycle** of the graph of $y = \sin x$. Once we know the shape of the graph, we need only plot the points for the quadrantal angles 0, $\dfrac{\pi}{2}$, π, $\dfrac{3\pi}{2}$, and also 2π, to draw one cycle of the graph of $y = \sin x$.

From $x = 2\pi$ to $x = 4\pi$, the values of y repeat their values on the unit circle and we have a second cycle identical to the first cycle:

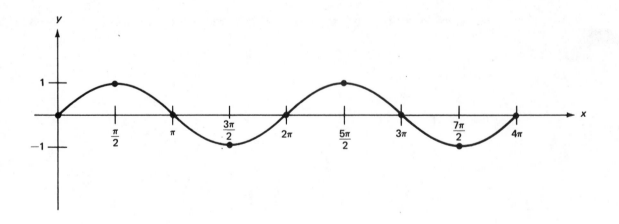

The graph continues to repeat in cycles to 6π, 8π, and so on.

EXAMPLE 9.1.   Draw three cycles of the graph of $y = \sin x$ starting from $x = 0$.

Solution.   Starting from $x = 0$, the first cycle ends at $x = 2\pi$, the second at $x = 4\pi$, and the third at $x = 6\pi$. Therefore, the graph of the three cycles is

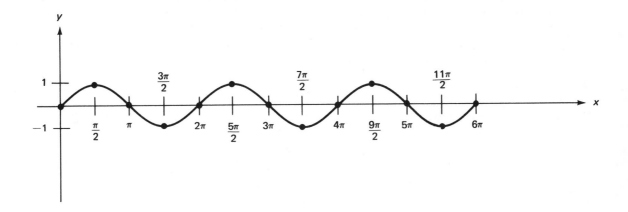

To the left of the origin, x takes on values of negative angles. From -2π to 0, we have a cycle identical to the cycle from 0 to 2π:

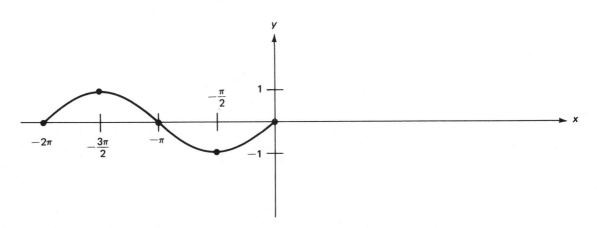

EXAMPLE 9.2.   Draw two cycles of $y = \sin x$ starting from $x = -2\pi$.

Solution.   We have one cycle from $x = -2\pi$ to $x = 0$, and the second from $x = 0$ to $x = 2\pi$:

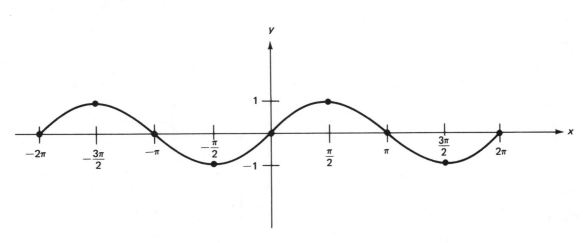

Recall from algebra that a **function** has exactly one value of y for each value of x. The relation $y = \sin x$ has this property. Therefore, we call $y = \sin x$ a **trigonometric function**.

Also, recall that the **domain** of a function is the set of all x-values for which a y-value is defined. Since the cycles of $y = \sin x$ continue indefinitely to both the right and left, and for each x-value there is a y-value, the domain of $y = \sin x$ is the set of all real numbers.

Finally, recall that the **range** of a function is the set of all y-values. Observe that y is never more than 1 nor less than -1. Therefore, the range of $y = \sin x$ is the set of all y-values from -1 to 1. We write this set as

$$-1 \leqslant y \leqslant 1.$$

We make one more important observation about the function $y = \sin x$. Observe that, for example,

$$\sin \frac{\pi}{6} = \frac{1}{2} \text{ and } \sin\left(-\frac{\pi}{6}\right) = -\frac{1}{2},$$

$$\sin \frac{\pi}{4} = \frac{1}{\sqrt{2}} \text{ and } \sin\left(-\frac{\pi}{4}\right) = -\frac{1}{\sqrt{2}},$$

$$\sin \frac{\pi}{3} = \frac{\sqrt{3}}{2} \text{ and } \sin\left(-\frac{\pi}{3}\right) = -\frac{\sqrt{3}}{2},$$

$$\sin \frac{\pi}{2} = 1 \text{ and } \sin\left(-\frac{\pi}{2}\right) = -1,$$

and so on. In general,

$$\sin(-x) = -\sin x$$

for all values of x. Functions with this property are called **odd** functions [because functions $f(x) = x^n$, where n is an odd integer, have this property].

In Section 1.2, we said that the name "sine" comes either from a Latin word meaning "fold in a garment," or from a connection with a chord of a circle. The cycles of the graph resemble folds in a piece of draped material. A chord of a circle is a line segment with its endpoints on the circle. The line on the unit circle which represents y is half of a chord.

The trigonometric function $y = \cos x$ is similar to $y = \sin x$. However, returning to the unit circle, we recall that $\cos \theta = x$:

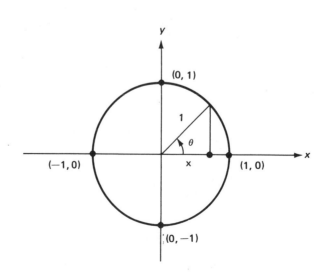

You should be careful not to confuse x in this interpretation with x as the angle in $y = \cos x$.

As θ goes from 0 to $\dfrac{\pi}{2}$, the value of x on the unit circle decreases from 1 to 0. From $\dfrac{\pi}{2}$ to π, x decreases through negative values from 0 to -1. From π to $\dfrac{3\pi}{2}$, x increases from -1 to 0, and from $\dfrac{3\pi}{2}$ to 2π, x increases from 0 to 1.

We interpret these values as the relation $y = \cos x$, where now, x is the angle in radians and y is the value of the cosine function. Other values of $y = \cos x$ are:

x	0	$\dfrac{\pi}{6}$	$\dfrac{\pi}{4}$	$\dfrac{\pi}{3}$	$\dfrac{\pi}{2}$	$\dfrac{2\pi}{3}$	$\dfrac{3\pi}{4}$	$\dfrac{5\pi}{6}$	π	$\dfrac{7\pi}{6}$	$\dfrac{5\pi}{4}$	$\dfrac{4\pi}{3}$	$\dfrac{3\pi}{2}$	$\dfrac{5\pi}{3}$	$\dfrac{7\pi}{4}$	$\dfrac{11\pi}{6}$	2π
y	1	.9	.7	.5	0	$-.5$	$-.7$	$-.9$	-1	$-.9$	$-.7$	$-.5$	0	.5	.7	.9	1

We plot these values:

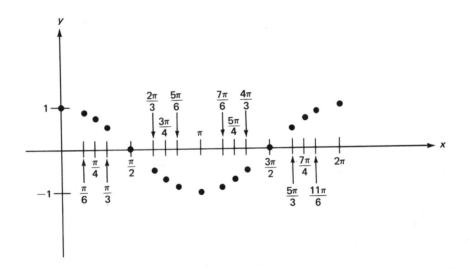

and connect them by a smooth curve to construct one cycle of the graph of $y = \cos x$:

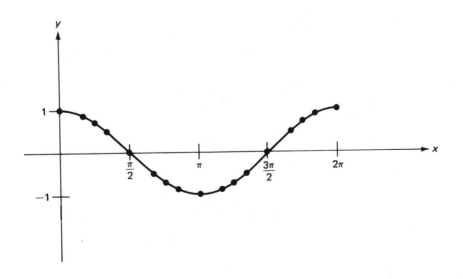

The graph has the same general shape as the graph of $y = \sin x$, but starts and ends at $y = 1$ when $x = 0$ and $x = 2\pi$. Again, to draw one cycle of the graph we need only plot the points for the quadrantal angles 0, $\dfrac{\pi}{2}$, π, $\dfrac{3\pi}{2}$, and also 2π.

EXAMPLE 9.3.   Draw two cycles of the graph of $y = \cos x$ starting from $x = 0$.

Solution.   We draw one cycle from $x = 0$ to $x = 2\pi$, and another identical cycle from $x = 2\pi$ to $x = 4\pi$. The graph is

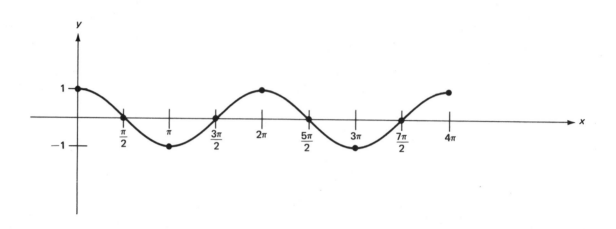

EXAMPLE 9.4.   Draw two cycles of the graph of $y = \cos x$ starting from $x = -2\pi$.

Solution.   The cycle of $y = \cos x$ from $x = -2\pi$ to $x = 0$ is identical to the cycle from $x = 0$ to $x = 2\pi$. Thus the graph is

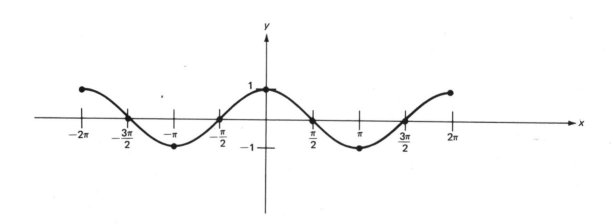

Since $y = \cos x$ continues indefinitely to the right and left, and there is a y-value for every x-value, the domain of $y = \cos x$ is all real numbers. However, y is never greater than 1 nor less than -1, so the range of $y = \cos x$ is

$$-1 \leqslant y \leqslant 1.$$

Moreover, observe that, for example,

$$\cos \frac{\pi}{6} = \frac{\sqrt{3}}{2} \text{ and } \cos\left(-\frac{\pi}{6}\right) = \frac{\sqrt{3}}{2},$$

$$\cos \frac{\pi}{4} = \frac{1}{\sqrt{2}} \text{ and } \cos\left(-\frac{\pi}{4}\right) = \frac{1}{\sqrt{2}},$$

$$\cos \frac{\pi}{3} = \frac{1}{2} \text{ and } \cos\left(-\frac{\pi}{3}\right) = \frac{1}{2},$$

and also,

$$\cos \frac{2\pi}{3} = -\frac{1}{2} \text{ and } \cos\left(-\frac{2\pi}{3}\right) = -\frac{1}{2},$$

$$\cos \frac{3\pi}{4} = -\frac{1}{\sqrt{2}} \text{ and } \cos\left(-\frac{3\pi}{4}\right) = -\frac{1}{\sqrt{2}},$$

$$\cos \frac{5\pi}{6} = -\frac{\sqrt{3}}{2} \text{ and } \cos\left(-\frac{5\pi}{6}\right) = -\frac{\sqrt{3}}{2},$$

$$\cos \pi = -1 \text{ and } \cos(-\pi) = -1,$$

and so on. In general,

$$\cos(-x) = \cos x$$

for all values of x. Functions with this property are called **even** functions [because functions $f(x) = x^n$, where n is an even integer, have this property].

Exercise 9.1

1. Draw four cycles of the graph of $y = \sin x$ starting from $x = 0$.

2. Draw three cycles of the graph of $y = \sin x$ starting from $x = -2\pi$.

3. Draw two cycles of the graph of $y = \sin x$ starting from $x = -4\pi$.

4. Draw four cycles of the graph of $y = \sin x$ starting from $x = -4\pi$.

5. Draw three cycles of the graph of $y = \cos x$ starting from $x = 0$.

6. Draw four cycles of the graph of $y = \cos x$ starting from $x = 0$.

7. Draw two cycles of the graph of $y = \cos x$ starting from $x = -4\pi$.

8. Draw four cycles of the graph of $y = \cos x$ starting from $x = -4\pi$.

9. State the domain of $y = \sin x$.   10. State the domain of $y = \cos x$.

11. State the range of $y = \sin x$.   12. State the range of $y = \cos x$.

The Functions $y = \tan x$ and $y = \cot x$

The trigonometric function $y = \tan x$ is very different from the functions $y = \sin x$ and $y = \cos x$. To interpret $\tan \theta$ in terms of (x, y), we use a different diagram on the unit circle:

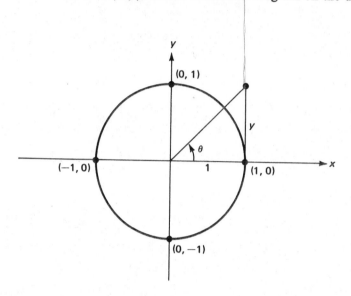

Observe that, in this diagram,

$$\tan \theta = \frac{y}{x}$$

$$= \frac{y}{1}$$

$$= y.$$

Also, observe that the line which represents y is half of the geometric tangent line to the circle at the point $(1, 0)$.

As θ goes from 0 to $\frac{\pi}{2}$, y starts from 0 and increases indefinitely. Recall from Section 3.4 that $\tan \frac{\pi}{2}$ is undefined. When $\theta = \frac{\pi}{2}$, the terminal side of θ and the line on the unit circle representing y are parallel. Therefore, the line has no endpoint, and so y has no real number value.

To interpret this part of the function $y = \tan x$, where x is the angle in radians, we observe that

$$\tan \frac{\pi}{6} = \frac{1}{\sqrt{3}} \approx .6,$$

$$\tan \frac{\pi}{4} = 1,$$

$$\tan \frac{\pi}{3} = \sqrt{3} \approx 1.7,$$

and

$$\tan \frac{\pi}{2} \text{ is undefined.}$$

Therefore, as x goes from 0 toward $\frac{\pi}{2}$, y increases indefinitely. We draw a dashed line at $x = \frac{\pi}{2}$, where y is undefined. The graph from 0 to $\frac{\pi}{2}$ is

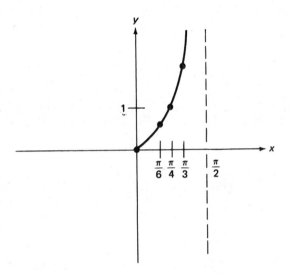

The dashed line is not part of the graph. The graph gets closer and closer to the line without touching it. Recall from algebra that such a line is called an **asymptote.** The line $x = \frac{\pi}{2}$ is an asymptote of the graph of $y = \tan x$.

Next we consider θ from 0 to $-\frac{\pi}{2}$. We use the negative half of the tangent line to the unit circle at the point (1, 0):

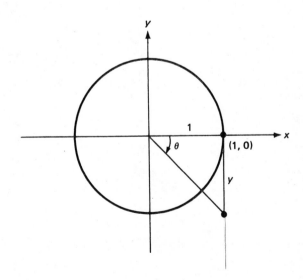

As θ goes from 0 to $-\dfrac{\pi}{4}$, y decreases from 0 to -1. Then y continues to decrease until at $\theta = -\dfrac{\pi}{2}$, $\tan \theta$ is undefined. Therefore, the graph of $y = \tan x$ from $-\dfrac{\pi}{2}$ to $\dfrac{\pi}{2}$ is

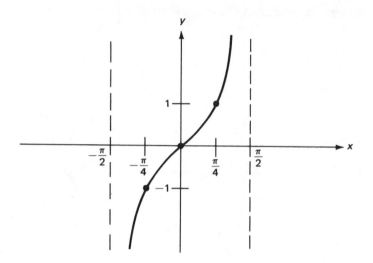

When θ goes from $\dfrac{\pi}{2}$ to π on the unit circle, we extend the terminal side of θ to meet the negative half of the tangent line at $(1, 0)$:

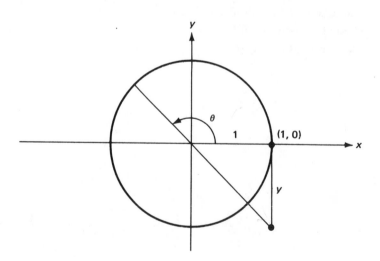

Similarly, from π to $\dfrac{3\pi}{2}$ we extend the terminal side of θ to meet the positive half of the tangent

line at $(1, 0)$. The graph of $y = \tan x$ from $-\dfrac{\pi}{2}$ to $\dfrac{3\pi}{2}$ is

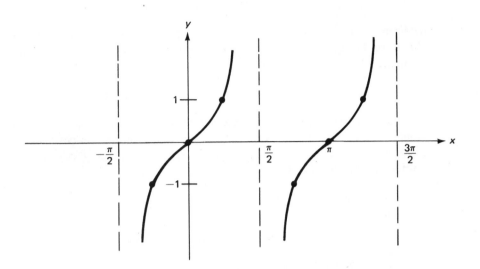

Observe that the graph repeats between each pair of asymptotes.

EXAMPLE 9.5.   Draw the graph of $y = \tan x$ from $x = 0$ to $x = 2\pi$.

Solution.   We have the part of the graph from $x = 0$ to $x = \dfrac{3\pi}{2}$. From $x = \dfrac{3\pi}{2}$ to 2π, the graph is the same as the graph from $x = -\dfrac{\pi}{2}$ to 0. Thus the graph is

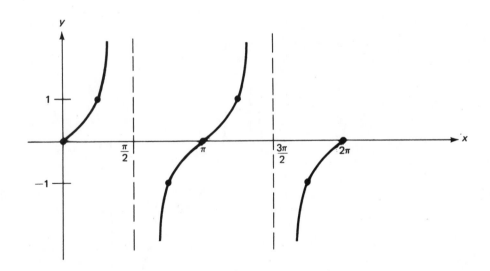

The graph of $y = \tan x$ continues indefinitely to both the right and left. However, there are no y-values defined when $x = \dfrac{\pi}{2}, \dfrac{3\pi}{2}, \dfrac{5\pi}{2}$, and so on, and $x = -\dfrac{\pi}{2}, -\dfrac{3\pi}{2}, -\dfrac{5\pi}{2}$, and so on.

Therefore, the domain of the function $y = \tan x$ is the set of all real numbers except those numbers for which a y-value is not defined. We may write the domain as

$$x \neq \pm \frac{\pi}{2}, \pm \frac{3\pi}{2}, \pm \frac{5\pi}{2}, \ldots .$$

Between each pair of asymptotes, the graph continues indefinitely upward and downward. Therefore, the graph includes all real numbers y. The range of $y = \tan x$ is the set of all real numbers. Also, observe that $y = \tan x$ is an odd function; that is,

$$\tan(-x) = -\tan x.$$

To draw the graph of $y = \cot x$, we could use another line tangent to the unit circle. You should find a tangent line to the unit circle which represents $\cot \theta$ in the (x, y) interpretation. However, we can also use the reciprocal relationship between $\tan \theta$ and $\cot \theta$ by which we defined $\cot \theta$ in Section 1.2:

$$\cot \theta = \frac{1}{\tan \theta}.$$

Then, for every x,

$$y = \cot x = \frac{1}{\tan x}.$$

In particular, when $\tan x = 0$,

$$y = \cot x = \frac{1}{\tan x} = \frac{1}{0},$$

which is undefined. For example, $y = \cot x$ is undefined at $x = 0$ and $x = \pi$. Similarly, when $\tan x$ is undefined, $\cot x = 0$. For example

$$y = \cot \frac{\pi}{2} = 0.$$

We recall that

$$y = \cot \frac{\pi}{4} = 1,$$

and

$$y = \cot \frac{3\pi}{4} = -1.$$

The graph of $y = \cot x$ from $x = 0$ to $x = \pi$ is

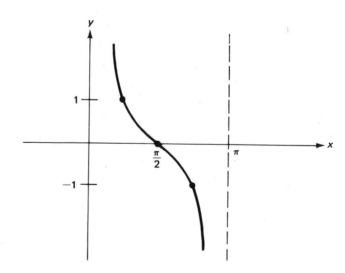

Observe that the y-axis, or the line $x = 0$, is an asymptote, and also the line $x = \pi$.

EXAMPLE 9.6.   Draw the graph of $y = \cot x$ from $x = -\dfrac{\pi}{2}$ to $x = \dfrac{5\pi}{2}$.

Solution.   There are asymptotes at $x = 0$, $x = \pi$, and $x = 2\pi$. The value is 0 at $x = -\dfrac{\pi}{2}$, $x = \dfrac{\pi}{2}$, $x = \dfrac{3\pi}{2}$, and $x = \dfrac{5\pi}{2}$. Therefore, the graph is

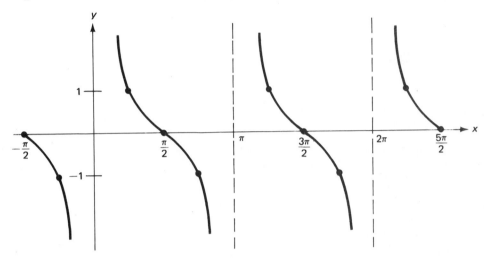

Observe that the domain of $y = \cot x$ is all real numbers except those where a y-value is not defined, or

$$x \neq 0,\ \pm\pi,\ \pm 2\pi,\ \dots .$$

The range of $y = \cot x$ is the set of all real numbers. Finally, $y = \cot x$ is also an odd function; that is,

$$\cot(-x) = -\cot x.$$

Exercise 9.2

1.   Draw the graph of $y = \tan x$ from $x = -\dfrac{\pi}{2}$ to $x = \dfrac{5\pi}{2}$.

2.   Draw the graph of $y = \tan x$ from $x = -\dfrac{\pi}{2}$ to $x = \dfrac{7\pi}{2}$.

3.   Draw the graph of $y = \tan x$ from $x = 0$ to $x = 3\pi$.

4.   Draw the graph of $y = \tan x$ from $x = -\pi$ to $x = \pi$.

5.   Draw the graph of $y = \cot x$ from $x = 0$ to $x = 2\pi$.

6.   Draw the graph of $y = \cot x$ from $x = -\pi$ to $x = \pi$.

7.   Draw the graph of $y = \cot x$ from $x = -\dfrac{\pi}{2}$ to $x = \dfrac{7\pi}{2}$.

8.   Draw the graph of $y = \cot x$ from $x = -\dfrac{3\pi}{2}$ to $x = \dfrac{3\pi}{2}$.

9.   State the domain of $y = \tan x$.   10.   State the domain of $y = \cot x$.

11.   State the range of $y = \tan x$.   12.   State the range of $y = \cot x$.

The Functions $y = \sec x$ and $y = \csc x$

To graph the function $y = \sec x$, we use the definition from Section 1.2,

$$\sec \theta = \frac{1}{\cos \theta}.$$

Therefore,

$$y = \sec x = \frac{1}{\cos x}.$$

In particular, whenever $\cos x = 0$, $y = \sec x$ is undefined. We start by drawing the graph of $y = \cos x$:

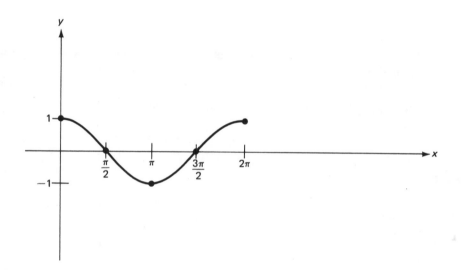

We see that $\cos x = 0$ when $x = \dfrac{\pi}{2}$ and $x = \dfrac{3\pi}{2}$. Therefore, $y = \sec x$ is undefined and has asymptotes at these points. Also, at $x = 0$,

$$y = \sec 0 = \frac{1}{\cos 0} = \frac{1}{1} = 1,$$

and at $x = \pi$,

$$y = \sec \pi = \frac{1}{\cos \pi} = \frac{1}{-1} = -1.$$

Thus we can use the graph of $y = \cos x$ to draw the graph of $y = \sec x$:

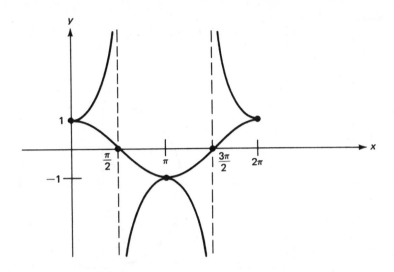

EXAMPLE 9.7.   On one set of axes, draw the graphs of $y = \cos x$ and $y = \sec x$ from $x = -\dfrac{\pi}{2}$ to $x = \dfrac{5\pi}{2}$.

Solution.   We draw the graph of $y = \cos x$ from $x = -\dfrac{\pi}{2}$ to $x = \dfrac{5\pi}{2}$. Then, since $y = \sec x$ is undefined whenever $\cos x = 0$, we draw the asymptotes and fill in the graph of $y = \sec x$:

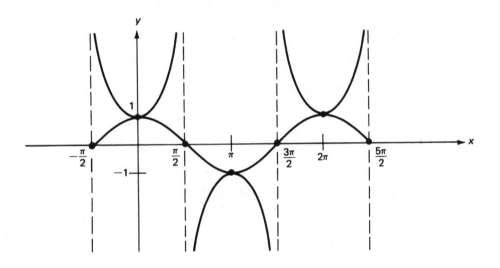

Observe that the domain of $y = \sec x$ is

$$x \neq \pm \frac{\pi}{2}, \ \pm \frac{3\pi}{2}, \ \pm \frac{5\pi}{2}, \ \ldots \ .$$

The y-values are always greater than or equal to 1, or less than or equal to -1. The range of $y = \sec x$ is

$$y \geq 1 \text{ or } y \leq -1.$$

The function $y = \sec x$, like $y = \cos x$, is an even function; that is,

$$\sec(-x) = \sec x.$$

On the unit circle, we may represent $\sec \theta$ by r in the (x, y) and r interpretation:

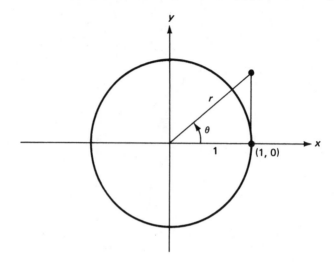

$$\sec \theta = \frac{r}{x}$$

$$= \frac{r}{1}$$

$$= r.$$

Observe that $r = 1$ when $\theta = 0$, and r increases until it is undefined when $\theta = \frac{\pi}{2}$. In geometry, a secant line crosses a circle at two points. The line representing r is half of a secant line of the circle.

To draw the graph of the function $y = \csc x$, we use the definition from Section 1.2,

$$\csc \theta = \frac{1}{\sin \theta}.$$

Therefore,

$$y = \csc x = \frac{1}{\sin x}.$$

Whenever $\sin x = 0, y = \csc x$ is undefined:

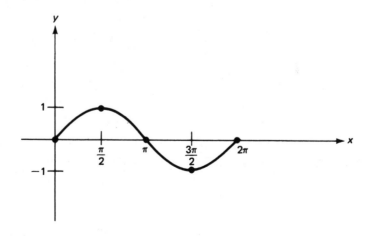

Since $\sin x = 0$ at $x = 0$, $x = \pi$, and $x = 2\pi$, $y = \csc x$ is undefined and has asymptotes at these points. Also, at $x = \dfrac{\pi}{2}$,

$$y = \csc \frac{\pi}{2} = \frac{1}{\sin \dfrac{\pi}{2}} = \frac{1}{1} = 1,$$

and at $x = \dfrac{3\pi}{2}$,

$$y = \csc \frac{3\pi}{2} = \frac{1}{\sin \dfrac{3\pi}{2}} = \frac{1}{-1} = -1.$$

Using the graph of $y = \sin x$ to draw the graph of $y = \csc x$, the graph is

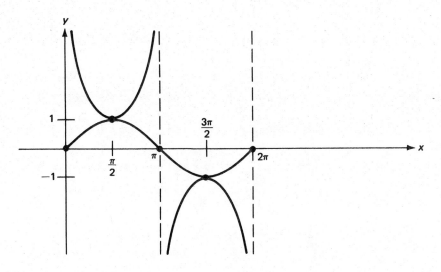

EXAMPLE 9.8.   On one set of axes, draw the graphs of $y = \sin x$ and $y = \csc x$ from $x = -\dfrac{3\pi}{2}$ to $x = \dfrac{3\pi}{2}$.

Solution.   We draw the graph of $y = \sin x$ from $x = -\dfrac{3\pi}{2}$ to $x = \dfrac{3\pi}{2}$, then draw the asymptotes and fill in the graph of $y = \csc x$:

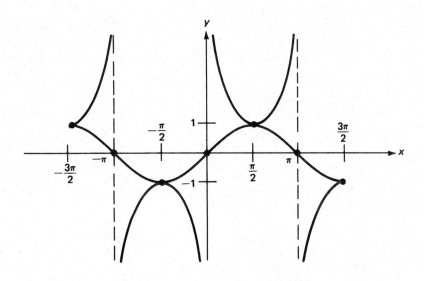

Observe that the domain of $y = \csc x$ is

$$x \neq 0, \pm\pi, \pm 2\pi, \ldots$$

and the range of $y = \csc x$ is

$$y \geqslant 1 \text{ or } y \leqslant -1.$$

The function $y = \csc x$, like $y = \sin x$, is an odd function; that is,

$$\csc(-x) = -\csc x.$$

You should find a line on the unit circle which represents $\csc \theta$ in the (x, y) and r interpretation.

Exercise 9.3

1. On one set of axes, draw the graphs of $y = \cos x$ and $y = \sec x$ from $x = -\dfrac{3\pi}{2}$ to $x = \dfrac{5\pi}{2}$.

2. On one set of axes, draw the graphs of $y = \cos x$ and $y = \sec x$ from $x = 0$ to $x = 4\pi$.

3. On one set of axes, draw the graphs of $y = \sin x$ and $y = \csc x$ from $x = 0$ to $x = 4\pi$.

4. On one set of axes, draw the graphs of $y = \sin x$ and $y = \csc x$ from $x = -\dfrac{5\pi}{2}$ to $x = \dfrac{5\pi}{2}$.

5. State the domain of $y = \sec x$.   6. State the domain of $y = \csc x$.

7. State the range of $y = \sec x$.   8. State the range of $y = \csc x$.

Self-test

1. Draw three cycles of the graph of $y = \cos x$ starting from $x = -2\pi$.

2. Draw the graph of $y = \tan x$ from $x = -\dfrac{3\pi}{2}$ to $x = \dfrac{3\pi}{2}$.

3. On one set of axes, draw the graphs of $y = \sin x$ and $y = \csc x$ from $x = -2\pi$ to $x = 2\pi$.

4. State the range of $y = \csc x$.

4._____

5. State the domain of $y = \tan x$.

5._____

10 General Sine Waves

In this unit you will learn how to analyze and graph the general form of the sine function. This function describes a common physical occurrence called simple harmonic motion. Examples of simple harmonic motion are the swing of a pendulum and the voltage of alternating current.

OBJECTIVES

When you have finished this unit you should be able to:

1. Draw the graph of functions of the form $y = A \sin x$ and $y = A \cos x$.
2. Draw the graph of functions of the form $y = \sin Bx$ and $y = \cos Bx$.
3. Draw the graph of functions of the form $y = \sin(x - C)$ and $y = \cos(x - C)$.
4. State the amplitude, period, phase shift, and draw the graph of functions of the form $y = A \sin(Bx - D)$.

Section
10.1

Amplitude

Recall from Section 9.1 the first cycle of the graph of $y = \sin x$:

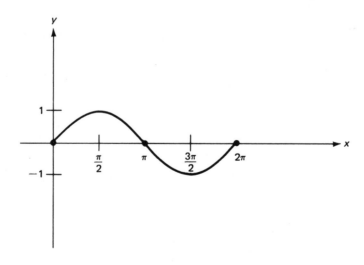

Now, consider the function $y = 2 \sin x$. Every value of the original graph is multiplied by 2. When $\sin x = 0$, $y = 2 \sin x$ is also 0. That is,

$$y = 2 \sin 0 = 2(0) = 0,$$

$$y = 2 \sin \pi = 2(0) = 0,$$

and

$$y = 2 \sin 2\pi = 2(0) = 0.$$

When $x = \dfrac{\pi}{2}$,

$$y = 2 \sin \frac{\pi}{2} = 2(1) = 2.$$

When $x = \dfrac{3\pi}{2}$,

$$y = 2 \sin \frac{3\pi}{2} = 2(-1) = -2.$$

One cycle of the graph of $y = 2 \sin x$ is

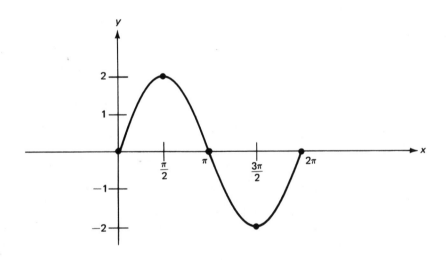

The range of $y = 2 \sin x$ is

$$-2 \leqslant y \leqslant 2.$$

We say that the **amplitude** of the function $y = 2 \sin x$ is 2.

In general, the function $y = A \sin x$ has values between A and $-A$. The range of the function $y = A \sin x$, when $A > 0$, is

$$-A \leqslant y \leqslant A.$$

We call $|A|$ the amplitude of the function $y = A \sin x$.

EXAMPLE 10.1. Draw one cycle of the graph of $y = 4 \sin x$.

Solution.   The amplitude is

$$|4| = 4.$$

When $\sin x = 0$, $y = 4 \sin x$ is also 0. Thus

$$y = 4 \sin 0 = 4(0) = 0,$$

$$y = 4 \sin \pi = 4(0) = 0,$$

and

$$y = 4 \sin 2\pi = 4(0) = 0.$$

When $x = \dfrac{\pi}{2}$,

$$y = 4 \sin \frac{\pi}{2} = 4(1) = 4.$$

When $x = \dfrac{3\pi}{2}$,

$$y = 4 \sin \frac{3\pi}{2} = 4(-1) = -4.$$

One cycle of the graph is

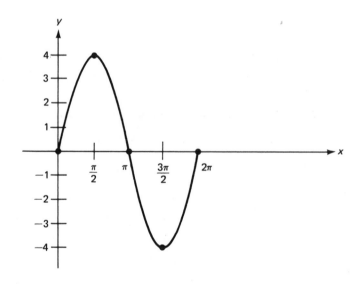

EXAMPLE 10.2.   Draw one cycle of the graph of $y = \dfrac{2}{3} \sin x$.

Solution.   The amplitude is

$$\left|\frac{2}{3}\right| = \frac{2}{3}.$$

Again, when $\sin x = 0$, $y = \dfrac{2}{3} \sin x$ is also 0. These are the points where $x = 0$, π, and 2π. When

$$x = \frac{\pi}{2},$$

$$y = \frac{2}{3} \sin \frac{\pi}{2} = \frac{2}{3}(1) = \frac{2}{3}.$$

When $x = \frac{3\pi}{2}$,

$$y = \frac{2}{3} \sin \frac{3\pi}{2} = \frac{2}{3}(-1) = -\frac{2}{3}.$$

One cycle of the graph is

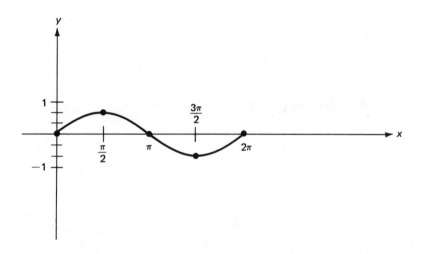

If A is negative, the graph of $y = A \sin x$ reverses with respect to the x-axis. That is, the first half of the cycle has negative values, and the second half of the cycle has positive values. Since the amplitude is defined to be the absolute value of A, however, the amplitude of $y = A \sin x$ is positive regardless of whether A is positive or negative.

EXAMPLE 10.3.   Draw one cycle of the graph of $y = -2 \sin x$.

Solution.   The amplitude is

$$|-2| = 2.$$

When $\sin x = 0$, $y = -2 \sin x$ is also 0. These are the points where $x = 0$, π, and 2π. However, when $x = \frac{\pi}{2}$,

$$y = -2 \sin \frac{\pi}{2} = -2(1) = -2,$$

and when $x = \frac{3\pi}{2}$,

$$y = -2 \sin \frac{3\pi}{2} = -2(-1) = 2.$$

Therefore, one cycle of the graph is

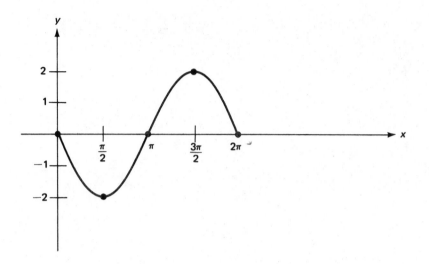

Now, recall the first cycle of the graph of $y = \cos x$:

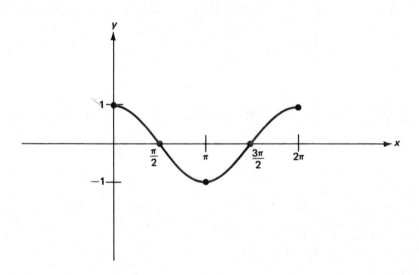

The graph of the function $y = A \cos x$ has values between A and $-A$. We say that $|A|$ is the amplitude of $y = A \cos x$.

EXAMPLE 10.4.   Draw one cycle of the graph of $y = 2 \cos x$.

Solution.   The amplitude is

$$|2| = 2.$$

When $\cos x = 0, y = 2 \cos x$ is also 0. Thus

$$y = 2 \cos \frac{\pi}{2} = 2(0) = 0$$

and

$$y = 2 \cos \frac{3\pi}{2} = 2(0) = 0.$$

When $x = 0$,

$$y = 2 \cos 0 = 2(1) = 2,$$

when $x = \pi$,

$$y = 2 \cos \pi = 2(-1) = -2,$$

and when $x = 2\pi$,

$$y = 2 \cos 2\pi = 2(1) = 2.$$

Therefore, one cycle of the graph is

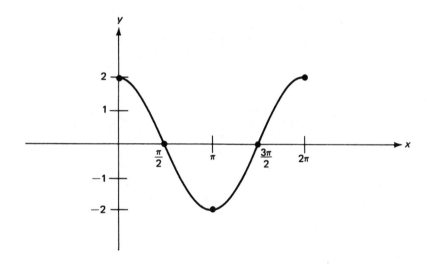

If A is negative, the graph of $y = A \cos x$ begins and ends at a negative number.

EXAMPLE 10.5.   Draw one cycle of the graph of $y = -\dfrac{1}{3} \cos x$.

Solution.   The amplitude is

$$\left| -\frac{1}{3} \right| = \frac{1}{3}.$$

Again, when $\cos x = 0$, $y = -\dfrac{1}{3} \cos x$ is also 0. These are the points where $x = \dfrac{\pi}{2}$ and $x = \dfrac{3\pi}{2}$. However, when $x = 0$,

$$y = -\frac{1}{3} \cos 0 = -\frac{1}{3}(1) = -\frac{1}{3}.$$

When $x = \pi$,

$$y = -\frac{1}{3} \cos \pi = -\frac{1}{3}(-1) = \frac{1}{3}.$$

When $x = 2\pi$,

$$y = -\frac{1}{3}\cos 2\pi = -\frac{1}{3}(1) = -\frac{1}{3}.$$

Therefore, one cycle of the graph is

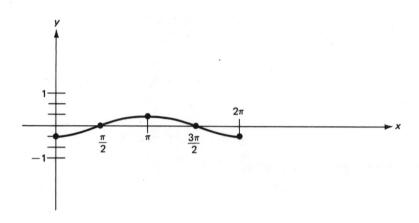

Exercise 10.1

Draw one cycle of the graph:

1. $y = 3 \sin x$

2. $y = 6 \sin x$

3. $y = \frac{1}{2} \sin x$

4. $y = \frac{3}{2} \sin x$

5. $y = -3 \sin x$

6. $y = -\frac{1}{3} \sin x$

7. $y = \frac{1}{3} \cos x$

8. $y = \frac{5}{2} \cos x$

9. $y = -\frac{1}{2} \cos x$

10. $y = -2 \cos x$

11. $y = -\sin x$

12. $y = -\cos x$

Section 10.2 — Period

Consider the function $y = \sin 2x$. Observe that this function is not the same as the function $y = 2 \sin x$. The function $y = \sin 2x$ has amplitude 1.

We will calculate some points of the graph of $y = \sin 2x$. When $x = 0$,

$$y = \sin 2(0) = \sin 0 = 0.$$

When $x = \frac{\pi}{2}$,

$$y = \sin 2\left(\frac{\pi}{2}\right) = \sin \pi = 0.$$

When $x = \pi$,

$$y = \sin 2(\pi) = \sin 2\pi = 0.$$

So far, all our values are 0. However, if we try $x = \dfrac{\pi}{4}$,

$$y = \sin 2\left(\frac{\pi}{4}\right) = \sin \frac{\pi}{2} = 1,$$

and if $x = \dfrac{3\pi}{4}$,

$$y = \sin 2\left(\frac{3\pi}{4}\right) = \sin \frac{3\pi}{2} = -1.$$

The first cycle of the graph of $y = \sin 2x$ is

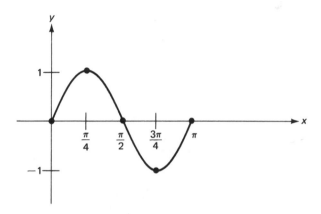

Observe that we complete the first cycle of $y = \sin 2x$ in π units instead of 2π units. We say that the **period** of the function $y = \sin 2x$ is π. In general, the period of a repeating function is the length of the interval covered by one cycle of the graph of the function.

There are two cycles of the graph of $y = \sin 2x$ in the same interval as one cycle of the graph of $y = \sin x$. The period of $y = \sin x$ is 2π. The period of $y = \sin 2x$ is π, or $\dfrac{2\pi}{2}$. In general, the graph of $y = Bx$, where $B > 0$, has B cycles in the same interval as one cycle of the graph of $y = \sin x$. Therefore, each cycle has length $\dfrac{2\pi}{B}$. The period of the function $y = \sin Bx$, where $B > 0$, is given by

$$\frac{2\pi}{B}.$$

EXAMPLE 10.6.   Draw one cycle of the graph of $y = \sin 3x$.

Solution.   Since

$$B = 3,$$

there are three cycles of the graph in 2π units. The period is

$$\frac{2\pi}{B} = \frac{2\pi}{3}.$$

Therefore, the first cycle ends at $x = \dfrac{2\pi}{3}$. Now,

$$\frac{1}{2}\left(\frac{2\pi}{3}\right) = \frac{\pi}{3},$$

so the middle of the cycle is at $x = \dfrac{\pi}{3}$. Also,

$$\frac{1}{2}\left(\frac{\pi}{3}\right) = \frac{\pi}{6},$$

so the graph increases to its maximum value of 1 at $x = \dfrac{\pi}{6}$. Finally, the point halfway between $\dfrac{\pi}{3}$ and $\dfrac{2\pi}{3}$ is $\dfrac{\pi}{2}$, so the graph decreases to its minimum value of -1 at $x = \dfrac{\pi}{2}$. One cycle of the graph is

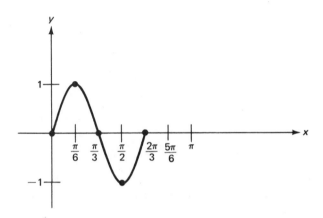

EXAMPLE 10.7.   Draw one cycle of the graph of $y = \sin \dfrac{1}{2} x$.

Solution.   Since

$$B = \frac{1}{2},$$

there is $\dfrac{1}{2}$ of a cycle of the graph in 2π units. The period is

$$\frac{2\pi}{B} = \frac{2\pi}{\dfrac{1}{2}} = 2\pi\left(\frac{2}{1}\right) = 4\pi.$$

The first cycle ends at $x = 4\pi$. Then the middle of the first cycle is at $x = 2\pi$, the graph rises to 1 at $x = \pi$, and decreases to -1 at $x = 3\pi$. One cycle of the graph is

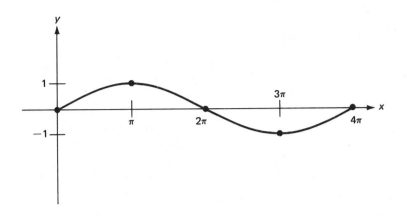

Similarly, the period of the function $y = \cos Bx$ is given by

$$\frac{2\pi}{B}.$$

EXAMPLE 10.8.   Draw one cycle of the graph of $y = \cos 2x$.

Solution.   The period is

$$\frac{2\pi}{B} = \frac{2\pi}{2} = \pi.$$

The first cycle ends at $x = \pi$. Then the middle of the first cycle is at $x = \frac{\pi}{2}$. Remember that the values at the beginning and end of a cycle of the graph of the cosine are 1, and the value at the middle of a cycle of the graph of the cosine is -1. Thus, when $x = 0$ and when $x = \pi$,

$$y = \cos 2x = 1,$$

and when $x = \frac{\pi}{2}$,

$$y = \cos 2x = -1.$$

Then,

$$\frac{1}{2}\left(\frac{\pi}{2}\right) = \frac{\pi}{4},$$

so the graph decreases to 0 at $x = \frac{\pi}{4}$. Also, the point halfway between $\frac{\pi}{2}$ and π is $\frac{3\pi}{4}$, so the graph increases again to 0 at $x = \dfrac{3\pi}{4}$. One cycle of the graph is

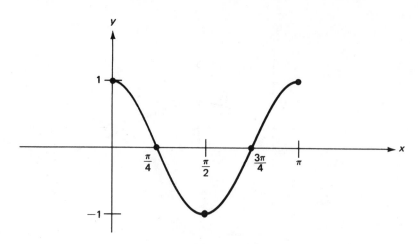

EXAMPLE 10.9.   Draw one cycle of the graph of $y = \cos \dfrac{2}{3} x$.

Solution.   The period is

$$\frac{2\pi}{B} = \frac{2\pi}{\dfrac{2}{3}} = 2\pi\left(\frac{3}{2}\right) = 3\pi.$$

Therefore, the first cycle ends at $x = 3\pi$, and the value is 1. The middle of the cycle is at $x = \dfrac{3\pi}{2}$, and the value is -1. Since

$$\frac{1}{2}\left(\frac{3\pi}{2}\right) = \frac{3\pi}{4},$$

the graph decreases to 0 at $x = \dfrac{3\pi}{4}$. The point halfway between $\dfrac{3\pi}{2}$ and 3π is $\dfrac{9\pi}{4}$, so the graph increases again to 0 at $x = \dfrac{9\pi}{4}$. One cycle of the graph is

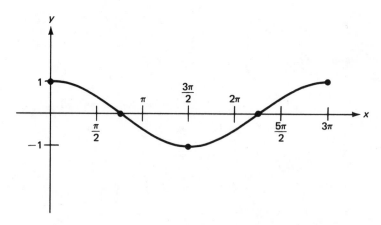

The graphs in the preceding examples all have amplitude 1. We may put together an amplitude A other than 1 as in Section 10.1, and a period other than 2π.

EXAMPLE 10.10.   Draw one cycle of the graph of $y = 3 \sin 3x$.

Solution.   The amplitude is

$$|A| = |3| = 3,$$

and the period is

$$\frac{2\pi}{B} = \frac{2\pi}{3}.$$

The graph of the sine is 0 at the beginning and end of a cycle, so the value is 0 at $x = 0$ and at $x = \frac{2\pi}{3}$. Also, the value is 0 at the middle of the cycle, $x = \frac{\pi}{3}$. Since

$$\frac{1}{2}\left(\frac{\pi}{3}\right) = \frac{\pi}{6},$$

the graph increases until it reaches its maximum value at $x = \frac{\pi}{6}$, and since the amplitude is 3, the value is 3 at $x = \frac{\pi}{6}$. The minimum value is -3 at $x = \frac{\pi}{2}$. One cycle of the graph is

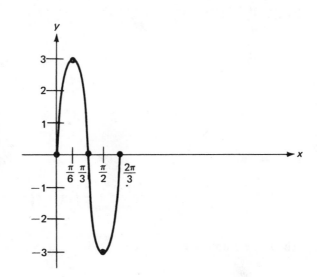

Exercise 10.2

Draw one cycle of the graph:

1. $y = \sin 4x$

2. $y = \sin \frac{1}{4} x$

3. $y = \sin \frac{1}{3} x$

4. $y = \sin \frac{2}{3} x$

5. $y = \sin \frac{3}{2} x$

6. $y = \sin \frac{2}{5} x$

7. $y = \cos 3x$

8. $y = \cos \frac{1}{3} x$

9. $y = \cos \frac{1}{2} x$

10. $y = \cos \frac{3}{2} x$

11. $y = 2 \sin 2x$

12. $y = \frac{1}{2} \sin \frac{1}{2} x$

13. $y = \frac{1}{3} \cos 3x$

14. $y = \frac{3}{2} \cos \frac{2}{3} x$

15. $y = -\sin 3x$

16. $y = -3 \cos \frac{1}{3} x$

Section
10.3

Phase Shift

Consider the function $y = \sin\left(x - \frac{\pi}{2}\right)$. The function has amplitude 1 and period 2π like the function $y = \sin x$. However, the function $y = \sin\left(x - \frac{\pi}{2}\right)$ clearly is not the same as the function $y = \sin x$.

We will calculate some points of the graph of $y = \sin\left(x - \frac{\pi}{2}\right)$. When $x = 0$,

$$y = \sin\left(0 - \frac{\pi}{2}\right) = \sin\left(-\frac{\pi}{2}\right) = -1.$$

When $x = \frac{\pi}{2}$,

$$y = \sin\left(\frac{\pi}{2} - \frac{\pi}{2}\right) = \sin(0) = 0.$$

When $x = \pi$,

$$y = \sin\left(\pi - \frac{\pi}{2}\right) = \sin\left(\frac{\pi}{2}\right) = 1.$$

When $x = \frac{3\pi}{2}$,

$$y = \sin\left(\frac{3\pi}{2} - \frac{\pi}{2}\right) = \sin(\pi) = 0,$$

and when $x = 2\pi$,

$$y = \sin\left(2\pi - \frac{\pi}{2}\right) = \sin\left(\frac{3\pi}{2}\right) = -1.$$

One cycle of the graph of $y = \sin\left(x - \frac{\pi}{2}\right)$ is

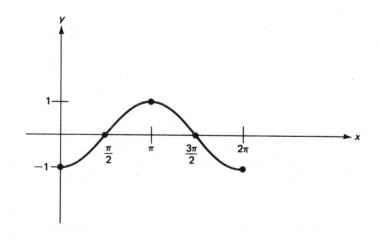

If we calculate one more point, for $x = \dfrac{5\pi}{2}$, we have the graph

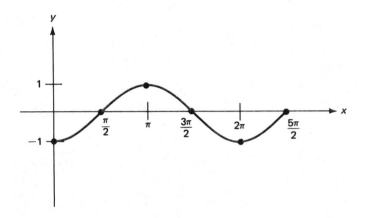

Observe that the graph from $x = \dfrac{\pi}{2}$ to $x = \dfrac{5\pi}{2}$ is one cycle of the sine graph moved to the right so that the cycle begins at $x = \dfrac{\pi}{2}$. We say that the graph has a **phase shift** of $\dfrac{\pi}{2}$.

The phase shift of $y = \sin\left(x - \dfrac{\pi}{2}\right)$ is a shift such that a cycle begins at $x = \dfrac{\pi}{2}$ which is the same as the cycle of $y = \sin x$ which begins at $x = 0$. If we write

$$x - \frac{\pi}{2} = 0,$$

we have

$$x = \frac{\pi}{2}.$$

In general, the graph of $y = \sin(x - C)$ is the same as the graph of $y = \sin x$ with a cycle beginning at $x = C$. The phase shift of the function $y = \sin(x - C)$ is C.

EXAMPLE 10.11. Draw two cycles of the graph of $y = \sin(x - \pi)$.

Solution. The phase shift is π. Thus one cycle begins at π. If we first draw the graph of $y = \sin x$, we can then move each point π units to right:

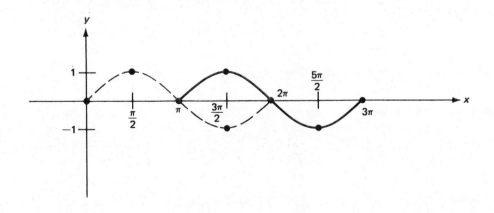

The cycle we have drawn does not cross the y-axis; therefore, we draw a second cycle to the left of the first cycle. The graph is

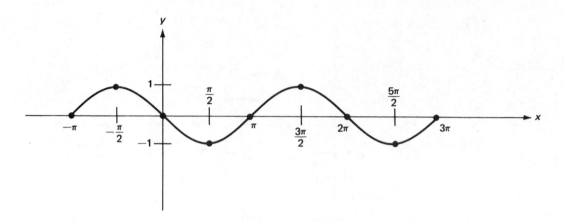

Similarly, the phase shift of the function $y = \cos(x - C)$ is C.

EXAMPLE 10.12.   Draw two cycles of the graph of $y = \cos\left(x - \dfrac{\pi}{2}\right)$.

Solution.   The phase shift is $\dfrac{\pi}{2}$. We draw the graph of $y = \cos x$ and move each point $\dfrac{\pi}{2}$ units to the right. Then, we draw another cycle to the left of the first cycle. The graph is

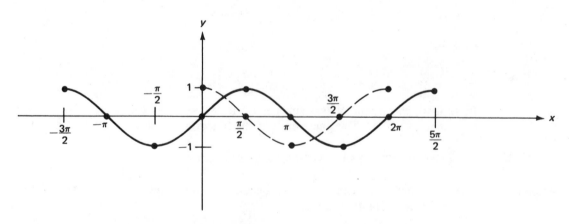

Observe that the resulting graph is the same as the graph of $y = \sin x$.

We may also have a phase shift to the left, when C is negative.

EXAMPLE 10.13.   Draw two cycles of the graph of $y = \sin\left(x + \dfrac{\pi}{2}\right)$.

Solution.   If we write

$$x + \frac{\pi}{2} = 0,$$

we have

$$x = -\frac{\pi}{2}.$$

The phase shift is $-\dfrac{\pi}{2}$. We draw the graph of $y = \sin x$ and move each point $\dfrac{\pi}{2}$ units to the left. Since the resulting graph crosses the y-axis, we draw the second cycle to the right of the first cycle. The graph is

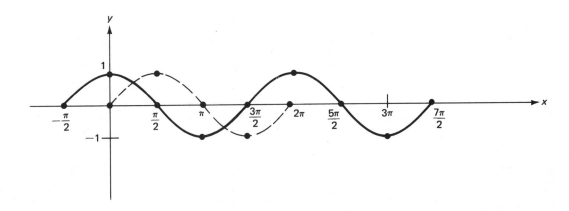

Observe that the graph is the same as the graph of $y = \cos x$. The graph of $y = \cos x$ is simply the graph of $y = \sin x$ with a phase shift. Therefore, in the next section, we need only consider the general sine function.

EXAMPLE 10.14.   Draw one cycle of the graph of $y = \sin\left(x + \dfrac{\pi}{3}\right)$.

Solution.   The phase shift is $-\dfrac{\pi}{3}$. Therefore, we must mark the x-axis in units of $\dfrac{\pi}{3}$ rather than $\dfrac{\pi}{2}$. The graph is

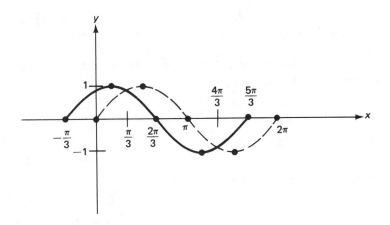

Observe that $\dfrac{\pi}{2}$ is halfway between $\dfrac{\pi}{3}$ and $\dfrac{2\pi}{3}$. It is a common error to place $\dfrac{\pi}{3}$ halfway between 0 and $\dfrac{\pi}{2}$. Clearly $\dfrac{\pi}{4}$, not $\dfrac{\pi}{3}$, is half of $\dfrac{\pi}{2}$.

EXAMPLE 10.15.   Draw one cycle of the graph of $y = \sin\left(x + \dfrac{\pi}{4}\right)$.

Solution.   The phase shift is $-\dfrac{\pi}{4}$, and the graph is

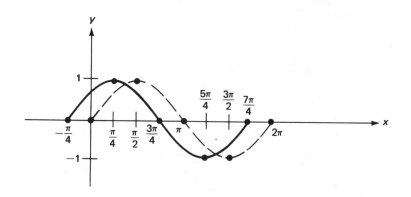

Exercise 10.3

Draw two cycles of the graph:

1. $y = \sin\left(x - \dfrac{\pi}{6}\right)$ 2. $y = \sin\left(x - \dfrac{\pi}{4}\right)$ 3. $y = \cos\left(x - \dfrac{\pi}{3}\right)$ 4. $y = \cos\left(x - \dfrac{2\pi}{3}\right)$

5. $y = \sin\left(x + \dfrac{3\pi}{2}\right)$ 6. $y = \sin\left(x + \dfrac{5\pi}{4}\right)$

Draw one cycle of the graph:

7. $y = \sin(x + \pi)$ 8. $y = \cos(x + \pi)$ 9. $y = \sin\left(x + \dfrac{3\pi}{4}\right)$ 10. $y = \sin\left(x + \dfrac{4\pi}{3}\right)$

Section 10.4   The General Sine Function

The general sine function is a function of the form $y = A \sin(Bx - D)$. The amplitude of the general sine function is

$$|A|,$$

and the period is given by

$$\frac{2\pi}{B}.$$

To find the phase shift, we write

$$Bx - D = 0$$
$$Bx = D$$
$$x = \frac{D}{B}.$$

The phase shift is given by

$$\frac{D}{B}.$$

We do not have to remember this formula to find the phase shift. However, it is important to remember that the phase shift is not simply D.

EXAMPLE 10.16.   Draw two cycles of the graph of $y = 2 \sin(2x - \pi)$.

Solution.   The amplitude is 2, and the period is π. To find the phase shift, we may write

$$y = 2 \sin(2x - \pi)$$

$$y = 2 \sin\left(2x - \frac{2\pi}{2}\right)$$

$$y = 2 \sin 2\left(x - \frac{\pi}{2}\right).$$

The phase shift is $\frac{\pi}{2}$. Observe that this method gives the same result as

$$\frac{D}{B} = \frac{\pi}{2}.$$

To draw the graph, it is easiest first to draw the graph of $y = 2 \sin 2x$, with amplitude 2 and period π:

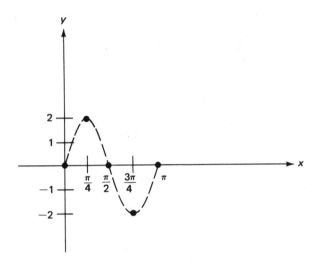

Now, we move each point $\frac{\pi}{2}$ to the right, and put a second cycle to the left of the first cycle. The graph is

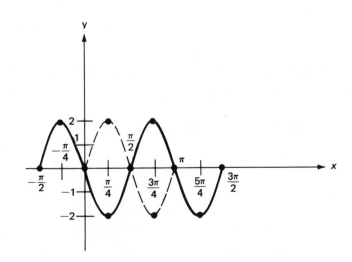

When the phase shift involves $\frac{\pi}{3}$ or $\frac{\pi}{6}$, we must be sure to mark the x-axis in appropriate units.

EXAMPLE 10.17.   Draw one cycle of the graph of $y = \frac{1}{2} \sin\left(3x + \frac{\pi}{2}\right)$.

Solution.   The amplitude is $\frac{1}{2}$, and the period is $\frac{2\pi}{3}$. To find the phase shift, we write

$$y = \frac{1}{2} \sin\left(3x + \frac{\pi}{2}\right)$$

$$y = \frac{1}{2} \sin\left(3x + \frac{3\pi}{6}\right)$$

$$y = \frac{1}{2} \sin 3\left(x + \frac{\pi}{6}\right).$$

The phase shift is $-\frac{\pi}{6}$. We must mark the x-axis in units of $\frac{\pi}{6}$. First, we draw the graph of $y = \frac{1}{2} \sin 3x$, with amplitude $\frac{1}{2}$ and period $\frac{2\pi}{3}$. Then, we move each point $\frac{\pi}{6}$ to the left: The graph is

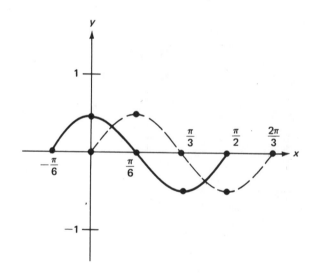

If B is a fraction less than 1, we may factor a common factor from the denominator of $Bx - D$ to find the phase shift.

EXAMPLE 10.18.   Draw one cycle of the graph of $y = 3 \sin\left(\frac{1}{3}x + \frac{\pi}{6}\right)$.

Solution.   The amplitude is 3, and the period is

$$\frac{2\pi}{\frac{1}{3}} = 2\pi\left(\frac{3}{1}\right) = 6\pi.$$

To find the phase shift, we write

$$y = 3 \sin\left(\frac{1}{3}x + \frac{\pi}{6}\right)$$

$$y = 3 \sin \frac{1}{3}\left(x + \frac{\pi}{2}\right).$$

The phase shift is $-\frac{\pi}{2}$. First, we draw the graph of $y = 3 \sin \frac{1}{3}x$, with amplitude 3 and period 6π. Then, we move each point $\frac{\pi}{2}$ to the left. The graph is

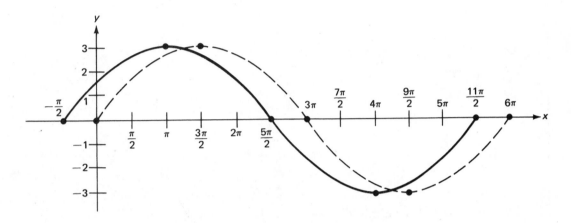

Exercise 10.4

Draw two cycles of the graph:

1. $y = 3 \sin(2x - 2\pi)$

2. $y = 2 \sin(3x - \pi)$

3. $y = \frac{1}{2} \sin\left(2x + \frac{\pi}{2}\right)$

4. $y = \frac{1}{3} \sin\left(3x + \frac{\pi}{2}\right)$

Draw one cycle of the graph:

5. $y = 2 \sin(2x + \pi)$

6. $y = \frac{1}{2} \sin\left(2x + \frac{\pi}{3}\right)$

7. $y = \frac{1}{3} \sin\left(\frac{1}{2}x + \frac{\pi}{6}\right)$

8. $y = 3 \sin\left(\frac{1}{3}x + \frac{\pi}{2}\right)$

Draw two cycles of the graph:

9. $y = 3 \sin\left(\frac{1}{3}x - \frac{\pi}{3}\right)$

10. $y = \frac{1}{2} \sin\left(\frac{1}{2}x - \pi\right)$

11. $y = \sin\left(\frac{2}{3}x - \frac{\pi}{3}\right)$

12. $y = \sin\left(\frac{3}{2}x - \frac{\pi}{2}\right)$

Self-test

1. Draw two cycles of the graph of $y = \sin\left(x - \dfrac{3\pi}{4}\right)$.

2. Draw one cycle of the graph of $y = \cos\dfrac{1}{4}x$.

3. Draw one cycle of the graph of $y = \dfrac{3}{2}\cos x$.

4. Find the amplitude, period, and phase shift of $y = 3\sin(3x + \pi)$.

5. Draw two cycles of the graph of $y = 3\sin(3x + \pi)$.

Unit 11 Cumulative Review

INTRODUCTION

In this unit you should make certain you remember all the material in all of the preceding units.

OBJECTIVE

When you have finished this unit you should be able to demonstrate that you can fulfill every objective of each preceding unit.

To prepare for this unit you should review the Self-Tests for Units 1 through 10. Do each problem of each Self-Test over again. If you cannot do a problem, or even if you have the slightest difficulty, you should:

1. Find out from the answer section the Objective for the unit to which the problem relates.
2. Review all the material in the section which has the same number as the objective, and redo all the Exercises for the section.
3. Try the Self-Test for the unit again.

Repeat these steps until you can do each problem in each Self-Test for each of Units 1 through 10 easily and accurately.

Self-test

1. Use a special triangle to find:
   a. sec 60°
   b. tan(−120°)

1a. _____

1b. _____

2. Use a unit point to find:
   a. cot 180°
   b. sin 630°

2a. _____

2b. _____

For problems 3–7, use Table II.

3. If cos θ = .8718, find all angles θ which are positive but less than 360°.

3. _____

4. If $a = 16$, $b = 8$, and $c = 10$ are sides of a triangle, find the largest angle to the nearest degrees and tens of minutes.

4. _____

5. If $\gamma = 40°50'$, $a = 120$, and $c = 150$, find all possible angles α to the nearest degrees and tens of minutes.

5. _____

6. A point on a cliff over the ocean is a vertical distance of 52 feet above the water. There is a boat on the water 22.4 feet from the base of the cliff. What is the angle of depression from the point on the cliff to the boat?

6. _____

7. Two forces combine to form a resultant. One force is 11.5 pounds and makes an angle of $50°30'$ with the resultant. If the second force is 8.5 pounds, what angle must it make with the resultant?

7. _____

8. An angle of $24°10'$ at the center of a circle subtends an arc on the circle. If the radius of the circle is 20 inches, how long is the arc?

8. _____

9. Draw the graph of $y = \cot x$ from $x = -\pi$ to $x = \pi$.

10. Draw two cycles of the graph of $y = 3 \sin(3x - \pi)$.

Unit 12

Basic Trigonometric Identities

INTRODUCTION

There are many useful relationships among the trigonometric functions. One group of relationships is called the trigonometric identities. Using the trigonometric identities, you can change a trigonometric expression into a simpler or more useful form for a given situation. In this unit you will learn six basic trigonometric identities. You will practice using these identities to change trigonometric expressions from one form to another by proving identities.

OBJECTIVES

When you have finished this unit you should be able to:

1. Use the six basic identities to write an expression in terms of sin x and cos x.
2. Prove identities involving the six basic identities.

Section 12.1

The Basic Identities

An **identity** is an equation which is true for all allowable values of the variable or variables. For example, the equation

$$x + 2 = 2 + x$$

is an identity because it is true for all values of x. The equation

$$\frac{xy}{x} = y$$

is an identity because it is true for all values of x and y except $x = 0$, which is not an allowable value of x. However, the equation

$$x^2 = 3x$$

is *not* an identity because it is true only for the values $x = 0$ and $x = 3$.

The trigonometric functions have values which are real numbers, and therefore they may be used in any algebraic identity. For example, the equation

$$\sin x + 2 = 2 + \sin x$$

is an identity. Also,

$$\frac{\sin x \cos x}{\sin x} = \cos x$$

is an identity, but any x such that $\sin x = 0$ is not an allowable value of x.

In addition to identities based on familiar algebraic manipulations, there are identities which are unique to the trigonometric functions. In this section, we will learn six such identities. We already know the first three. We know that $\csc x$ is the reciprocal of $\sin x$, $\sec x$ is the reciprocal of $\cos x$, and $\cot x$ is the reciprocal of $\tan x$. These are the **reciprocal identities**:

$$\csc x = \frac{1}{\sin x}$$

$$\sec x = \frac{1}{\cos x}$$

$$\cot x = \frac{1}{\tan x}.$$

From each reciprocal identity we can derive two related identities:

$$\sin x \csc x = 1$$

$$\cos x \sec x = 1$$

$$\tan x \cot x = 1,$$

and

$$\sin x = \frac{1}{\csc x}$$

$$\cos x = \frac{1}{\sec x}$$

$$\tan x = \frac{1}{\cot x}.$$

EXAMPLE 12.1.   Write $\sec x \csc x$ in terms of $\sin x$ and $\cos x$.

Solution.   Using the reciprocal identities

$$\sec x = \frac{1}{\cos x}$$

and

$$\csc x = \frac{1}{\sin x},$$

we have

$$\sec x \csc x = \frac{1}{\cos x} \cdot \frac{1}{\sin x}.$$

Then, ordinary algebra gives

$$\frac{1}{\cos x} \cdot \frac{1}{\sin x} = \frac{1}{\cos x \sin x}.$$

EXAMPLE 12.2.   Write $\dfrac{\sec x}{\csc x}$ in terms of $\sin x$ and $\cos x$.

Solution.   We write

$$\frac{\sec x}{\csc x} = \sec x \cdot \frac{1}{\csc x}.$$

Then, a reciprocal identity gives

$$\sec x = \frac{1}{\cos x},$$

and a related identity gives

$$\frac{1}{\csc x} = \sin x.$$

Therefore,

$$\frac{\sec x}{\csc x} = \sec x \cdot \frac{1}{\csc x}$$

$$= \frac{1}{\cos x} \cdot \sin x$$

$$= \frac{\sin x}{\cos x}.$$

So far, we have no way to write $\tan x$ and $\cot x$ in terms of $\sin x$ and $\cos x$. For this purpose, we use the next two identities, which we will call the **ratio identities**:

$$\tan x = \frac{\sin x}{\cos x}$$

$$\cot x = \frac{\cos x}{\sin x}.$$

The ratio identities are very easy to derive. Recall that, on the unit circle, $\sin \theta = y$ and $\cos \theta = x$. Therefore,

$$\tan \theta = \frac{y}{x} = \frac{\sin \theta}{\cos \theta}$$

and

$$\cot \theta = \frac{x}{y} = \frac{\cos \theta}{\sin \theta}.$$

Interpreting x as the angle in place of θ, we have the ratio identities.

Using the ratio and reciprocal identities, we can write any expression involving any of the trigonometric functions in terms of just $\sin x$ and $\cos x$.

EXAMPLE 12.3.   Write $\sec x \cot x$ in terms of $\sin x$ and $\cos x$.

Solution.   Using a reciprocal identity,

$$\sec x = \frac{1}{\cos x},$$

and using a ratio identity,

$$\cot x = \frac{\cos x}{\sin x}.$$

Therefore,

$$\sec x \cot x = \frac{1}{\cos x} \cdot \frac{\cos x}{\sin x}.$$

Now, using techniques from algebra, we can divide out a common factor from the numerator and denominator:

$$\frac{1}{\cancel{\cos x}} \cdot \frac{\cancel{\cos x}}{\sin x} = \frac{1}{\sin x}.$$

It is possible for all the factors to divide out, so that neither $\sin x$ nor $\cos x$ appears in the solution.

EXAMPLE 12.4.   Write $\cos x \tan x \csc x$ in terms of $\sin x$ and $\cos x$.

Solution.   Using a reciprocal and a ratio identity,

$$\csc x = \frac{1}{\sin x}$$

and

$$\tan x = \frac{\sin x}{\cos x}.$$

Therefore,

$$\cos x \tan x \csc x = \cancel{\cos x} \cdot \frac{\cancel{\sin x}}{\cancel{\cos x}} \cdot \frac{1}{\cancel{\sin x}}$$

$$= 1.$$

Finally, we have an identity which relates $\sin x$ and $\cos x$. It is called the **Pythagorean identity** because it is derived using the Pythagorean theorem. Recall that, on the unit circle,

$$x^2 + y^2 = 1.$$

Therefore,

$$(\cos \theta)^2 + (\sin \theta)^2 = 1,$$

or

$$(\sin \theta)^2 + (\cos \theta)^2 = 1.$$

As an example of the Pythagorean identity, consider a 30° angle in the 30–60–90 special triangle:

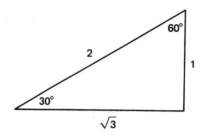

We see that

$$\sin 30° = \frac{1}{2}$$

and

$$\cos 30° = \frac{\sqrt{3}}{2}.$$

Then,

$$(\sin 30°)^2 + (\cos 30°)^2 = \left(\frac{1}{2}\right)^2 + \left(\frac{\sqrt{3}}{2}\right)^2$$

$$= \frac{1}{4} + \frac{3}{4}$$

$$= 1.$$

To avoid constant use of parentheses, it is usual to write $(\sin \theta)^2$ as $\sin^2 \theta$. Thus we agree that

$$\sin^2 \theta = (\sin \theta)^2.$$

For example,

$$\sin^2 30° = (\sin 30°)^2$$

$$= \left(\frac{1}{2}\right)^2$$

$$= \frac{1}{4}.$$

You must be careful to distinguish between $\sin^2 \theta$ and $\sin \theta^2$. For example,

$$\sin^2 30° = \left(\frac{1}{2}\right)^2 = \frac{1}{4},$$

but,

$$\sin 30°^2 = \sin 900°,$$

which you find has a value of 0, using coterminal and reference angles.

Using the notation above, the Pythagorean identity for $\sin x$ and $\cos x$ is

$$\sin^2 x + \cos^2 x = 1.$$

There are two very useful identities related to the Pythagorean identity. Subtracting $\cos^2 x$ from both sides of the Pythagorean identity, we have

$$\sin^2 x = 1 - \cos^2 x.$$

Subtracting $\sin^2 x$ from both sides, we have

$$\cos^2 x = 1 - \sin^2 x.$$

EXAMPLE 12.5.   Write $\sec x - \sin x \tan x$ in terms of $\sin x$ and $\cos x$.

Solution.   First, we write $\sec x$ and $\tan x$ in terms of $\sin x$ and $\cos x$:

$$\sec x - \sin x \tan x = \frac{1}{\cos x} - \sin x \cdot \frac{\sin x}{\cos x}.$$

Now, $\sin x \cdot \sin x = (\sin x)^2$, which we write as $\sin^2 x$:

$$\frac{1}{\cos x} - \sin x \cdot \frac{\sin x}{\cos x} = \frac{1}{\cos x} - \frac{\sin^2 x}{\cos x}.$$

Writing the difference of the two numerators over the common denominator $\cos x$,

$$\frac{1}{\cos x} - \frac{\sin^2 x}{\cos x} = \frac{1 - \sin^2 x}{\cos x}.$$

Finally, a related identity of the Pythagorean identity gives $1 - \sin^2 x = \cos^2 x$, thus

$$\frac{1 - \sin^2 x}{\cos x} = \frac{\cos^2 x}{\cos x},$$

and recalling the meaning of $\cos^2 x$,

$$\frac{\cos^2 x}{\cos x} = \frac{(\cos x)^2}{\cos x}$$

$$= \frac{\cos x \cos x}{\cos x}$$

$$= \cos x.$$

EXAMPLE 12.6.   Write $\tan x + \cot x$ in terms of $\sin x$ and $\cos x$.

Solution.   Using the ratio formulas,

$$\tan x + \cot x = \frac{\sin x}{\cos x} + \frac{\cos x}{\sin x}.$$

To combine these terms, recall from algebra that we must use a common denominator:

$$\frac{\sin x}{\cos x} + \frac{\cos x}{\sin x} = \frac{\sin x \sin x}{\sin x \cos x} + \frac{\cos x \cos x}{\sin x \cos x}$$

$$= \frac{\sin^2 x}{\sin x \cos x} + \frac{\cos^2 x}{\sin x \cos x}$$

$$= \frac{\sin^2 x + \cos^2 x}{\sin x \cos x}.$$

Using the Pythagorean identity, $\sin^2 x + \cos^2 x = 1$, we have

$$\frac{\sin^2 x + \cos^2 x}{\sin x \cos x} = \frac{1}{\sin x \cos x}.$$

EXAMPLE 12.7.   Write $\sec^2 x - 1$ in terms of $\sin x$ and $\cos x$.

Solution.   Since

$$\sec x = \frac{1}{\cos x},$$

we may write

$$\sec^2 x = \left(\frac{1}{\cos x}\right)^2$$

$$= \frac{1^2}{(\cos x)^2}$$

$$= \frac{1}{\cos^2 x}.$$

Therefore,

$$\sec^2 x - 1 = \frac{1}{\cos^2 x} - 1.$$

Using a common denominator,

$$\frac{1}{\cos^2 x} - 1 = \frac{1}{\cos^2 x} - \frac{\cos^2 x}{\cos^2 x}$$

$$= \frac{1 - \cos^2 x}{\cos^2 x}$$

$$= \frac{\sin^2 x}{\cos^2 x}.$$

EXAMPLE 12.8.   Write $\dfrac{\tan x}{\sec x - \cos x}$ in terms of $\sin x$ and $\cos x$.

Solution.   Since

$$\tan x = \frac{\sin x}{\cos x}$$

and

$$\sec x = \frac{1}{\cos x},$$

we have

$$\frac{\tan x}{\sec x - \cos x} = \frac{\dfrac{\sin x}{\cos x}}{\dfrac{1}{\cos x} - \cos x}.$$

The resulting expression is a complex fraction. There is a very efficient way to simplify complex fractions. We multiply each term in the numerator and denominator by the least common denominator of all of the fractions in the complex fraction. In this complex fraction, the least common denominator is simply cos x. Therefore,

$$\frac{\dfrac{\sin x}{\cos x}}{\dfrac{1}{\cos x} - \cos x} = \frac{\cos x \cdot \dfrac{\sin x}{\cos x}}{\cos x \cdot \dfrac{1}{\cos x} - \cos x \cdot \cos x}$$

$$= \frac{\sin x}{1 - \cos^2 x}$$

$$= \frac{\sin x}{\sin^2 x}$$

$$= \frac{1}{\sin x}.$$

In some cases, we can simplify an expression by replacing $\sin^2 x$ by $1 - \cos^2 x$ or $\cos^2 x$ by $1 - \sin^2 x$.

EXAMPLE 12.9.   Write $(\sec x - \tan x)^2$ in terms of $\sin x$ and $\cos x$.

Solution.   First, we must work out $(\sec x - \tan x)^2$. Remember that we must be careful not to forget the middle term:

$$(\sec x - \tan x)^2 = (\sec x - \tan x)(\sec x - \tan x)$$

$$= \sec^2 x - \sec x \tan x - \tan x \sec x + \tan^2 x$$

$$= \sec^2 x - 2 \sec x \tan x + \tan x^2.$$

Writing the resulting expression in terms of $\sin x$ and $\cos x$,

$$\sec^2 x - 2 \sec x \tan x + \tan^2 x = \frac{1}{\cos^2 x} - 2 \cdot \frac{1}{\cos x} \cdot \frac{\sin x}{\cos x} + \frac{\sin^2 x}{\cos^2 x}$$

$$= \frac{1}{\cos^2 x} - \frac{2 \sin x}{\cos^2 x} + \frac{\sin^2 x}{\cos^2 x}.$$

We have a common denominator, so

$$\frac{1}{\cos^2 x} - \frac{2 \sin x}{\cos^2 x} + \frac{\sin^2 x}{\cos^2 x} = \frac{1 - 2 \sin x + \sin^2 x}{\cos^2 x}.$$

Now, we can simplify this expression by replacing $\cos^2 x$ by $1 - \sin^2 x$ and factoring:

$$\frac{1 - 2 \sin x + \sin^2 x}{\cos^2 x} = \frac{1 - 2 \sin x + \sin^2 x}{1 - \sin^2 x}$$

$$= \frac{(1 - \sin x)(1 - \sin x)}{(1 + \sin x)(1 - \sin x)}$$

$$= \frac{1 - \sin x}{1 + \sin x}.$$

This expression cannot be reduced further because the numerator and denominator cannot be factored further. You must *never* divide out *addends* such as sin x in the final expression.

Exercise 12.1

Write in terms of sin x and cos x:

1. $\sin x \sec x$
2. $\cos x \csc x$
3. $\dfrac{\csc x}{\sec x}$

4. $\dfrac{\cos x}{\sec x}$
5. $\tan x \csc x$
6. $\dfrac{\sec x}{\tan x}$

7. $\sin x \cot x \sec x$
8. $\cos x \tan x \cot x \sec x$
9. $\tan x \sec x \cot x$

10. $\cot x \csc x \sec x$
11. $\csc x - \cos x \cot x$
12. $\sec^2 x - \tan^2 x$

13. $\sec^2 x + \csc^2 x$
14. $\csc x \tan x + \sec x \cot x$
15. $\csc^2 x - 1$

16. $\tan^2 x + 1$
17. $\dfrac{\cot x}{\csc x - \sin x}$
18. $\dfrac{\tan^2 x}{\sec^2 x - 1}$

19. $\dfrac{\tan x + \cot x}{\sec x + \csc x}$
20. $\dfrac{\tan x + \cot x}{\sec x \csc x}$
21. $(\csc x - \cot x)^2$

22. $\dfrac{(\sec x - 1)^2}{\tan^2 x}$
23. $\sin x \sec^2 x - \tan^2 x$
24. $\cos x \csc^2 x + \cot^2 x$

Section 12.2 Proving Identities

To prove an identity, we must show that the expressions on each side of the equality are the same. We may use the six basic identities and their variations, and also algebraic manipulations such as those we used in the preceding section. Often, the best way to prove an identity is to write the expression on each side of the equality in terms of sine and cosine.

EXAMPLE 12.10. Prove that $\sin x \cot x = \cos x$ is an identity.

Solution.   We write $\sin x \cot x$ in terms of $\sin x$ and $\cos x$:

$$\sin x \cot x \overset{?}{=} \cos x$$

$$\cancel{\sin x} \cdot \frac{\cos x}{\cancel{\sin x}} \overset{?}{=} \cos x$$

$$\cos x = \cos x.$$

Since the expressions on each side of the equality are the same, the identity is proved.

EXAMPLE 12.11.   Prove that $\sec^2 x - \tan^2 x = 1$ is an identity.

Solution.   First, we write $\sec^2 x - \tan^2 x$ in terms of $\sin x$ and $\cos x$:

$$\sec^2 x - \tan^2 x \overset{?}{=} 1$$

$$\frac{1}{\cos^2 x} - \frac{\sin^2 x}{\cos^2 x} \overset{?}{=} 1$$

$$\frac{1 - \sin^2 x}{\cos^2 x} \overset{?}{=} 1.$$

Then, we use a variation of the Pythagorean identity to replace $1 - \sin^2 x$ by $\cos^2 x$:

$$\frac{\cos^2 x}{\cos^2 x} \overset{?}{=} 1$$

$$1 = 1,$$

and the identity is proved.

We must never use any algebraic manipulation which involves both sides of the equality. For example, in proving the identity above, it is a common error to multiply both sides of the equality by $\cos^2 x$ to eliminate the fractions. We cannot use this kind of operation in proving an identity because we do not know that the two sides are equal. Indeed, that is precisely what we are to prove.

EXAMPLE 12.12.   Prove that $\dfrac{\cos x}{1 + \sin x} = \dfrac{1 - \sin x}{\cos x}$ is an identity.

Solution.   It is a common error simply to multiply across so we have $\cos^2 x$ on the left-hand side and $(1 + \sin x)(1 - \sin x)$ on the right-hand side. However, this would assume that the two sides are equal, which is what we must prove.

To prove the identity, we multiply the numerator and denominator on one side by the conjugate of the binomial. Conjugates are two binomials of the form $a + b$ and $a - b$. Thus $1 - \sin x$ is the conjugate of $1 + \sin x$, and vice versa. Multiplying the numerator and the denominator on the left-hand side by $1 - \sin x$,

$$\frac{\cos x}{1 + \sin x} \overset{?}{=} \frac{1 - \sin x}{\cos x}$$

$$\frac{\cos x(1 - \sin x)}{(1 + \sin x)(1 - \sin x)} \overset{?}{=} \frac{1 - \sin x}{\cos x}$$

$$\frac{\cos x(1 - \sin x)}{1 - \sin^2 x} \overset{?}{=} \frac{1 - \sin x}{\cos x}.$$

Then, using a variation of the Pythagorean identity,

$$\frac{\cos x(1 - \sin x)}{\cos^2 x} \overset{?}{=} \frac{1 - \sin x}{\cos x}.$$

Dividing out the common factor $\cos x$,

$$\frac{1 - \sin x}{\cos x} = \frac{1 - \sin x}{\cos x},$$

and the identity is proved. You should prove the identity by multiplying the numerator and the denominator on the right-hand side by $1 + \sin x$.

We may use basic trigonometric identities and algebraic manipulations on each side of an identity if the operations are not those that involve both sides of the equality. That is, we may reduce each side independently to the same expression.

EXAMPLE 12.13.   Prove that $\dfrac{\sin x}{1 - \cos x} = \csc x + \cot x$ is an identity.

Solution.   On the left-hand side we multiply the numerator and the denominator by $1 + \cos x$, while on the right-hand side we write the expression in terms of $\sin x$ and $\cos x$:

$$\frac{\sin x}{1 - \cos x} \overset{?}{=} \csc x + \cot x$$

$$\frac{\sin x(1 + \cos x)}{(1 - \cos x)(1 + \cos x)} \overset{?}{=} \frac{1}{\sin x} + \frac{\cos x}{\sin x}$$

$$\frac{\sin x(1 + \cos x)}{1 - \cos^2 x} \overset{?}{=} \frac{1 + \cos x}{\sin x}$$

$$\frac{\sin x(1 + \cos x)}{\sin^2 x} \overset{?}{=} \frac{1 + \cos x}{\sin x}$$

$$\frac{1 + \cos x}{\sin x} = \frac{1 + \cos x}{\sin x}.$$

Each side of the equality has been reduced independently to another expression. Since the two resulting expressions are the same, the identity is proved.

There are often several ways to prove an identity. It is usually helpful to write each expression in terms of $\sin x$ and $\cos x$, but you will still have to find appropriate basic identities and algebraic tools.

EXAMPLE 12.14.   Prove that $\dfrac{\cos^2 x}{\sin x - \sin^2 x} = \dfrac{1 + \sin x}{\sin x}$ is an identity.

Solution.   Both sides are already written in terms of $\sin x$ and $\cos x$. However, observe that the right-hand side is in terms of $\sin x$ only. We may use a variation of the Pythagorean identity to write the left-hand side in terms of $\sin x$ only, and then observe that the numerator and denominator can be factored and reduced algebraically:

$$\frac{\cos^2 x}{\sin x - \sin^2 x} \stackrel{?}{=} \frac{1 + \sin x}{\sin x}$$

$$\frac{1 - \sin^2 x}{\sin x - \sin^2 x} \stackrel{?}{=} \frac{1 + \sin x}{\sin x}$$

$$\frac{(1 + \sin x)(1 - \sin x)}{\sin x (1 - \sin x)} \stackrel{?}{=} \frac{1 + \sin x}{\sin x}$$

$$\frac{1 + \sin x}{\sin x} = \frac{1 + \sin x}{\sin x}.$$

Alternatively, we may prove the identity by multiplying the numerator and the denominator on the right-hand side by $1 - \sin x$:

$$\frac{\cos^2 x}{\sin x - \sin^2 x} \stackrel{?}{=} \frac{1 + \sin x}{\sin x}$$

$$\frac{\cos^2 x}{\sin x - \sin^2 x} \stackrel{?}{=} \frac{(1 + \sin x)(1 - \sin x)}{\sin x (1 - \sin x)}$$

$$\frac{\cos^2 x}{\sin x - \sin^2 x} \stackrel{?}{=} \frac{1 - \sin^2 x}{\sin x - \sin^2 x}$$

$$\frac{\cos^2 x}{\sin x - \sin^2 x} = \frac{\cos^2 x}{\sin x - \sin^2 x}.$$

There are two other identities related to the Pythagorean identity which can be useful. To derive the first, we start with the Pythagorean identity and divide each term by $\cos^2 x$:

$$\sin^2 x + \cos^2 x = 1$$

$$\frac{\sin^2 x}{\cos^2 x} + \frac{\cos^2 x}{\cos^2 x} = \frac{1}{\cos^2 x}$$

$$\tan^2 x + 1 = \sec^2 x.$$

Observe that since the two sides of the first equality are known to be equal, we may divide both sides of the equality. To derive the second identity, we divide both sides of the Pythagorean identity by $\sin^2 x$:

$$\sin^2 x + \cos^2 x = 1$$

$$\frac{\sin^2 x}{\sin^2 x} + \frac{\cos^2 x}{\sin^2 x} = \frac{1}{\sin^2 x}$$

$$1 + \cot^2 x = \csc^2 x$$

or

$$\cot^2 x + 1 = \csc^2 x.$$

The two identities are:

$$\tan^2 x + 1 = \sec^2 x$$

$$\cot^2 x + 1 = \csc^2 x.$$

These identities are more useful in the form

$$\tan^2 x = \sec^2 x - 1$$

$$\cot^2 x = \csc^2 x - 1.$$

EXAMPLE 12.15.   Prove that $\dfrac{\tan x}{\sec x + 1} = \sec x \cot x - \cot x$ is an identity.

Solution.   We may multiply the numerator and the denominator on the left-hand side by $\sec x - 1$ and use the new variation of the Pythagorean identity:

$$\frac{\tan x}{\sec x + 1} \overset{?}{=} \sec x \cot x - \cot x$$

$$\frac{\tan x(\sec x - 1)}{(\sec x + 1)(\sec x - 1)} \overset{?}{=} \sec x \cot x - \cot x$$

$$\frac{\tan x(\sec x - 1)}{\sec^2 x - 1} \overset{?}{=} \sec x \cot x - \cot x$$

$$\frac{\tan x(\sec x - 1)}{\tan^2 x} \overset{?}{=} \sec x \cot x - \cot x$$

$$\frac{\sec x - 1}{\tan x} \overset{?}{=} \sec x \cot x - \cot x$$

There are several ways to complete this proof. One is to write the right-hand side in terms of $\sec x$ and $\tan x$:

$$\frac{\sec x - 1}{\tan x} \overset{?}{=} \sec \cdot \frac{1}{\tan x} - \frac{1}{\tan x}$$

$$\frac{\sec x - 1}{\tan x} \overset{?}{=} \frac{\sec x}{\tan x} - \frac{1}{\tan x}$$

$$\frac{\sec x - 1}{\tan x} = \frac{\sec x - 1}{\tan x},$$

and the identity is proved.

It is not necessary to know the new variations of the Pythagorean identity to prove the identity above. We can also prove the identity by writing each expression in terms of $\sin x$ and $\cos x$ and using the algebraic techniques of Section 12.1:

$$\frac{\tan x}{\sec x + 1} \overset{?}{=} \sec x \cot x - \cot x$$

$$\frac{\dfrac{\sin x}{\cos x}}{\dfrac{1}{\cos x} + 1} \overset{?}{=} \frac{1}{\cos x} \cdot \frac{\cos x}{\sin x} - \frac{\cos x}{\sin x}$$

$$\frac{\cos x\left(\dfrac{\sin x}{\cos x}\right)}{\cos x\left(\dfrac{1}{\cos x}\right) + \cos x} \overset{?}{=} \frac{1}{\sin x} - \frac{\cos x}{\sin x}$$

$$\frac{\sin x}{1 + \cos x} \overset{?}{=} \frac{1 - \cos x}{\sin x}.$$

We may now use the method in the preceding examples:

$$\frac{\sin x(1 - \cos x)}{(1 + \cos x)(1 - \cos x)} \overset{?}{=} \frac{1 - \cos x}{\sin x}$$

$$\frac{\sin x(1 - \cos x)}{1 - \cos^2 x} \overset{?}{=} \frac{1 - \cos x}{\sin x}$$

$$\frac{\sin x(1 - \cos x)}{\sin^2 x} \overset{?}{=} \frac{1 - \cos x}{\sin x}$$

$$\frac{1 - \cos x}{\sin x} = \frac{1 - \cos x}{\sin x}.$$

Exercise 12.2

Prove that the equality is an identity. (Since there may be several correct methods of proof for some identities, answers for this Exercise cannot be given. Consult your instructor to find out if your method of proof is correct.)

1. $\cos x \tan x = \sin x$

2. $\sec x \cot x \sin x = 1$

3. $\dfrac{\sin x}{\tan x} = \cos x$

4. $\dfrac{\cot x}{\cos x} = \csc x$

5. $\csc^2 x - \cot^2 x = 1$

6. $\cot^2 x - \dfrac{\cos^2 x}{\tan^2 x} = \cos^2 x$

7. $\sin^2 x + \dfrac{\sin^2 x}{\tan^2 x} = 1$

8. $\sin^2 x \cot^2 x + \cos^2 x \tan^2 x = 1$

9. $\tan x + \cot x = \sec x \csc x$

10. $\sec^2 x + \csc^2 x = \sec^2 x \csc^2 x$

11. $\dfrac{1 + \csc x}{1 - \csc x} = \dfrac{\sin x + 1}{\sin x - 1}$

12. $\dfrac{1 + \tan x}{1 + \cot x} = \dfrac{\sin x}{\cos x}$

13. $\dfrac{\cos x}{1 - \sin x} = \dfrac{1 + \sin x}{\cos x}$

14. $\dfrac{\sin x}{1 + \cos x} = \csc x - \cot x$

15. $\dfrac{1}{1 + \sin x} = \sec^2 x - \sin x \sec^2 x$

16. $\dfrac{1 + \sin x}{1 - \sin x} = \sec^2 x + 2 \sec x \tan x + \tan^2 x$

17. $\dfrac{\sin^2 x}{\cos x + \cos^2 x} = \dfrac{1 - \cos x}{\cos x}$

18. $\dfrac{\cos^2 x}{\sin x - \sin^2 x} = \csc x + 1$

19. $\dfrac{\cot x}{\csc x + 1} = \csc x \tan x - \tan x$

20. $\dfrac{\sin^2 x}{\sec x - 1} = \cos x + \cos^2 x$

21. $\sin^4 x - \cos^4 x = \sin^2 x - \cos^2 x$

22. $\sec^4 x - 1 = \dfrac{\sin^2 x + \sin^2 x \cos^2 x}{\cos^4 x}$

23. $\tan^4 x + 2 \tan^2 x + 1 = \sec^4 x$

24. $\tan^4 x + \tan^2 x = \sec^4 x - \sec^2 x$

Summary of Identities

Reciprocal Identities

$$\csc x = \frac{1}{\sin x}$$

$$\sec x = \frac{1}{\cos x}$$

$$\cot x = \frac{1}{\tan x}$$

Variations

$$\sin x \csc x = 1$$

$$\cos x \sec x = 1$$

$$\tan x \cot x = 1$$

and

$$\sin x = \frac{1}{\csc x}$$

$$\cos x = \frac{1}{\sec x}$$

$$\tan x = \frac{1}{\cot x}$$

Ratio Identities

$$\tan x = \frac{\sin x}{\cos x}$$

$$\cot x = \frac{\cos x}{\sin x}$$

Pythagorean Identity

$$\sin^2 x + \cos^2 x = 1$$

Variations

$$\sin^2 x = 1 - \cos^2 x$$

$$\cos^2 x = 1 - \sin^2 x$$

and

$$\tan^2 x + 1 = \sec^2 x$$

$$\cot^2 x + 1 = \csc^2 x$$

and

$$\tan^2 x = \sec^2 x - 1$$

$$\cot^2 x = \csc^2 x - 1$$

Self-test

Write in terms of $\sin x$ and $\cos x$:

1. $\dfrac{\cot x}{\csc x}$

2. $\dfrac{\tan x}{\tan x + \cot x}$

1._____

2._____

Prove that the equality is an identity:

3. $\cos x \tan x \csc x = 1$

4. $\cos x + \sin x \tan x = \sec x$

5. $\dfrac{\tan x}{\sec x - 1} = \csc x + \cot x$

Unit 13   Conditional Equations

INTRODUCTION

You know from algebra courses that equation solving is one of the most important tools of mathematics. In this unit you will learn how to solve equations involving the trigonometric functions. To solve such equations, you will use the trigonometric identities and also techniques from algebra.

OBJECTIVES

When you have finished this unit you should be able to:

1. Solve equations which are linear in one trigonometric function.
2. Solve equations which are quadratic in one trigonometric function.
3. Solve equations involving trigonometric functions which are reciprocal functions or cofunctions.

Section 13.1   Linear Equations

We saw in Unit 12 that an identity is an equation which is true for all allowable values of the variable. A **conditional equation** is an equation which is true for just some values of the variable. For example,

$$x^2 = 3x$$

is a conditional equation which is true for $x = 0$ and $x = 3$.

Recall from algebra that an equation is a linear equation if it involves variables to the first power only. The equation above is not a linear equation. The equation

$$2x + 3 = 4$$

is a linear equation. An equation is linear in a trigonometric function if it involves the trigonometric function to the first power only. The equation

$$2 \sin x + 3 = 4$$

is a linear equation in sin x.

To solve a linear equation in one trigonometric function, we first solve for the function. Since the two sides of a conditional equation are given as equal, we may use usual techniques of algebra involving both sides of the equation.

EXAMPLE 13.1.   Solve $2 \sin x + 3 = 4$ for all values of x such that $0° \leqslant x < 360°$.

Solution.   First, we solve for $\sin x$:

$$2 \sin x + 3 = 4$$

$$2 \sin x = 1$$

$$\sin x = \frac{1}{2}.$$

Now, we must find all values of x where x is positive or $0°$ but less than $360°$. Recall that $\frac{1}{2}$ is the sine of a special angle which is an angle of the 30–60–90 special triangle:

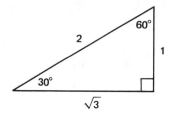

We see that

$$\sin 30° = \frac{1}{2}.$$

Since $\sin x$ is also positive in the second quadrant,

$$\sin 150° = \frac{1}{2}.$$

The solutions between $0°$ and $360°$ are $30°$ and $150°$. Observe that there are an infinite number of other solutions coterminal with $30°$ and also an infinite number of other solutions coterminal with $150°$. However, these solutions are more than $360°$ or less than $0°$. Throughout this unit, we will solve equations for all values of x such that $0° \leqslant x < 360°$.

EXAMPLE 13.2.   Solve $\sqrt{2} \cos x + 1 = 0$.

Solution.   Solving for $\cos x$,

$$\sqrt{2} \cos x + 1 = 0$$

$$\sqrt{2} \cos x = -1$$

$$\cos x = -\frac{1}{\sqrt{2}}.$$

Again we have a special angle, but it is an angle of the 45–45–90 special triangle:

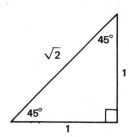

Therefore,

$$\cos 45° = \frac{1}{\sqrt{2}}.$$

Since cos x is negative, the solutions are in the second and third quadrants:

$$\cos 135° = -\frac{1}{\sqrt{2}}$$

and

$$\cos 225° = -\frac{1}{\sqrt{2}}.$$

The solutions are 135° and 225°.

EXAMPLE 13.3.   Solve $3 \cos x + 2 = 5$.

Solution.   Solving for cos x,

$$3 \cos x + 2 = 5$$

$$3 \cos x = 3$$

$$\cos x = 1.$$

In this case we have a quadrantal angle. Using a unit point and $\cos \theta = x$, we have the diagram

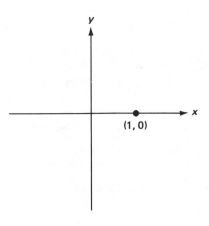

Therefore,

$$\cos 0° = 1.$$

Since there is no other unit point with $x = 1$, the only solution is 0°. Observe that we have used x in two ways, as the variable in the equation and as a coordinate of a unit point. It should be clear that these are two very different meanings. You should be able to tell from the context which use of x is meant.

The solution of an equation involving a trigonometric function need not be a special or quadrantal angle. If it is not, we use Table II or a scientific calculator.

EXAMPLE 13.4.   Solve $4 \tan x - 3 = 8$.

Solution.   Solving for $\tan x$,

$$4 \tan x - 3 = 8$$

$$4 \tan x = 11$$

$$\tan x = 2.75.$$

Using Table II, we find

$$x = 70°$$

to the nearest degrees and tens of minutes. Since $\tan x$ is also positive in the third quadrant,

$$x = 250°.$$

The solutions are 70° and 250°.

A trigonometric function may appear more than once in an equation. We use the techniques of algebra to solve for the trigonometric function, treating it just like a variable in an algebraic equation.

EXAMPLE 13.5.   Solve $4 \csc x + 2 = \csc x - 2$.

Solution.   Solving for $\csc x$,

$$4 \csc x + 2 = \csc x - 2$$

$$3 \csc x = -4$$

$$\csc x = -1.333.$$

Using Table II, the reference angle is

$$x' = 48°40'$$

to the nearest degrees and tens of minutes. Since $\csc x$ is negative in the third and fourth quadrants, the solutions are 228°40′ and 311°20′.

We may check such solutions by substitution, just as in an algebraic equation. To check 228°40′, we write

$$4 \csc x + 2 = \csc x - 2$$

$$4 \csc 228°40' + 2 \overset{?}{=} \csc 228°40' - 2.$$

Using reference angles and Table II, we find that $\csc 228°40' = -1.332$, so

$$4(-1.332) + 2 \overset{?}{=} -1.332 - 2$$

$$-3.328 \approx -3.332.$$

The two sides may not be exactly equal because of our approximations in using Table II. You should check the solution 311°20′.

It is possible for a conditional equation to have no solution.

EXAMPLE 13.6.   Solve $2 \sin x - 3 = 0$.

Solution.

$$2 \sin x - 3 = 0$$

$$2 \sin x = 3$$

$$\sin x = \frac{3}{2}.$$

But we know that $\sin x$ is never greater than 1, so there is no x such that $\sin x = \frac{3}{2}$. Therefore, the equation has no solution.

Exercise 13.1

Solve for all values of x such that $0° \leqslant x < 360°$:

1. $2 \cos x - 1 = 0$
2. $2 \sin x + \sqrt{3} = 0$
3. $2 \tan x + 2 = 0$

4. $\sqrt{3} \sec x - 2 = 0$
5. $\sqrt{3} \sin x - 2 = 0$
6. $2 \csc x + 1 = 0$

7. $2 \csc x - 3 = 1$
8. $3 \cos x + 6 = 2$
9. $4 \sin x + 5 = 1$

10. $\cot x + 2 = 2$
11. $5 \sec x - 6 = 5$
12. $9 \cos x + 11 = 3$

13. $5 \sin x + 2 = 2 \sin x + 4$
14. $3 \cot x + 12 = 5 - 3 \cot x$
15. $\frac{1}{3} \tan x + \frac{1}{2} = \frac{1}{3}$

16. $\frac{3}{5} \sin x + \frac{1}{2} = \sin x - \frac{2}{5}$

Section 13.2

Quadratic Equations

Recall from algebra that a **quadratic equation** involves variables to the second power. The equation

$$x^2 = 3x$$

is a quadratic equation. An equation is quadratic in a trigonometric function if it involves the square of the trigonometric function. The equation

$$\sin^2 x = 3 \sin x$$

is a quadratic equation in $\sin x$.

Some quadratic equations can be solved by factoring. To use this method, the equation must be in the standard form

$$ax^2 + bx + c = 0,$$

where all the nonzero terms are on one side and 0 is on the other side. Then we may be able to factor the expression of nonzero terms, and we may set each factor equal to 0. Remember that to use this method, one side of the equation must be 0.

EXAMPLE 13.7.   Solve $\sin^2 x = 3 \sin x$.

Solution.   We collect all the nonzero terms on the left-hand side of the equation:

$$\sin^2 x = 3 \sin x$$

$$\sin^2 x - 3 \sin x = 0.$$

Now, we may factor the expression of nonzero terms:

$$\sin x(\sin x - 3) = 0.$$

Setting each factor equal to zero,

$$\sin x = 0 \text{ and } \sin x - 3 = 0$$

$$\sin x = 0 \text{ and } \sin x = 3.$$

The first equation gives the solutions

$$x = 0°$$

$$x = 180°.$$

The second equation has no solutions since $\sin x$ cannot be greater than 1. Therefore, the solutions are 0° and 180°.

It is a common error to divide both sides by a common factor such as $\sin x$. You must *never* divide out a factor involving a variable unless you are sure the factor cannot be 0. The factor $\sin x$ can be 0, and indeed $\sin x = 0$ gives the solutions to the equation above. Were you to divide out $\sin x$, it would appear that the equation had no solutions. In general, if you divide out a common factor involving a variable from an equation, you will lose some solutions.

When the expression of nonzero terms in the standard form of a quadratic equation is a trinomial, we may be able to factor it into two binomial factors.

EXAMPLE 13.8.   Solve $2 \cos^2 x + \cos x = 1$.

Solution.   It is a common error simply to factor the expression on the left-hand side as we did before, and set each factor equal to 1. Remember that all the nonzero terms must be on one side, with 0 on the other side of the equation. We collect the nonzero terms on the left-hand side:

$$2 \cos^2 x + \cos x = 1$$

$$2 \cos^2 x + \cos x - 1 = 0.$$

The expression of nonzero terms can be factored into two binomial factors:

$$(\cos x + 1)(2 \cos x - 1) = 0.$$

Setting each factor equal to zero,

$$\cos x + 1 = 0 \text{ and } 2 \cos x - 1 = 0$$

$$\cos x = -1 \text{ and } \cos x = \frac{1}{2}.$$

The first equation gives the solution

$$x = 180°.$$

The second equation gives the solutions

$$x = 60°$$
$$x = 300°.$$

The solutions are 180°, 60°, and 300°.

The solutions may be checked by substitution. For 180°, we use a unit point to check that

$$\cos 180° = -1.$$

Then,

$$2\cos^2 x + \cos x = 1$$
$$2\cos^2 180° + \cos 180° \overset{?}{=} 1$$
$$2(-1)^2 + (-1) \overset{?}{=} 1$$
$$2(1) - 1 \overset{?}{=} 1$$
$$2 - 1 = 1.$$

To check 60°, we use a special triangle to check that

$$\cos 60° = \frac{1}{2}.$$

Then,

$$2\cos^2 x + \cos = 1$$
$$2\cos^2 60° + \cos 60° \overset{?}{=} 1$$
$$2\left(\frac{1}{2}\right)^2 + \frac{1}{2} \overset{?}{=} 1$$
$$2\left(\frac{1}{4}\right) + \frac{1}{2} \overset{?}{=} 1$$
$$\frac{1}{2} + \frac{1}{2} = 1.$$

You should use a reference angle to check 300°.

We have seen an example of an equation quadratic in a trigonometric function which has two solutions, and one with three solutions. Clearly such an equation can have no solutions, if neither of the equations resulting from the factors has a solution. We may also have one solution or four solutions.

EXAMPLE 13.9. Solve $2\sin^2 x - \sin x - 3 = 0$.

Solution. We factor the expression of nonzero terms and set each factor equal to zero:

$$2\sin^2 x - \sin x - 3 = 0$$
$$(\sin x + 1)(2\sin x - 3) = 0$$
$$\sin x + 1 = 0 \text{ and } 2\sin x - 3 = 0$$
$$\sin x = -1 \text{ and } \sin x = \frac{3}{2}.$$

The first equation gives the solution

$$x = 270°.$$

The second equation has no solution. Therefore, 270° is the only solution. You should check this solution.

EXAMPLE 13.10. Solve $\tan^2 x + \tan x = 0$.

Solution. In this case we have a common factor $\tan x$. Remember that you must not divide out the common factor if it involves a variable. We factor the expression of nonzero terms and set each factor equal to zero:

$$\tan^2 x + \tan x = 0$$

$$\tan x(\tan x + 1) = 0$$

$$\tan x = 0 \text{ and } \tan x + 1 = 0$$

$$\tan x = 0 \text{ and } \tan x = -1.$$

The first equation gives the solutions

$$x = 0°$$

$$x = 180°.$$

The second equation gives the solutions

$$x = 135°$$

$$x = 315°.$$

Therefore, there are four solutions: 0°, 180°, 135°, and 315°. You should check these solutions.

There is a useful method for solving equations of the form

$$ax^2 + c = 0.$$

We solve for x^2,

$$x^2 = -\frac{c}{a},$$

and take the square root, remembering that the square root can be positive or negative:

$$x = \pm\sqrt{-\frac{c}{a}}.$$

To have a real solution, either a or c must be negative. Otherwise, there is no real solution.

EXAMPLE 13.11. Solve $4\cos^2 x - 3 = 0$.

Solution. Using the method above, with $\cos x$ in place of x,

$$4\cos^2 x - 3 = 0$$

$$\cos^2 x = \frac{3}{4}$$

$$\cos x = \pm\sqrt{\frac{3}{4}}$$

$$\cos x = \pm\frac{\sqrt{3}}{2}.$$

Thus we have two equations,

$$\cos x = \frac{\sqrt{3}}{2} \text{ and } \cos x = -\frac{\sqrt{3}}{2}.$$

The first equation gives the solutions

$$x = 30°$$
$$x = 330°.$$

The second equation gives the solutions

$$x = 150°$$
$$x = 210°.$$

The solutions are 30°, 330°, 150°, and 210°. You should check these solutions.

If we cannot solve a quadratic equation by factoring, and it is not the special form above, we must use the quadratic formula. You may recall from algebra that, for a quadratic equation in standard form

$$ax^2 + bx + c = 0,$$

the quadratic formula is

$$x = \frac{-b \pm \sqrt{b^2 - 4ac}}{2a}.$$

EXAMPLE 13.12. Solve $\sin^2 x + 3 \sin x = 2$.

Solution. Remember that we must write the equation in standard form, with all of the nonzero terms on one side and 0 on the other:

$$\sin^2 x + 3 \sin x - 2 = 0.$$

A little experimenting will convince you that the expression of nonzero terms cannot be factored into integral factors. Therefore, we use the quadratic formula, with $\sin x$ in place of x, and $a = 1$, $b = 3$, and $c = -2$:

$$\sin x = \frac{-b \pm \sqrt{b^2 - 4ac}}{2a}$$

$$\sin x = \frac{-3 \pm \sqrt{3^2 - 4(1)(-2)}}{2(1)}$$

$$\sin x = \frac{-3 \pm \sqrt{9 + 8}}{2}$$

$$\sin x = \frac{-3 \pm \sqrt{17}}{2}.$$

To find any possible solutions, we use Table I or a calculator to find $\sqrt{17} \approx 4.123$. Then,

$$\frac{-3 + \sqrt{17}}{2} \approx \frac{-3 + 4.123}{2} = .5615,$$

and

$$\frac{-3 - \sqrt{17}}{2} \approx \frac{-3 - 4.123}{2} = -3.5615.$$

Therefore, we have

$$\sin x = .5615 \text{ and } \sin x = -3.5615.$$

The first equation gives the solutions $34°10'$ and $145°50'$ to the nearest degrees and tens of minutes. The second equation gives no further solutions since $\sin x$ cannot be less than -1. To check $34°10'$, we find that

$$\sin 34°10' = .5616,$$

and so

$$\sin^2 x + 3 \sin x = 2$$

$$\sin^2 34°10' + 3 \sin 34°10' \overset{?}{=} 2$$

$$(.5616)^2 + 3(.5616) \overset{?}{=} 2$$

$$2.0002 \approx 2.$$

You should check $145°50'$.

Exercise 13.2

Solve for all values of x such that $0° \leqslant x < 360°$:

1. $\cos^2 x = 3 \cos x$

2. $\tan^2 x - \tan x = 0$

3. $3 \sin^2 x + \sin x = 0$

4. $2 \sec^2 x = 3 \sec x$

5. $2 \sin^2 x + \sin x = 1$

6. $\sec^2 x + \sec x = 2$

7. $2 \cos^2 x + \cos x - 3 = 0$

8. $2 \csc^2 x - 3 \csc x + 1 = 0$

9. $2 \sec^2 x + 2 = 5 \sec x$

10. $\sqrt{3} \tan^2 x - 4 \tan x + \sqrt{3} = 0$

11. $2 \sin^2 x + \sin x - 6 = 0$

12. $2 \csc^2 x - \csc x = 0$

13. $2 \cos^2 x = 1$

14. $4 \sin^2 x - 3 = 0$

15. $\sin^2 x + \sin x = 1$

16. $\sec^2 x - \sec x = 1$

17. $\tan^2 x + 2 \tan x = 1$

18. $2 \cos^2 x = 2 \cos x + 3$

19. $\sin^2 x + \sin x = 3$

20. $\cos^2 x + \cos x + 1 = 0$

Section 13.3   Equations Related to Quadratic Equations

In the two preceding sections, we have used only the techniques of algebra to solve an equation for the trigonometric function. We can use the trigonometric identities along with algebraic techniques to solve types of equations involving more than one trigonometric function.

First, we consider a type of equation which is linear in two trigonometric functions which are reciprocal functions.

EXAMPLE 13.13.   Solve $2 \sin x - 3 \csc x = 1$.

Solution.   We use the reciprocal identity

$$\csc x = \frac{1}{\sin x}$$

to write the equation as

$$2 \sin x - 3 \csc x = 1$$

$$2 \sin x - 3\left(\frac{1}{\sin x}\right) = 1$$

$$2 \sin x - \frac{3}{\sin x} = 1.$$

Now we have an equation involving a rational expression. To solve such an equation we multiply each term, which is the same as multiplying each side, by the least common denominator. Multiplying each term by $\sin x$,

$$\sin x(2 \sin x) - \sin x\left(\frac{3}{\sin x}\right) = 1(\sin x)$$

$$2 \sin^2 x - 3 = \sin x.$$

The resulting equation is a quadratic equation in $\sin x$, which we solve for $\sin x$ using algebraic techniques:

$$2 \sin^2 x - \sin x - 3 = 0$$

$$(\sin x + 1)(2 \sin x - 3) = 0$$

$$\sin x + 1 = 0 \text{ and } 2 \sin x - 3 = 0$$

$$\sin x = -1 \text{ and } \sin x = \frac{3}{2}.$$

The first equation has one solution

$$x = 270°,$$

and the second equation has no solution. Therefore, the solution is $270°$. To check this solution, we must check in the original equation:

$$2 \sin x - 3 \csc x = 1$$

$$2 \sin 270° - 3 \csc 270° \stackrel{?}{=} 1$$

$$2(-1) - 3(-1) \stackrel{?}{=} 1$$

$$-2 + 3 = 1.$$

A second type of equation is quadratic in one or both of two trigonometric functions which are cofunctions. We use the Pythagorean identity or its variations.

EXAMPLE 13.14.   Solve $\sin^2 x + 2 \cos^2 x = 2$.

Solution.   The equation is quadratic in both $\sin x$ and $\cos x$. We may use either

$$\sin^2 x = 1 - \cos^2 x$$

or

$$\cos^2 x = 1 - \sin^2 x.$$

If we choose the first, we derive a quadratic equation in $\cos^2 x$:

$$\sin^2 x + 2 \cos^2 x = 2$$

$$1 - \cos^2 x + 2 \cos^2 x = 2$$

$$1 + \cos^2 x = 2$$

$$\cos^2 x = 1.$$

Remember that we must allow for both the positive and negative square roots of 1:

$$\cos x = \pm 1$$

$$\cos x = 1 \text{ and } \cos x = -1.$$

From the first equation

$$x = 0°,$$

and from the second equation

$$x = 180°.$$

The solutions are $0°$ and $180°$. You should check these solutions in the original equation.

EXAMPLE 13.15.   Solve $\sin x - \cos^2 x + 1 = 0$.

Solution.   Because we do not have $\sin^2 x$, we must use

$$\cos^2 x = 1 - \sin^2 x.$$

We must also be careful in using parentheses for the expression $1 - \sin^2 x$:

$$\sin x - \cos^2 x + 1 = 0$$

$$\sin x - (1 - \sin^2 x) + 1 = 0$$

$$\sin x - 1 + \sin^2 x + 1 = 0$$

$$\sin^2 x + \sin x = 0.$$

Remember that we must not divide out the common factor $\sin x$. We solve the equation by factoring:

$$\sin x(\sin x + 1) = 0$$

$$\sin x = 0 \text{ and } \sin x + 1 = 0$$

$$\sin x = 0 \text{ and } \sin x = -1.$$

From the first equation

$$x = 0°$$

$$x = 180°,$$

and from the second equation

$$x = 270°.$$

The solutions are $0°$, $180°$, and $270°$. You should check these solutions in the original equation. Also, observe that $90°$ is not a solution, and does not check in the original equation.

EXAMPLE 13.16.   Solve $\tan^2 x + \sec x + 1 = 0$.

Solution.   There are two ways to solve this equation. You may remember, or derive, the variation of the Pythagorean identity

$$\tan^2 x = \sec^2 x - 1.$$

Then,

$$\tan^2 x + \sec x + 1 = 0$$

$$\sec^2 x - 1 + \sec x + 1 = 0$$

$$\sec^2 x + \sec x = 0$$

$$\sec x(\sec x + 1) = 0$$

$$\sec x = 0 \text{ and } \sec x + 1 = 0$$

$$\sec x = 0 \text{ and } \sec x = -1.$$

The first equation has no solution, and from the second equation

$$x = 180°.$$

Therefore, the only solution is $180°$. You should check this solution.

We may also solve the equation by writing the equation in terms of $\sin x$ and $\cos x$:

$$\tan^2 x + \sec x + 1 = 0$$

$$\frac{\sin^2 x}{\cos^2 x} + \frac{1}{\cos x} + 1 = 0.$$

Now, we must multiply every term by the least common denominator $\cos^2 x$:

$$\cos^2 x\left(\frac{\sin^2 x}{\cos^2 x}\right) + \cos^2 x\left(\frac{1}{\cos x}\right) + 1(\cos^2 x) = 0(\cos^2 x)$$

$$\sin^2 x + \cos x + \cos^2 x = 0.$$

We must use $\sin^2 x = 1 - \cos^2 x$, so we have

$$1 - \cos^2 x + \cos x + \cos^2 x = 0$$

$$1 + \cos x = 0$$

$$\cos x = -1.$$

The only solution is $180°$.

As in Section 13.2, we may encounter equations which we cannot solve by factoring. As before, we use the quadratic formula.

EXAMPLE 13.17.   Solve $2 \sin^2 x + 2 \cos x = 1$.

Solution.   Using $\sin^2 x = 1 - \cos^2 x$, we derive a quadratic equation in $\cos x$:

$$2 \sin^2 x + 2 \cos x = 1$$

$$2(1 - \cos^2 x) + 2 \cos x = 1$$

$$2 - 2 \cos^2 x + 2 \cos x = 1$$

$$0 = 2 \cos^2 x - 2 \cos x - 1.$$

We use the quadratic formula with $\cos x$ in place of x:

$$\cos x = \frac{-b \pm \sqrt{b^2 - 4ac}}{2a}$$

$$\cos x = \frac{-(-2) \pm \sqrt{(-2)^2 - 4(2)(-1)}}{2(2)}$$

$$\cos x = \frac{2 \pm \sqrt{4 + 8}}{4}$$

$$\cos x = \frac{2 \pm \sqrt{12}}{4}$$

The radical expression may be reduced, but this is not necessary since we will approximate it by a decimal. Using $\sqrt{12} \approx 3.464$,

$$\frac{2 + \sqrt{12}}{4} \approx \frac{2 + 3.464}{4} = 1.366,$$

so there is no solution from this part.

$$\frac{2 - \sqrt{12}}{4} \approx \frac{2 - 3.464}{4} = -0.3660.$$

The solutions for

$$\cos x = -0.3660$$

are 111°30′ and 248°30′ to the nearest degrees and tens of minutes. Checking 111°30′,

$$2 \sin^2 x + 2 \cos x = 1$$

$$2 \sin^2 111°30′ + 2 \cos 111°30′ \overset{?}{=} 1$$

$$2(.9304)^2 + 2(-.3665) \overset{?}{=} 1$$

$$.9983 \approx 1.$$

You should check 248°30′.

Another type of equation involving $\sin x$ and $\cos x$ poses an interesting problem in using the variations of the Pythagorean identity.

EXAMPLE 13.18.   Solve $\sin x + 2 \cos x = 1$.

Solution.   Neither $\sin x$ nor $\cos x$ is squared, so we cannot use the Pythagorean identity. However, since

$$\sin^2 x = 1 - \cos^2 x,$$

we may use

$$\sin x = \pm \sqrt{1 - \cos^2 x} \, .$$

Then,

$$\sin x + 2 \cos x = 1$$

$$\pm \sqrt{1 - \cos^2 x} + 2 \cos x = 1.$$

We have an equation involving a radical expression. Recall from algebra that to solve such an equation we must isolate the radical and then square each side:

$$\pm \sqrt{1 - \cos^2 x} = 1 - 2 \cos x$$

$$\left(\pm \sqrt{1 - \cos^2 x}\right)^2 = (1 - 2 \cos x)^2.$$

Squaring either the positive or negative root gives $1 - \cos^2 x$. Also remember that, in squaring the right-hand side, we must be careful not to forget the middle term. We have

$$1 - \cos^2 x = 1 - 4 \cos x + 4 \cos^2 x$$

$$0 = 5 \cos^2 x - 4 \cos x$$

$$0 = \cos x (5 \cos x - 4)$$

$$\cos x = 0 \text{ and } 5 \cos x - 4 = 0$$

$$\cos x = 0 \text{ and } \cos x = \frac{4}{5}.$$

The possible solutions are 90°, 270°, 36°50′, and 323°10′. But note that when we square both sides of the equation we lose minus signs. We may have extraneous solutions belonging to related equations. We must check every solution. For $x = 90°$,

$$\sin x + 2 \cos x = 1$$

$$\sin 90° + 2 \cos 90° \overset{?}{=} 1$$

$$1 + 2(0) = 1.$$

Therefore, 90° is a solution. For $x = 270°$,

$$\sin x + 2 \cos x = 1$$

$$\sin 270° + 2 \cos 270° \overset{?}{=} 1$$

$$-1 + 2(0) \neq 1.$$

Therefore, 270° is not a solution. A related equation is the original equation with one or more signs changed. Observe that 270° is a solution to the related equation

$$-\sin x + 2 \cos x = 1.$$

For $x = 36°50'$,

$$\sin x + 2 \cos x = 1$$

$$\sin 36°50' + 2 \cos 36°50' \overset{?}{=} 1$$

$$.5995 + 2(.8004) \overset{?}{=} 1$$

$$2.2 \neq 1.$$

Therefore, 36°50' is not a solution. Observe that 36°50' is also a solution to the related equation

$$-\sin x + 2 \cos x = 1.$$

For $x = 323°10'$,

$$\sin x + 2 \cos x = 1$$

$$\sin 323°10' + 2 \cos 323°10' \overset{?}{=} 1$$

$$-.5995 + 2(.8004) \overset{?}{=} 1$$

$$1.001 \approx 1.$$

Therefore, 323°10' is a solution. The solutions of the original equation are 90° and 323°10'.

Exercise 13.3

Solve for all values of x such that $0° \leqslant x < 360°$:

1. $\sin x - 2 \csc x = 1$ 2. $2 \cos x - \sec x + 1 = 0$ 3. $\cos x + \sec x = 2$

4. $\tan x + \cot x = 2$ 5. $\cos^2 x - \sin^2 x + 1 = 0$ 6. $\cos^2 x - \sin^2 x = 1$

7. $\sin^2 x - 3\cos^2 x = 0$

8. $\sec^2 x + 2\tan^2 x = 2$

9. $\sin^2 x - \cos x = 1$

10. $2\cos^2 x + \sin x = 2$

11. $\sin^2 x + \cos x + 1 = 0$

12. $2\sin^2 x + \cos x = 1$

13. $\sin x - \cos^2 x = 1$

14. $\cos x - 2\sin^2 x + 1 = 0$

15. $\tan^2 x + \sec x = 1$

16. $2\cot^2 x + \csc x + 1 = 0$

17. $\cos^2 x - \sin x = 0$

18. $2\sin^2 x + \cos x = 0$

19. $3\sin x + \cos^2 x + 1 = 0$

20. $\sin^2 x + \cos x + 2 = 0$

21. $\sin x + \cos x = 1$

22. $\cos x - \sin x = 1$

23. $\cos x - 2\sin x = 1$

24. $\sin x + \cos x = 2$

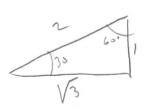

$$\frac{\sin^2 x + \cos^2 x = 1}{\cos^2 x \qquad \cos^2 x \quad \cos^2 x}$$

$$\tan^2 x + 1 = \sec^2 x$$

Self-test

Solve for all values of x such that $0° \leqslant x < 360°$:

1. $2 \cos x + 5 = 4$

                                                       1._____

2. $\sin^2 x - \cos^2 x = 0$

                                                       2._____

3. $2 \sin x - \csc x + 1 = 0$

                                                       3._____

4. $2 \cos^2 x + \cos x = 2$

                                                       4._____

5. $\cos^2 x = - \cos x$

                                                       5._____

Unit 14

Other Identities

INTRODUCTION

In the preceding units you have seen that trigonometric identities are essential, along with algebraic techniques, in simplifying expressions and solving equations involving trigonometric functions. In this unit you will learn another basic type of identity dealing with the sum or difference of two angles. From these identities you can derive other identities, in particular, identities for twice an angle and for half an angle. You will use these identities to prove other identities and solve other types of conditional equations.

OBJECTIVES

When you have finished this unit you should be able to:

1. Prove identities involving $\sin(\theta \pm x)$, $\cos(\theta \pm x)$, $\tan(\theta \pm x)$, and $\cot(\theta \pm x)$, where θ is a quadrantal angle.
2. Prove identities and solve conditional equations involving $\sin 2x$ and $\cos 2x$.
3. Prove identities and solve conditional equations involving $\sin\dfrac{x}{2}$ and $\cos\dfrac{x}{2}$.

Section 14.1

The Sum and Difference Identities

So far, our identities have involved one angle x. Now, we will consider a type of identity involving two angles. We will call the two angles α and β. First, we consider $\sin(\alpha + \beta)$. It is a common error to think that $\sin(\alpha + \beta)$ must be the same as $\sin\alpha + \sin\beta$. It is easy to show that this is not the case. For example, suppose $\alpha = 60°$ and $\beta = 30°$. Then,

$$\sin(\alpha + \beta) = \sin(60° + 30°)$$
$$= \sin 90°$$
$$= 1.$$

But,

$$\sin\alpha + \sin\beta = \sin 60° + \sin 30°$$
$$= \frac{\sqrt{3}}{2} + \frac{1}{2}$$
$$= \frac{\sqrt{3} + 1}{2}$$
$$\approx 1.366.$$

Therefore, in general, $\sin(\alpha + \beta) \neq \sin\alpha + \sin\beta$.

We should observe that the type of proof above is a proof by **counterexample**. A counterexample shows that something is *not* true. If we want to prove that something is, in general, true, then we must prove it for all allowable values. We did this when we proved identities for all allowable values of x using sin x, cos x, and so on. If we want to prove that something is not true, then we need to find just one instance for which it is not true. To do this we may use a counterexample.

EXAMPLE 14.1.   Use $\alpha = 60°$ and $\beta = 30°$ to prove by counterexample that $\sin(\alpha - \beta) \neq \sin \alpha - \sin \beta$.

Solution.

$$\sin(\alpha - \beta) = \sin(60° - 30°)$$
$$= \sin 30°$$
$$= \frac{1}{2},$$

and

$$\sin \alpha - \sin \beta = \sin 60° - \sin 30°$$
$$= \frac{\sqrt{3}}{2} - \frac{1}{2}$$
$$= \frac{\sqrt{3} - 1}{2}$$
$$\approx .366.$$

Therefore, $\sin(\alpha - \beta) \neq \sin \alpha - \sin \beta$.

The identities for $\sin(\alpha + \beta)$ and $\sin(\alpha - \beta)$ are given by two of the **sum and difference identities**.

Sum and Difference Identities for the Sine:

$$\sin(\alpha + \beta) = \sin \alpha \cos \beta + \cos \alpha \sin \beta$$
$$\sin(\alpha - \beta) = \sin \alpha \cos \beta - \cos \alpha \sin \beta$$

Observe that these identities are the same except for the minus sign. Although a specific example does not constitute a proof that something is true in general, an example can be enlightening. Again suppose $\alpha = 60°$ and $\beta = 30°$. Then,

$$\sin \alpha \cos \beta + \cos \alpha \sin \beta = \sin 60° \cos 30° + \cos 60° \sin 30°$$
$$= \frac{\sqrt{3}}{2} \cdot \frac{\sqrt{3}}{2} + \frac{1}{2} \cdot \frac{1}{2}$$
$$= \frac{3}{4} + \frac{1}{4}$$
$$= 1,$$

which is indeed the same as $\sin(60° + 30°)$. You should show that the difference identity also works in the case that $\alpha = 60°$ and $\beta = 30°$. The proofs of these identities are given at the end of this unit.

We use the sum and difference identities to prove another type of identity.

EXAMPLE 14.2.   Prove that $\sin(\pi - x) = \sin x$ is an identity.

Solution.   We use the difference identity with $\alpha = \pi$ and $\beta = x$:

$$\sin(\alpha - \beta) = \sin \alpha \cos \beta - \cos \alpha \sin \beta$$

$$\sin(\pi - x) = \sin \pi \cos x - \cos \pi \sin x.$$

Now, recall from Section 8.3 that, since π radians $= 180°$,

$$\sin \pi = \sin 180° = 0$$

and

$$\cos \pi = \cos 180° = -1.$$

Then,

$$\sin(\pi - x) \overset{?}{=} \sin x$$

$$\sin \pi \cos x - \cos \pi \sin x \overset{?}{=} \sin x$$

$$(0)(\cos x) - (-1)(\sin x) \overset{?}{=} \sin x$$

$$0 + (1)(\sin x) \overset{?}{=} \sin x$$

$$\sin x = \sin x.$$

We have similar identities for $\cos(\alpha + \beta)$ and $\cos(\alpha - \beta)$.

Sum and Difference Identities for the Cosine:

$$\cos(\alpha + \beta) = \cos \alpha \cos \beta - \sin \alpha \sin \beta$$

$$\cos(\alpha - \beta) = \cos \alpha \cos \beta + \sin \alpha \sin \beta$$

Observe that these identities are the same except for the minus sign, and that the minus sign is in the identity for the sum of the angles. The proofs of these identities are given at the end of this unit.

We may also use the sum and difference identities for the cosine to prove the new type of identity.

EXAMPLE 14.3.   Prove that $\cos\left(\dfrac{\pi}{2} + x\right) = -\sin x$ is an identity.

Solution.   Using $\alpha = \dfrac{\pi}{2}$ and $\beta = x$,

$$\cos\left(\frac{\pi}{2} + x\right) \overset{?}{=} -\sin x$$

$$\cos \frac{\pi}{2} \cos x - \sin \frac{\pi}{2} \sin x \overset{?}{=} -\sin x$$

$$(0)(\cos x) - (1)(\sin x) \overset{?}{=} -\sin x$$

$$0 - \sin x \overset{?}{=} -\sin x$$

$$-\sin x = -\sin x.$$

We do not need sum and difference identities for the other trigonometric functions because each of the other trigonometric functions can be written in terms of the sine and cosine. For example, recall the ratio identity from Section 12.1:

$$\tan x = \frac{\sin x}{\cos x}.$$

If $x = \alpha + \beta$, we have

$$\tan(\alpha + \beta) = \frac{\sin(\alpha + \beta)}{\cos(\alpha + \beta)}.$$

EXAMPLE 14.4.   Prove that $\tan(\pi - x) = -\tan x$ is an identity.

Solution.   If $\alpha = \pi$ and $\beta = x$,

$$\tan(\pi - x) \overset{?}{=} -\tan x$$

$$\frac{\sin(\pi - x)}{\cos(\pi - x)} \overset{?}{=} -\frac{\sin x}{\cos x}$$

$$\frac{\sin \pi \cos x - \cos \pi \sin x}{\cos \pi \cos x + \sin \pi \sin x} \overset{?}{=} -\frac{\sin x}{\cos x}$$

$$\frac{(0)(\cos x) - (-1)(\sin x)}{(-1)(\cos x) + (0)(\sin x)} \overset{?}{=} -\frac{\sin x}{\cos x}$$

$$\frac{0 + \sin x}{-\cos x + 0} \overset{?}{=} -\frac{\sin x}{\cos x}$$

$$-\frac{\sin x}{\cos x} = -\frac{\sin x}{\cos x}.$$

Exercise 14.1

(Answers for proofs are not given.)

Use $\alpha = 60°$ and $\beta = 30°$ to prove by counterexample:

1. $\cos(\alpha + \beta) \neq \cos \alpha + \cos \beta$

2. $\cos(\alpha - \beta) \neq \cos \alpha - \cos \beta$

3. $\tan(\alpha + \beta) \neq \tan \alpha + \tan \beta$

4. $\tan(\alpha - \beta) \neq \tan \alpha - \tan \beta$

Prove that the equality is an identity:

5. $\sin(\pi + x) = -\sin x$

6. $\sin\left(\frac{3\pi}{2} - x\right) = -\cos x$

7. $\cos(\pi + x) = -\cos x$

8. $\cos\left(\frac{3\pi}{2} + x\right) = \sin x$

9. $\tan(\pi + x) = \tan x$

10. $\tan\left(\frac{\pi}{2} - x\right) = \cot x$

11. $\cot(\pi - x) = -\cot x$

12. $\cot\left(\frac{\pi}{2} + x\right) = -\tan x$

13. $\sin(\alpha + \beta) \sin \alpha + \cos(\alpha + \beta) \cos \alpha = \cos \beta$

14. $\cos(\alpha - \beta) \cos \beta - \sin(\alpha - \beta) \sin \beta = \cos \alpha$

15. $\tan(\alpha + \beta) = \dfrac{\tan \alpha + \tan \beta}{1 - \tan \alpha \tan \beta}$

16. $\cot(\alpha + \beta) = \dfrac{\cot \alpha \cot \beta - 1}{\cot \beta + \cot \alpha}$

Section 14.2   The Double-Angle Identities

We use the identities for the sum of two angles to derive another set of identities. Consider the identity

$$\sin(\alpha + \beta) = \sin \alpha \cos \beta + \cos \alpha \sin \beta,$$

and let $\alpha = x$ and also $\beta = x$. Then, we have

$$\sin(x + x) = \sin x \cos x + \cos x \sin x$$
$$= \sin x \cos x + \sin x \cos x$$
$$= 2 \sin x \cos x.$$

But, $\sin(x + x) = \sin 2x$, therefore

$$\sin 2x = 2 \sin x \cos x.$$

To illustrate this identity, we will use it to find $\sin 120°$. Since $120° = 2(60°)$,

$$\sin 120° = \sin 2(60°)$$
$$= 2 \sin 60° \cos 60°$$
$$= 2 \cdot \frac{\sqrt{3}}{2} \cdot \frac{1}{2}$$
$$= \frac{\sqrt{3}}{2}.$$

You can use a reference angle to verify this result.
Similarly, consider the identity

$$\cos(\alpha + \beta) = \cos \alpha \cos \beta - \sin \alpha \sin \beta.$$

If we let $\alpha = x$ and also $\beta = x$, then

$$\cos(x + x) = \cos x \cos x - \sin x \sin x$$
$$= \cos^2 x - \sin^2 x.$$

But, $\cos(x + x) = \cos 2x$, therefore

$$\cos 2x = \cos^2 x - \sin^2 x.$$

To illustrate this identity, we will find cos 120°. Since 120° = 2(60°),

$$\cos 120° = \cos 2(60°)$$

$$= \cos^2 60° - \sin^2 60°$$

$$= \left(\frac{1}{2}\right)^2 - \left(\frac{\sqrt{3}}{2}\right)^2$$

$$= \frac{1}{4} - \frac{3}{4}$$

$$= -\frac{1}{2}.$$

Again, you can use a reference angle to verify this result.

Since the two identities we have derived involve twice an angle, they are called the **double-angle identities**.

The Double-Angle Identities:

$$\sin 2x = 2 \sin x \cos x$$
$$\cos 2x = \cos^2 x - \sin^2 x$$

The double-angle identities are useful in proving other identities and in solving equations.

EXAMPLE 14.5.   Prove that $\sin 2x \sec^2 x = 2 \tan x$ is an identity.

Solution.   We use the double-angle identity to replace $\sin 2x$, and write both the expressions in terms of $\sin x$ and $\cos x$:

$$\sin 2x \sec^2 x \overset{?}{=} 2 \tan x$$

$$2 \sin x \cos x \cdot \frac{1}{\cos^2 x} \overset{?}{=} 2 \cdot \frac{\sin x}{\cos x}$$

$$\frac{2 \sin x \, \cancel{\cos x}}{\cos x \, \cancel{\cos x}} \overset{?}{=} \frac{2 \sin x}{\cos x}$$

$$\frac{2 \sin x}{\cos x} = \frac{2 \sin x}{\cos x}.$$

EXAMPLE 14.6.   Prove that $\dfrac{\cos 2x}{\cos^2 x} = 1 - \tan^2 x$ is an identity.

Solution.   We use the double-angle identity to replace $\cos 2x$, and write both the expressions in terms of $\sin x$ and $\cos x$:

$$\frac{\cos 2x}{\cos^2 x} \overset{?}{=} 1 - \tan^2 x$$

$$\frac{\cos^2 x - \sin^2 x}{\cos^2 x} \overset{?}{=} 1 - \frac{\sin^2 x}{\cos^2 x}.$$

There are two ways to complete the proof. We may separate the fractions on the left-hand side, or we may use a common denominator to combine the terms on the right-hand side. Using the first method,

$$\frac{\cos^2 x}{\cos^2 x} - \frac{\sin^2 x}{\cos^2 x} = 1 - \frac{\sin^2 x}{\cos^2 x}$$

$$1 - \frac{\sin^2 x}{\cos^2 x} = 1 - \frac{\sin^2 x}{\cos^2 x}.$$

You should complete the proof using the second method.

EXAMPLE 14.7.   Solve $\sin 2x = \cos x$ for all values of x such that $0° \leqslant x < 360°$.

Solution.   We use the double-angle identity to replace $\sin 2x$:

$$\sin 2x = \cos x$$

$$2 \sin x \cos x = \cos x.$$

Remember that we must not divide out a factor containing a variable. We collect the nonzero terms on one side and factor out the common factor $\cos x$:

$$2 \sin x \cos x - \cos x = 0$$

$$\cos x(2 \sin x - 1) = 0.$$

Since each factor contains only one trigonometric function, we can set each factor equal to 0 and solve for the trigonometric function:

$$\cos x = 0 \text{ and } 2 \sin x - 1 = 0$$

$$\cos x = 0 \text{ and } \sin x = \frac{1}{2}.$$

The solutions are 90°, 270°, 30°, and 150°.

We should check these solutions in the original equation. To check 90°,

$$\sin 2x = \cos x$$

$$\sin 2(90°) \overset{?}{=} \cos 90°$$

$$\sin 180° \overset{?}{=} \cos 90°$$

$$0 = 0.$$

To check 270°,

$$\sin 2x = \cos x$$

$$\sin 2(270°) \overset{?}{=} \cos 270°$$

$$\sin 540° \overset{?}{=} \cos 270°.$$

Recall from Section 4.1 that 540° is coterminal with 180°:

$$\sin 180° \overset{?}{=} \cos 270°$$

$$0 = 0.$$

To check 30°,

$$\sin 2x = \cos x$$

$$\sin 2(30°) \overset{?}{=} \cos 30°$$

$$\sin 60° \overset{?}{=} \cos 30°$$

$$\frac{\sqrt{3}}{2} = \frac{\sqrt{3}}{2}.$$

You should use a reference angle to check 150°.

To prove some identities and solve some equations involving cos 2x, there are two useful variations of the double-angle identity for the cosine. Sometimes we need an identity involving only sin x. Then, using the variation of the Pythagorean identity

$$\cos^2 x = 1 - \sin^2 x,$$

we may write

$$\cos 2x = \cos^2 x - \sin^2 x$$

$$\cos 2x = 1 - \sin^2 x - \sin^2 x$$

$$\cos 2x = 1 - 2\sin^2 x.$$

It is a common error to confuse the variation of the double-angle identity with the Pythagorean identity. Indeed, it is a common error to confuse cos 2x with cos² x. First observe that, for example,

$$\cos 2(60°) = \cos 120° = -\frac{1}{2},$$

whereas

$$\cos^2 60° = \left(\frac{1}{2}\right)^2 = \frac{1}{4}.$$

Then, observe that the variation of the double-angle identity is

$$\cos 2x = 1 - 2\sin^2 x$$

and the variation of the Pythagorean identity is

$$\cos^2 x = 1 - \sin^2 x.$$

To know which identity to use, you must be careful to observe whether you are replacing cos 2x or cos² x.

A second variation of the double-angle identity for the cosine is derived using the variation of the Pythagorean identity

$$\sin^2 x = 1 - \cos^2 x.$$

If we need an identity involving only cos x, we may write

$$\cos 2x = \cos^2 x - \sin^2 x$$

$$\cos 2x = \cos^2 x - (1 - \cos^2 x)$$

$$\cos 2x = \cos^2 x - 1 + \cos^2 x$$

$$\cos 2x = 2\cos^2 x - 1.$$

> **Variations of the Double-Angle Identity for cos 2x:**
>
> $$\cos 2x = 1 - 2\sin^2 x$$
> $$\cos 2x = 2\cos^2 x - 1$$

EXAMPLE 14.8. Prove that $\dfrac{\cos 2x}{\sin x} = \csc x - 2\sin x$ is an identity.

Solution.   Since the right-hand side will be in terms of sin x only, we use the variation of the double-angle identity for cos $2x$ which involves only sin x:

$$\frac{\cos 2x}{\sin x} \stackrel{?}{=} \csc x - 2\sin x$$

$$\frac{1 - 2\sin^2 x}{\sin x} \stackrel{?}{=} \frac{1}{\sin x} - 2\sin x.$$

Again, we may separate the fractions on the left-hand side or use a common denominator on the right-hand side. Using the first method,

$$\frac{1}{\sin x} - \frac{2\sin^2 x}{\sin x} \stackrel{?}{=} \frac{1}{\sin x} - 2\sin x$$

$$\frac{1}{\sin x} - 2\sin x = \frac{1}{\sin x} - 2\sin x.$$

EXAMPLE 14.9. Solve $\cos x + \cos 2x = 2$ for all values of x such that $0° \leqslant x < 360°$.

Solution.   To derive an equation quadratic in cos x only, we use the variation of the double-angle identity for cos $2x$ which involves only cos x:

$$\cos x + \cos 2x = 2$$

$$\cos x + 2\cos^2 x - 1 = 2$$

$$2\cos^2 x + \cos x - 3 = 0.$$

This equation can be solved by factoring:

$$(\cos x - 1)(2\cos x + 3) = 0$$

$$\cos x - 1 = 0 \text{ and } 2\cos x + 3 = 0$$

$$\cos x = 1 \text{ and } \cos x = -\frac{3}{2}.$$

There are no solutions from the second equation, so the only solution is 0°. You should check this solution.

We do not need identities for tan $2x$ and cot $2x$ because

$$\tan 2x = \frac{\sin 2x}{\cos 2x}$$

and

$$\cot 2x = \frac{\cos 2x}{\sin 2x}.$$

EXAMPLE 14.10.   Prove that $2 \cot 2x = \csc x \sec x - 2 \tan x$ is an identity

Solution.   We write $\cot 2x$ in terms of $\cos 2x$ and $\sin 2x$:

$$2 \cot 2x \overset{?}{=} \csc x \sec x - 2 \tan x$$

$$\frac{2 \cos 2x}{\sin 2x} \overset{?}{=} \csc x \sec x - 2 \tan x.$$

We cannot tell yet which form of $\cos 2x$ will be most convenient, so we will leave $\cos 2x$ and write the other parts in terms of $\sin x$ and $\cos x$:

$$\frac{2 \cos 2x}{2 \sin x \cos x} \overset{?}{=} \frac{1}{\sin x \cos x} - \frac{2 \sin x}{\cos x}$$

$$\frac{\cos 2x}{\sin x \cos x} \overset{?}{=} \frac{1}{\sin x \cos x} - \frac{2 \sin^2 x}{\sin x \cos x}$$

$$\frac{\cos 2x}{\sin x \cos x} \overset{?}{=} \frac{1 - 2 \sin^2 x}{\sin x \cos x}.$$

Now we see that clearly we need the variation of the double-angle identity for $\cos 2x$ in terms of $\sin x$ only:

$$\frac{1 - 2 \sin^2 x}{\sin x \cos x} = \frac{1 - 2 \sin^2 x}{\sin x \cos x}.$$

Exercise 14.2

(Answers for proofs are not given.)
Prove that the equality is an identity:

1. $\dfrac{\sin 2x}{2 \sin x} = \cos x$

2. $\sin 2x \csc x = 2 \cos x$

3. $\sin 2x \tan x = 2 \sin^2 x$

4. $\dfrac{2 \cos x}{\sin 2x} = \csc x$

5. $\dfrac{\cos 2x}{\cos x \sin x} = \cot x - \tan x$

6. $\dfrac{\cos 2x}{\sin^2 x} = \cot^2 x - 1$

7. $\dfrac{\cos 2x}{\cos x} = 2 \cos x - \sec x$

8. $\cos 2x \csc^2 x = \csc^2 x - 2$

9. $\cos 2x \cot x = \cot x - \sin 2x$

10. $\sin 2x - \tan x = \cos 2x \tan x$

11. $\dfrac{2}{1 + \cos 2x} = \sec^2 x$

12. $\dfrac{2}{1 - \cos 2x} = \csc^2 x$

13. $\dfrac{\sin 2x}{\cos 2x + 1} = \tan x$

14. $\dfrac{\sin 2x}{\cos 2x - 1} = - \cot x$

15. $\tan 2x = \dfrac{2 \tan x}{1 - \tan^2 x}$

16. $\cot 2x = \dfrac{\cot^2 x - 1}{2 \cot x}$

17. $2 \cot 2x = 2 \cot x - \sec x \csc x$

18. $2 \cot 2x = \cot x - \tan x$

19. $\sin 3x = 3 \sin x - 4 \sin^3 x$
     (Write $\sin 3x = \sin(2x + x)$ and use the identity for the sum of two angles.)

20. $\cos 3x = 4 \cos^3 x - 3 \cos x$

Solve for all values of x such that $0° \leqslant x < 360°$:

21. $\dfrac{1}{2} \sin 2x = \sin x$ 22. $\dfrac{1}{2} \sin 2x = -\cos x$

23. $\cos 2x + \cos x + 1 = 0$ 24. $\cos 2x + \sin x = 1$

25. $\cos 2x = \sin x$ 26. $\cos 2x = \cos x$

27. $\cos 2x = \cos^2 x$ 28. $\cos 2x = 2 \sin^2 x - 2$

29. $\cos 2x = \sin x - \sin^2 x$ 30. $\cos 2x - \cos^2 x = \cos x$

Section 14.3   The Half-Angle Identities

We derive another set of identities from the variations of the double-angle identity for the cosine. Recall from Section 14.2 that we derived the two variations

$$\cos 2x = 1 - 2 \sin^2 x$$

and

$$\cos 2x = 2 \cos^2 x - 1.$$

Starting with the first variation, and temporarily replacing x by α,

$$\cos 2\alpha = 1 - 2 \sin^2 \alpha$$

$$2 \sin^2 \alpha = 1 - \cos 2\alpha$$

$$\sin^2 \alpha = \frac{1 - \cos 2\alpha}{2}$$

$$\sin \alpha = \pm\sqrt{\frac{1 - \cos 2\alpha}{2}} \ .$$

Now, we replace α, not by x, but by $\dfrac{x}{2}$. Then,

$$\alpha = \frac{x}{2}$$

$$2\alpha = x.$$

Therefore,

$$\sin\left(\frac{x}{2}\right) = \pm\sqrt{\frac{1 - \cos x}{2}} \ .$$

Similarly, starting with the second variation of the double-angle identity, and using α,

$$\cos 2\alpha = 2\cos^2 \alpha - 1$$

$$1 + \cos 2\alpha = 2\cos^2 \alpha$$

$$\frac{1 + \cos 2\alpha}{2} = \cos^2 \alpha$$

$$\cos^2 \alpha = \frac{1 + \cos 2\alpha}{2}$$

$$\cos \alpha = \pm\sqrt{\frac{1 + \cos 2\alpha}{2}}$$

Now, replacing α by $\frac{x}{2}$ and 2α by x,

$$\cos\left(\frac{x}{2}\right) = \pm\sqrt{\frac{1 + \cos x}{2}}.$$

The two identities we have derived are the **half-angle identities**.

The Half-Angle Identities:

$$\sin\left(\frac{x}{2}\right) = \pm\sqrt{\frac{1 - \cos x}{2}}$$

$$\cos\left(\frac{x}{2}\right) = \pm\sqrt{\frac{1 + \cos x}{2}}$$

Alternate form:

$$\sin^2\left(\frac{x}{2}\right) = \frac{1 - \cos x}{2}$$

$$\cos^2\left(\frac{x}{2}\right) = \frac{1 + \cos x}{2}$$

Like the other identities, the half-angle identities are useful in proving identities and in solving equations.

EXAMPLE 14.11.   Prove that $\dfrac{2\sin^2\left(\dfrac{x}{2}\right)}{\cos x} = \sec x - 1$ is an identity.

Solution.   We use the variation of the half-angle identity

$$\sin^2\left(\frac{x}{2}\right) = \frac{1 - \cos x}{2}$$

Also, writing the right-hand side in terms of $\cos x$,

$$\frac{2\sin^2\left(\dfrac{x}{2}\right)}{\cos x} \overset{?}{=} \sec x - 1$$

$$\frac{2\left(\dfrac{1 - \cos x}{2}\right)}{\cos x} \overset{?}{=} \frac{1}{\cos x} - 1$$

$$\frac{1 - \cos x}{\cos x} \overset{?}{=} \frac{1}{\cos x} - 1$$

$$\frac{1}{\cos x} - \frac{\cos x}{\cos x} \overset{?}{=} \frac{1}{\cos x} - 1$$

$$\frac{1}{\cos x} - 1 = \frac{1}{\cos x} - 1.$$

EXAMPLE 14.12.   Solve $\cos \dfrac{x}{2} = \cos x + 1$ for all values of x such that $0° \leqslant x < 360°$.

Solution.   Using the half-angle identity $\cos\left(\dfrac{x}{2}\right) = \pm \sqrt{\dfrac{1 + \cos x}{2}}$,

we have

$$\cos\left(\frac{x}{2}\right) = \cos x + 1$$

$$\pm\sqrt{\frac{1 + \cos x}{2}} = \cos x + 1.$$

We square both sides, remembering the middle term on the right-hand side:

$$\left(\pm\sqrt{\frac{1 + \cos x}{2}}\right)^2 = (\cos x + 1)^2$$

$$\frac{1 + \cos x}{2} = \cos^2 x + 2 \cos x + 1$$

$$1 + \cos x = 2 \cos^2 x + 4 \cos x + 2$$

$$0 = 2 \cos^2 x + 3 \cos x + 1$$

$$0 = (\cos x + 1)(2 \cos x + 1)$$

$$\cos x + 1 = 0 \text{ and } 2 \cos x + 1 = 0$$

$$\cos x = -1 \text{ and } \cos x = -\frac{1}{2}.$$

The possible solutions are 180°, 120°, and 240°. But recall that, when we square both sides of an equation, we lose minus signs. Therefore, we may have extraneous solutions. To check 180°,

$$\cos\left(\frac{x}{2}\right) = \cos x + 1$$

$$\cos \frac{180°}{2} \overset{?}{=} \cos 180° + 1$$

$$\cos 90° \overset{?}{=} \cos 180° + 1$$

$$0 = -1 + 1.$$

To check 120°,

$$\cos\left(\frac{x}{2}\right) = \cos x + 1$$

$$\cos \frac{120°}{2} \overset{?}{=} \cos 120° + 1$$

$$\cos 60° \overset{?}{=} \cos 120° + 1$$

$$\frac{1}{2} = -\frac{1}{2} + 1.$$

To check 240°,

$$\cos\left(\frac{x}{2}\right) = \cos x + 1$$

$$\cos \frac{240°}{2} \overset{?}{=} \cos 240° + 1$$

$$\cos 120° \overset{?}{=} \cos 240° + 1$$

$$-\frac{1}{2} \neq -\frac{1}{2} + 1.$$

Thus 240° is an extraneous solution. The solutions are 180° and 120°.

Exercise 14.3

(Answers for proofs are not given. To avoid difficulty with signs, you may assume $\sin\left(\frac{x}{2}\right)$ and $\cos\left(\frac{x}{2}\right)$ are positive.)

$\frac{x}{2}$ in QI

Prove that the equality is an identity:

1. $\dfrac{2\cos^2\left(\frac{x}{2}\right)}{\cos x} = \sec x + 1$

2. $2\sin^2\left(\frac{x}{2}\right)\tan x = \tan x - \sin x$

3. $\sin\left(\frac{x}{2}\right)\cos\left(\frac{x}{2}\right) = \dfrac{\sin x}{2}$

4. $\dfrac{\sin^2\left(\frac{x}{2}\right)}{\cos^2\left(\frac{x}{2}\right)} = \dfrac{1 - \cos x}{1 + \cos x}$

5. $\tan\left(\frac{x}{2}\right) = \pm\sqrt{\dfrac{1 - \cos x}{1 + \cos x}}$

6. $\cot\left(\frac{x}{2}\right) = \pm\sqrt{\dfrac{1 + \cos x}{1 - \cos x}}$

7. $\tan\left(\frac{x}{2}\right) = \dfrac{\sin x}{1 + \cos x}$ $\left(\text{Multiply numerator and denominator of } \sqrt{\dfrac{1 - \cos x}{1 + \cos x}} \text{ by } 1 + \cos x.\right)$

8. $\tan\left(\frac{x}{2}\right) = \dfrac{1 - \cos x}{\sin x}$

Solve for all values of x such that $0° \leqslant x < 360°$:

9. $\cos^2\left(\frac{x}{2}\right) = \cos^2 x - 1$

10. $\sin^2\left(\frac{x}{2}\right) = \sin^2 x$

11. $\sin\left(\frac{x}{2}\right) = \cos x - 1$

12. $\sin\left(\frac{x}{2}\right) = \cos x$

Proofs of the Sum and Difference Identities

We begin by proving the identity for $\cos(\alpha - \beta)$. For simplicity, we will assume $\alpha > \beta$:

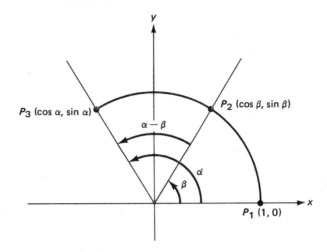

If the point P_2 is on the unit circle and also on the terminal side of β, then

$$\cos \beta = x$$

and also

$$\sin \beta = y.$$

Therefore, P_2 has coordinates $(\cos \beta, \sin \beta)$. Similarly, P_3 has coordinates $(\cos \alpha, \sin \alpha)$. Let $\overline{P_2P_3}$ represent the distance between P_2 and P_3. Using the distance formula,

$$\overline{P_2P_3} = \sqrt{(\cos \alpha - \cos \beta)^2 + (\sin \alpha - \sin \beta)^2}$$

$$\overline{P_2P_3}^2 = (\cos \alpha - \cos \beta)^2 + (\sin \alpha - \sin \beta)^2$$

$$= \cos^2 \alpha - 2 \cos \alpha \cos \beta + \cos^2 \beta + \sin^2 \alpha - 2 \sin \alpha \sin \beta + \sin^2 \beta.$$

But $\sin^2 \alpha + \cos^2 \alpha = 1$ and $\sin^2 \beta + \cos^2 \beta = 1$; therefore,

$$\overline{P_2P_3}^2 = 2 - 2 \cos \alpha \cos \beta - 2 \sin \alpha \sin \beta.$$

Now, if we move $\alpha - \beta$ into standard position, we have

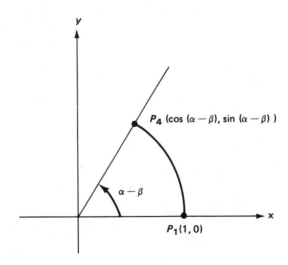

$$\overline{P_1P_4}^2 = [\cos(\alpha - \beta) - 1]^2 + [\sin(\alpha - \beta) - 0]^2$$

$$= \cos^2(\alpha - \beta) - 2\cos(\alpha - \beta) + 1 + \sin^2(\alpha - \beta).$$

But $\sin^2(\alpha - \beta) + \cos^2(\alpha - \beta) = 1$; therefore,

$$\overline{P_1P_4}^2 = 2 - 2\cos(\alpha - \beta).$$

Since $\alpha - \beta$ is the same size in each diagram,

$$\overline{P_2P_3} = \overline{P_1P_4}$$

$$\overline{P_2P_3}^2 = \overline{P_1P_4}^2,$$

so

$$2 - 2\cos\alpha\cos\beta - 2\sin\alpha\sin\beta = 2 - 2\cos(\alpha - \beta)$$

$$2\cos(\alpha - \beta) = 2\cos\alpha\cos\beta + 2\sin\alpha\sin\beta$$

$$\cos(\alpha - \beta) = \cos\alpha\cos\beta + \sin\alpha\sin\beta.$$

To prove the sum formula, we recall from Section 9.1 that $\sin(-x) = -\sin x$ and $\cos(-x) = \cos x$. Then,

$$\cos(\alpha + \beta) = \cos[\alpha - (-\beta)]$$

$$= \cos\alpha\cos(-\beta) + \sin\alpha\sin(-\beta)$$

$$= \cos\alpha\cos\beta + \sin\alpha(-\sin\beta)$$

$$= \cos\alpha\cos\beta - \sin\alpha\sin\beta.$$

To prove the sum and difference formulas for the sine, we recall from Section 1.4 that, since the sine and cosine are cofunctions, $\sin x = \cos(90° - x)$ and $\cos x = \sin(90° - x)$:

$$\sin(\alpha - \beta) = \cos[90° - (\alpha - \beta)]$$

$$= \cos[(90° - \alpha) + \beta]$$

$$= \cos(90° - \alpha)\cos\beta - \sin(90° - \alpha)\sin\beta$$

$$= \sin\alpha\cos\beta - \cos\alpha\sin\beta.$$

Finally,

$$\sin(\alpha + \beta) = \sin[\alpha - (-\beta)]$$

$$= \sin\alpha\cos(-\beta) - \cos\alpha\sin(-\beta)$$

$$= \sin\alpha\cos\beta + \cos\alpha\sin\beta.$$

A Note on Other Identities:

Recall that the double-angle and half-angle identities are easily derived from the sum and difference identities. There are many other identities we may derive from the sum and difference identities. We occasionally use the **product identities**. To derive the first product identity, we add together the sum and difference identities for the sine:

$$\sin(\alpha + \beta) = \sin \alpha \cos \beta + \cos \alpha \sin \beta$$

$$\sin(\alpha - \beta) = \sin \alpha \cos \beta - \cos \alpha \sin \beta$$

$$\overline{\sin(\alpha + \beta) + \sin(\alpha - \beta) = 2 \sin \alpha \cos \beta,}$$

therefore,

$$\sin \alpha \cos \beta = \frac{\sin(\alpha + \beta) + \sin(\alpha - \beta)}{2}.$$

Similarly, if we subtract the sum and difference identities for the sine,

$$\sin(\alpha + \beta) = \sin \alpha \cos \beta + \cos \alpha \sin \beta$$

$$-\sin(\alpha - \beta) = -\sin \alpha \cos \beta + \cos \alpha \sin \beta$$

$$\overline{\sin(\alpha + \beta) - \sin(\alpha - \beta) = 2 \cos \alpha \sin \beta,}$$

therefore,

$$\cos \alpha \sin \beta = \frac{\sin(\alpha + \beta) - \sin(\alpha - \beta)}{2}.$$

If we add the sum and difference identities for the cosine, we have

$$\cos(\alpha + \beta) = \cos \alpha \cos \beta - \sin \alpha \sin \beta$$

$$\cos(\alpha - \beta) = \cos \alpha \cos \beta + \sin \alpha \sin \beta$$

$$\overline{\cos(\alpha + \beta) + \cos(\alpha - \beta) = 2 \cos \alpha \cos \beta,}$$

therefore,

$$\cos \alpha \cos \beta = \frac{\cos(\alpha + \beta) + \cos(\alpha - \beta)}{2}.$$

Similarly, if we subtract the sum and difference identities for the cosine,

$$\cos(\alpha + \beta) = \cos \alpha \cos \beta - \sin \alpha \sin \beta$$

$$-\cos(\alpha - \beta) = -\cos \alpha \cos \beta - \sin \alpha \sin \beta$$

$$\overline{\cos(\alpha + \beta) - \cos(\alpha - \beta) = -2 \sin \alpha \sin \beta,}$$

therefore,

$$\sin \alpha \sin \beta = \frac{\cos(\alpha + \beta) - \cos(\alpha - \beta)}{-2}$$

or

$$\sin \alpha \sin \beta = \frac{\cos(\alpha - \beta) - \cos(\alpha + \beta)}{2}.$$

There are even more identities which can be derived from these identities.

Summary of Identities

Sum and Difference Identities

$$\sin(\alpha + \beta) = \sin \alpha \cos \beta + \cos \alpha \sin \beta$$

$$\sin(\alpha - \beta) = \sin \alpha \cos \beta - \cos \alpha \sin \beta$$

$$\cos(\alpha + \beta) = \cos \alpha \cos \beta - \sin \alpha \sin \beta$$

$$\cos(\alpha - \beta) = \cos \alpha \cos \beta + \sin \alpha \sin \beta$$

Double-Angle Identities

$$\sin 2x = 2 \sin x \cos x$$

$$\cos 2x = \cos^2 x - \sin^2 x$$

Variations

$$\cos 2x = 1 - 2 \sin^2 x$$

$$\cos 2x = 2 \cos^2 x - 1$$

Half-Angle Identities

$$\sin\left(\frac{x}{2}\right) = \pm\sqrt{\frac{1 - \cos x}{2}}$$

$$\cos\left(\frac{x}{2}\right) = \pm\sqrt{\frac{1 + \cos x}{2}}$$

Variations

$$\sin^2\left(\frac{x}{2}\right) = \frac{1 - \cos x}{2}$$

$$\cos^2\left(\frac{x}{2}\right) = \frac{1 + \cos x}{2}$$

Self-test

Prove that the equality is an identity:

1. $\cot\left(\dfrac{3\pi}{2} + x\right) = -\tan x$

2. $\dfrac{2 \cos 2x}{\sin 2x} = \cot x - \tan x$

3. $\cos^2\left(\dfrac{x}{2}\right) - \sin^2\left(\dfrac{x}{2}\right) = \cos x$

Solve for all values of x such that $0° \leqslant x < 360°$:

4. $\sin 2x = \sin x$

4. _____

5. $\cos 2x - \cos x = 2$

5. _____

Unit 15

Inverse Trigonometric Functions

INTRODUCTION

In the last several units you have interpreted the trigonometric ratios as functions. A function relates a variable x to a unique variable y. For every function, there is an inverse relation which interchanges the variables x and y. In this unit you will learn about the inverse relation. Then you will apply the inverse relation to three of the trigonometric functions, $y = \sin x$, $y = \cos x$, and $y = \tan x$.

OBJECTIVES

When you have finished this unit you should be able to:

1. Find the equation of the inverse relation for a linear or quadratic function and draw the graphs of both.
2. State the range of principal values and draw the graph of $y = \mathrm{Sin}^{-1}x$, $y = \mathrm{Cos}^{-1}x$, and $y = \mathrm{Tan}^{-1}x$.
3. Given a value of x, find the principal value of $y = \mathrm{Sin}^{-1}x$, $y = \mathrm{Cos}^{-1}x$, or $y = \mathrm{Tan}^{-1}x$.

Section 15.1 — Inverse Relations

In algebra you learned about functions, which relate two variables x and y. We give the definition of a function here.

> **Definition:** A relation in two variables x and y is a **function** if for each x-value there is just one y-value.

An example of a function is given by an equation such as $y = x + 1$, which represents a linear function. Its graph is

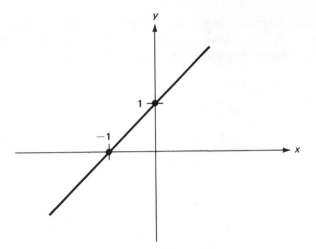

Its domain and range both are the set of all real numbers.

The relation given by $y = x^2$ is an example of a quadratic function. Its graph is

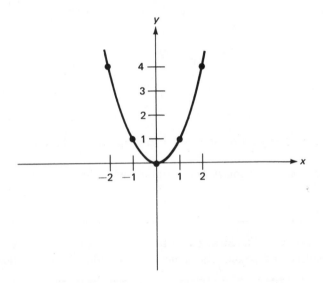

Its domain is the set of all real numbers, and its range is $y \geqslant 0$. Observe that a y-value can occur more than once. For example, when $x = 2$, $y = 4$, and also when $x = -2$, $y = 4$. For some y-values there are two different x-values. However, we must not have any x-values which have two different y-values. For each x-value there must be just one y-value.

A trigonometric relation such as $y = \sin x$ is a function:

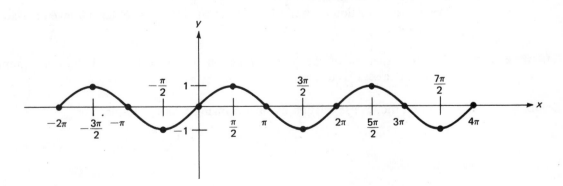

For each y-value, there are infinitely many x-values. For example, for $y = 0$, $x = 0$, $\pm \pi$, $\pm 2\pi$, ... However, for each x-value, there is just one y-value.

The relation given by $y^2 = x$ is not a function. The graph of this relation is

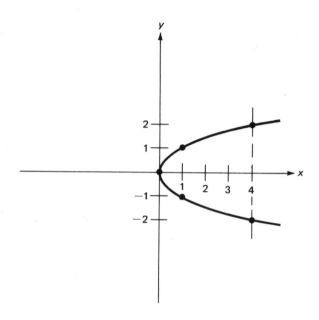

For each x-value, except $x = 0$, there are two y-values. For example, when $x = 4$, $y = 2$ and also $y = -2$.

For any function in two variables, we define another relation which can be formed from the function.

Definition: The **inverse** of a function is the relation in two variables obtained by reversing the first and second coordinates in the ordered pairs of the function.

For a function given by an equation in x and y, the inverse can be found by interchanging x and y in the equation of the function. When we interchange x and y, the domain of the function becomes the range of the inverse, and the range of the function becomes the domain of the inverse. The inverse of a function may or may not also be a function. When the inverse also is a function, we call it the **inverse function**.

Every linear function, except a function of the form $y = b$, has an inverse relation which is also a function.

EXAMPLE 15.1. Find the equation of the inverse function of $y = x + 1$, and draw the graphs of both functions on one set of axes.

Solution. First, we find the inverse relation by interchanging x and y. In place of

$$y = x + 1$$

we write

$$x = y + 1.$$

Now, we solve for y to write the inverse function

$$y = x - 1.$$

The graphs are

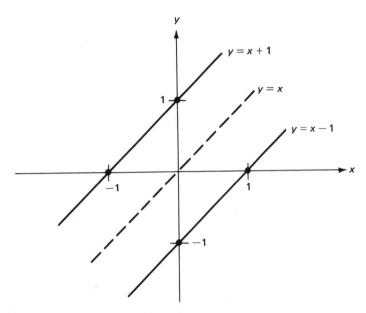

Observe that the graphs of $y = x + 1$ and its inverse $y = x - 1$ are symmetric with respect to the line $y = x$. The graph of an inverse relation or function is always the mirror image of the graph of the given function, with the line $y = x$ as the "mirror."

EXAMPLE 15.2.   Find the equation of the inverse function of $2x + 3y = 6$, and draw the graphs of both functions on one set of axes.

Solution.   In place of

$$2x + 3y = 6,$$

we write

$$2y + 3x = 6.$$

The graphs are

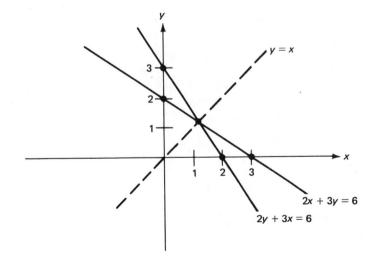

Observe that the lines are mirror images in the line $y = x$. You should also observe that the coordinates of each point reverse. For example, the y-intercept $(0, 2)$ becomes the x-intercept $(2, 0)$ of the inverse, and the x-intercept $(3, 0)$ becomes the y-intercept $(0, 3)$ of the inverse.

The inverse relation of a linear function of the form $y = b$ is not a function.

EXAMPLE 15.3.   Find the equation of the inverse relation of $y = 2$, and draw the graphs of the function and inverse relation on one set of axes.

Solution.   Starting with

$$y = 2,$$

we replace y by x to derive the equation

$$x = 2.$$

This is the equation of the inverse relation. The graphs are

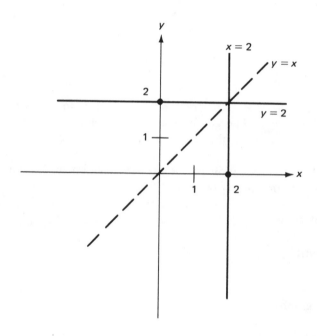

The graph of the inverse relation $x = 2$ is the mirror image of the graph of the function $y = 2$ in the line $y = x$. Recall that the range of the function is the domain of the inverse relation. The range of the function $y = 2$ is just the number 2, and the domain of the inverse relation $x = 2$ is just the number 2. The inverse relation is not a function because for the one value of x there are infinitely many y-values.

The inverse relation of a quadratic function is not a function.

EXAMPLE 15.4.   Find the equation of the inverse relation of $y = x^2$, and draw the function and inverse relation on one set of axes.

Solution.   We write

$$y = x^2,$$

and reversing x and y,

$$x = y^2$$

or

$$y^2 = x.$$

Solving for y, we may write

$$y = \pm\sqrt{x}\,.$$

Observe that for every value of x, except $x = 0$, there are two y-values, one positive and one negative. The graphs are

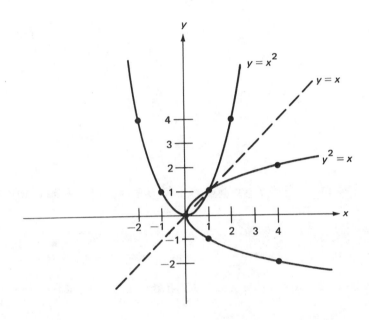

The range of the function $y = x^2$ is the set $y \geqslant 0$, and the domain of the inverse relation $y^2 = x$ is the set $x \geqslant 0$. The domain of $y = x^2$ is the set of all real numbers and the range of $y^2 = x$ is the set of all real numbers.

By restricting the domain of a quadratic function, we may find an inverse relation which is a function. When we do this, we restrict the range of the inverse function.

EXAMPLE 15.5.   Find the equation of the inverse relation of $y = x^2 - 4, x \geqslant 0$, and draw the graphs of both on one set of axes.

Solution.   We start with

$$y = x^2 - 4, x \geqslant 0.$$

Interchanging x and y, we have

$$x = y^2 - 4, y \geqslant 0.$$

Then, solving for y,

$$y^2 = x + 4, y \geqslant 0$$

$$y = \sqrt{x+4}\,.$$

We have only the positive square root because we have the restriction $y \geq 0$. The graphs are

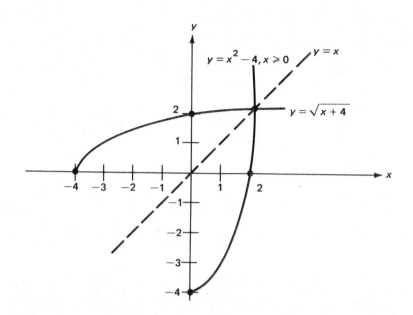

The range of the function $y = x^2 - 4$ is $y \geq -4$, and the domain of the inverse function $y = \sqrt{x + 4}$ is $x \geq -4$. The domain of $y = x^2 - 4$ is restricted to $x \geq 0$ and the range of $y = \sqrt{x + 4}$ is $y \geq 0$.

Exercise 15.1

Find the equation of the inverse relation, state whether or not it is a function, and draw both graphs on one set of axes:

1. $y = x - 2$ 2. $y = 2x + 1$ 3. $x + 2y = 4$

4. $3x - 4y = 12$ 5. $y = 3$ 6. $y = -1$

7. $y = x^2 - 1$ 8. $y = x^2 + 2$ 9. $y = x^2 - 1, x \geq 0$

10. $y = x^2 + 2, x \geq 0$ 11. $y = x^2 - 1, x \leq 0$ 12. $y = x^2 + 2, x \leq 0$

Section
15.2
The Inverse Trigonometric Functions

Consider the sine function $y = \sin x$. To find an inverse relation for $y = \sin x$ we interchange x and y in the equation. We have the inverse relation given by the equation

$$x = \sin y.$$

Recall from Section 9.1 that the range of $y = \sin x$ is $-1 \leq y \leq 1$. When we interchange x and y, we have $-1 \leq x \leq 1$. Therefore, the domain of $x = \sin y$ is $-1 \leq x \leq 1$. The range is all real numbers y.

However, there is no algebraic method by which the equation $x = \sin y$ can be solved for y. Therefore, we define an equation which is the solution for y of the equation $x = \sin y$.

Definition: The **inverse sine** $y = \sin^{-1}x$, where $-1 \leqslant x \leqslant 1$, is the solution for y of the equation $x = \sin y$.

The equation $y = \sin^{-1}x$ is also written $y = \arcsin x$.

To draw the graph of $y = \sin^{-1}x$, we start with the graph of $y = \sin x$. Two cycles of the graph starting from $x = -2\pi$ are

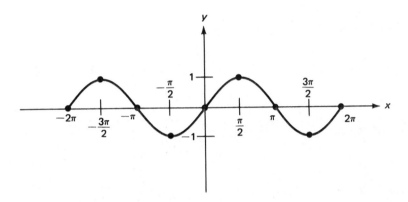

We reverse the coordinates of the ordered pairs to obtain the graph of $y = \sin^{-1}x$. For example, since $\left(\dfrac{\pi}{2}, 1\right)$ is an ordered pair of $y = \sin x$, $\left(1, \dfrac{\pi}{2}\right)$ is an ordered pair of $y = \sin^{-1}x$. Similarly, $(0, \pi)$ and $\left(-1, \dfrac{3\pi}{2}\right)$ are ordered pairs of $y = \sin^{-1}x$, and so on. Two cycles of the graph starting from $y = -2\pi$ are

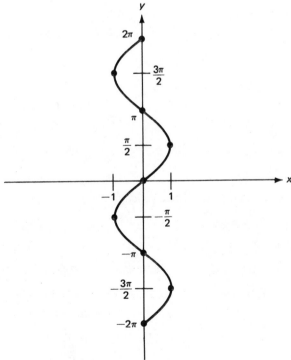

Observe that the domain of $y = \sin^{-1}x$ is $-1 \leqslant x \leqslant 1$. The graph may be continued indefinitely upward and downward, so the range is the set of all real numbers. For each value of x in the domain $-1 \leqslant x \leqslant 1$, there are infinitely many values of y because the cycles may be repeated indefinitely. Therefore, the relation $y = \sin^{-1}x$ is not a function.

In Section 15.1, we saw that by restricting the domain of a function, we may be able to restrict the range of the inverse relation so that it is also a function. Consider $y = \sin x$ restricted to $-\dfrac{\pi}{2} \leqslant x \leqslant \dfrac{\pi}{2}$:

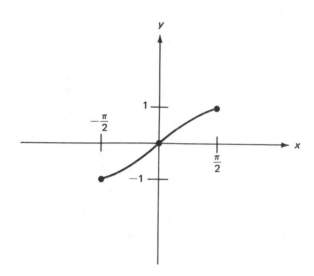

The graph of the inverse relation is

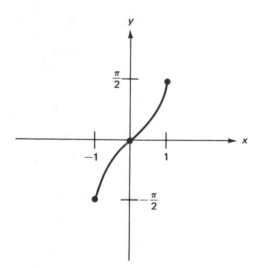

This inverse relation is a function. The restricted range of the inverse function is called the **range of principal values.** When we use the range of principal values, we write the equation of the function with a capital S:

$$y = \text{Sin}^{-1} x.$$

The range of principal values of $y = \text{Sin}^{-1} x$ is

$$-\frac{\pi}{2} \leqslant y \leqslant \frac{\pi}{2}.$$

Observe that the graph of $y = \text{Sin}^{-1} x$ is the mirror image of the graph of $y = \sin x$, $-\frac{\pi}{2} \leqslant x \leqslant \frac{\pi}{2}$, in the line $y = x$:

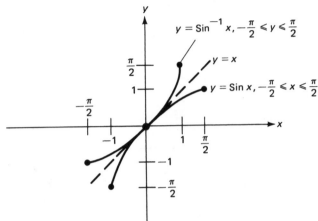

There are other restricted domains of $y = \sin x$ which would also restrict the range of the inverse relation so that it is a function. Only the range $-\frac{\pi}{2} \leqslant y \leqslant \frac{\pi}{2}$, however, is the range of principal values. This is the range which is most useful for applications to other areas of mathematics and science.

Now, we consider $y = \cos x$. The inverse relation is given by the equation

$$x = \cos y.$$

We define an equation which is the solution for y of the equation $x = \cos y$.

Definition: The **inverse cosine** $y = \cos^{-1} x$, where $-1 \leqslant x \leqslant 1$, is the solution for y of the equation $x = \cos y$.

The equation $y = \cos^{-1} x$ is also written $y = \arccos x$.

Two cycles of the graph of $y = \cos x$ starting from $x = -2\pi$ are

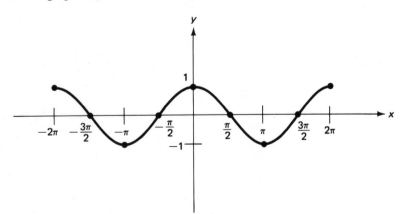

We reverse the coordinates of the ordered pairs to obtain the graph of $y = \cos^{-1} x$. For example, $(1, 0)$, $\left(0, \dfrac{\pi}{2}\right)$, $(-1, \pi)$, and so on, are ordered pairs of $y = \cos^{-1} x$. Two cycles of the graph starting from $y = -2\pi$ are

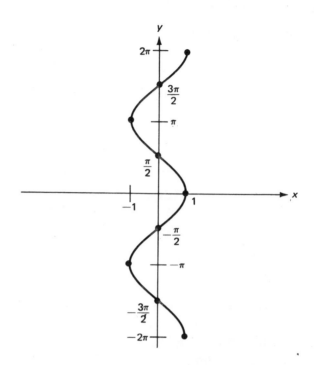

The domain of $y = \cos^{-1} x$ is $-1 \leqslant x \leqslant 1$. The graph may be continued indefinitely upward and downward, so the range is the set of all real numbers. For each value of x in the domain $-1 \leqslant x \leqslant 1$, there are infinitely many values of y, so the relation $y = \cos^{-1} x$ is not a function.

Observe that we cannot use the same restricted domain for $y = \cos x$ to restrict the range of $y = \cos^{-1} x$ so that it is a function. Were we to use $-\dfrac{\pi}{2} \leqslant y \leqslant \dfrac{\pi}{2}$ to restrict the range of $y = \cos^{-1} x$, we would have two values of y for each x such that $0 \leqslant x \leqslant 1$, and no value of y for each x such that $-1 \leqslant x < 0$. Therefore, we must find a different restricted domain, one which restricts the range of $y = \cos^{-1} x$ so that we have a function.

Consider $y = \cos x$ restricted to $0 \leqslant x \leqslant \pi$:

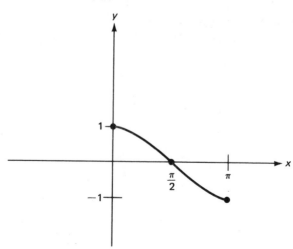

The graph of the inverse relation is

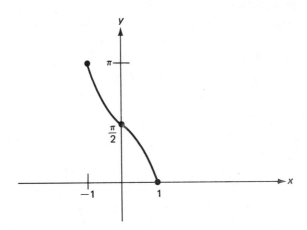

This inverse relation is a function. We write the equation of the function with a capital C:

$$y = \text{Cos}^{-1} x.$$

The range of principal values is

$$0 \leqslant y \leqslant \pi.$$

Observe that the graph of $y = \text{Cos}^{-1} x$ is the mirror image of the graph of $y = \cos x, 0 \leqslant x \leqslant \pi$, in the line $y = x$:

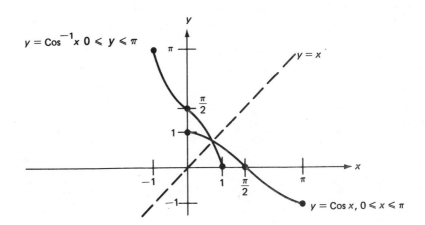

Finally, we consider $y = \tan x$. The inverse relation is given by the equation

$$x = \tan y.$$

We define an equation which is the solution for y of the equation $x = \tan y$.

Definition: The **inverse tangent** $y = \tan^{-1} x$ is the solution for y of the equation $x = \tan y$.

The equation $y = \tan^{-1} x$ is also written $y = \arctan x$.

The graph of $y = \tan x$ from $x = -\dfrac{3\pi}{2}$ to $x = \dfrac{3\pi}{2}$ is

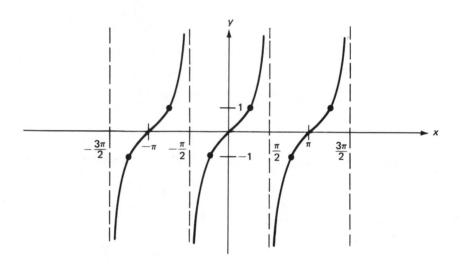

You should observe that, when we reverse the coordinates of the ordered pairs and draw the graph of $y = \tan^{-1} x$, the inverse relation from $y = -\dfrac{3\pi}{2}$ to $y = \dfrac{3\pi}{2}$ is clearly not a function.

A natural restriction for the domain of $y = \tan x$ is to a part of the graph between two asymptotes. Consider $y = \tan x$ restricted to $-\dfrac{\pi}{2} < x < \dfrac{\pi}{2}$:

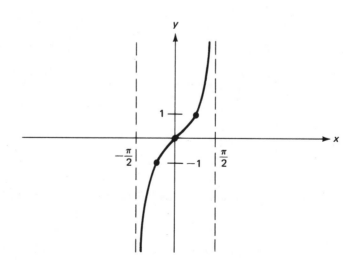

$\left(\text{Observe that the endpoints of } -\dfrac{\pi}{2} < x < \dfrac{\pi}{2} \text{ are not included because } y = \tan x \text{ is undefined at}\right.$

$\left. x = -\dfrac{\pi}{2} \text{ and } x = \dfrac{\pi}{2} . \right)$

The graph of the inverse relation is

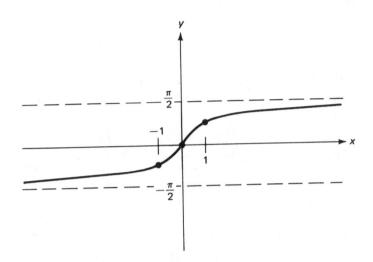

This inverse relation is a function. We write the equation of the function with a capital T:

$$y = \mathrm{Tan}^{-1} x.$$

The range of principal values is

$$-\frac{\pi}{2} < y < \frac{\pi}{2}.$$

You should show that the graph of $y = \mathrm{Tan}^{-1} x$ is the mirror image of the graph of $y = \tan x$ in the line $y = x$. Observe that the range of $y = \tan x$ is the set of all real numbers, and the domain of $y = \mathrm{Tan}^{-1} x$ is the set of all real numbers. Moreover, observe that the asymptotes of $y = \tan x$ are vertical lines but, when x and y are reversed, the asymptotes of $y = \mathrm{Tan}^{-1} x$ are horizontal lines.

It is possible to define inverse relations and ranges of principal values for inverse functions for $y = \cot x$, $y = \sec x$, and $y = \csc x$. However, these inverse functions are rarely encountered, so we will not discuss them here.

Exercise 15.2

1. State the range of principal values and draw the graph of the function $y = \mathrm{Sin}^{-1} x$.

2. State the range of principal values and draw the graph of the function $y = \mathrm{Cos}^{-1} x$.

3. State the range of principal values and draw the graph of the function $y = \mathrm{Tan}^{-1} x$.

4. If the range of principal values is given to be $0 < y < \pi$, draw the graph of the function $y = \mathrm{Cot}^{-1} x$.

Section
15.3

Values of the Inverse Trigonometric Functions

In solving equations in Units 13 and 14, we often faced the problem of finding x given a value of $\sin x$. For example, given

$$\sin x = \frac{1}{2},$$

we said $x = 30°$ and $x = 150°$. Recall that actually these are only the values of x such that $0° \leqslant x < 360°$. There are infinitely many angles x coterminal with $30°$ and $150°$ which have $\sin x = \frac{1}{2}$.

Using the ranges of principal values of the inverse trigonometric functions, we may solve the same problem of finding an angle given the value of a trigonometric function. However, we find just one angle which is not necessarily between $0°$ and $360°$, but is the one angle in the range of principal values. Also, we use radian measure so that the number we find can be interpreted not just as an angle but also as a quantity such as time.

Whenever we have an inverse trigonometric function of a positive number or zero, the value is between 0 and $\frac{\pi}{2}$. All of the trigonometric ratios are positive for angles between 0 and $\frac{\pi}{2}$, and $0 \leqslant y \leqslant \frac{\pi}{2}$ $\left(\text{not including } \frac{\pi}{2} \text{ for } y = \text{Tan}^{-1} x\right)$ is part of the range of principal values of each of the inverse trigonometric functions.

EXAMPLE 15.6.   Find the principal value of

    a. $\text{Sin}^{-1}\left(\dfrac{1}{2}\right)$ b. $\text{Cos}^{-1}(0)$ c. $\text{Tan}^{-1}(1)$

Solutions.   a.   If we write

$$y = \text{Sin}^{-1}\left(\frac{1}{2}\right),$$

then, since $y = \sin^{-1} x$ is the solution of $x = \sin y$,

$$\frac{1}{2} = \sin y.$$

We must find an angle y such that $\sin y = \frac{1}{2}$. Using the 30–60–90 special triangle,

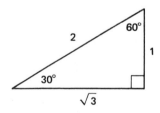

$$\sin 30° = \frac{1}{2}.$$

In radian measure

$$30° = 30\left(\frac{\pi}{180}\right) = \frac{\pi}{6}.$$

Since $\frac{\pi}{6}$ is in the range $0 < y < \frac{\pi}{2}$,

$$\text{Sin}^{-1}\left(\frac{1}{2}\right) = \frac{\pi}{6}.$$

b. If we write

$$y = \text{Cos}^{-1}(0),$$

then, since $y = \cos^{-1} x$ is the solution of $x = \cos y$,

$$0 = \cos y.$$

We must find an angle y such that $\cos y = 0$. Using a unit point,

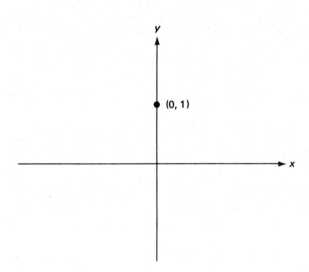

$$\cos 90° = 0.$$

In radian measure,

$$90° = 90\left(\frac{\pi}{180}\right) = \frac{\pi}{2},$$

and since $\frac{\pi}{2}$ is in the range $0 < y < \pi$,

$$\text{Cos}^{-1}(0) = \frac{\pi}{2}.$$

c. If we write

$$y = \text{Tan}^{-1}(1),$$

then, since $y = \tan^{-1} x$ is the solution of $x = \tan y$,

$$1 = \tan y.$$

We must find an angle y such that $\tan y = 1$. Using the 45–45–90 special triangle,

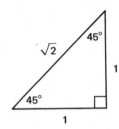

$$\tan 45° = 1.$$

In radian measure

$$45° = 45\left(\frac{\pi}{180}\right) = \frac{\pi}{4},$$

and since $\frac{\pi}{4}$ is in the range $-\frac{\pi}{2} < y < \frac{\pi}{2}$ $\left(\text{remember that } -\frac{\pi}{2} \text{ and } \frac{\pi}{2} \text{ are not included}\right.$

for $\left. y = \text{Tan}^{-1}x\right),$

$$\text{Tan}^{-1}(1) = \frac{\pi}{4}.$$

When we have an inverse trigonometric function of a negative number, we must use the range of principal values of the inverse trigonometric function to determine the appropriate value.

EXAMPLE 15.7.   Find the principal value of $\text{Sin}^{-1}\left(-\frac{1}{2}\right)$.

Solution.   From

$$y = \text{Sin}^{-1}\left(-\frac{1}{2}\right),$$

we have

$$-\frac{1}{2} = \sin y.$$

The reference angle for y is 30° or, in radians, $\frac{\pi}{6}$. We need an angle with reference angle $\frac{\pi}{6}$ and, in the range of principal values for $y = \text{Sin}^{-1} x$,

$$-\frac{\pi}{2} \leqslant y \leqslant \frac{\pi}{2}.$$

Clearly, we must use an angle y such that $-\frac{\pi}{2} \leqslant y < 0$. The angle in this range with reference angle $\frac{\pi}{6}$ is $-\frac{\pi}{6}$. Therefore,

$$\text{Sin}^{-1}\left(-\frac{1}{2}\right) = -\frac{\pi}{6}.$$

EXAMPLE 15.8.   Find the principal value of $\operatorname{Tan}^{-1}\left(-\dfrac{1}{\sqrt{3}}\right)$.

Solution.   From

$$y = \operatorname{Tan}^{-1}\left(-\frac{1}{\sqrt{3}}\right),$$

we have

$$-\frac{1}{\sqrt{3}} = \tan y.$$

Again the reference angle is 30°, or $\dfrac{\pi}{6}$. The range of principal values for $y = \operatorname{Tan}^{-1} x$ is

$$-\frac{\pi}{2} < y < \frac{\pi}{2},$$

so we must use an angle y such that $-\dfrac{\pi}{2} < y < 0$. Therefore,

$$\operatorname{Tan}^{-1}\left(-\frac{1}{\sqrt{3}}\right) = -\frac{\pi}{6}.$$

Since $y = \operatorname{Cos}^{-1} x$ has a different range of principal values, the case when x is negative is different from the preceding cases.

EXAMPLE 15.9.   Find the principal value of $\operatorname{Cos}^{-1}\left(-\dfrac{1}{\sqrt{2}}\right)$.

Solution.   From

$$y = \operatorname{Cos}^{-1}\left(-\frac{1}{\sqrt{2}}\right),$$

we have

$$-\frac{1}{\sqrt{2}} = \cos y.$$

The reference angle is 45°, or $\dfrac{\pi}{4}$. It is a common error to write the principal value as $-\dfrac{\pi}{4}$. However, recall that the range of principal values of $y = \operatorname{Cos}^{-1} x$ is

$$0 \leqslant y \leqslant \pi.$$

There are no negative angles in this range. We must use an angle y such that $\dfrac{\pi}{2} < y \leqslant \pi$. The angle in this range with reference angle $\dfrac{\pi}{4}$ is $\dfrac{3\pi}{4}$. Therefore,

$$\operatorname{Cos}^{-1}\left(-\frac{1}{\sqrt{2}}\right) = \frac{3\pi}{4}.$$

Exercise 15.3

Find the principal value in radians in terms of π:

1. $\operatorname{Sin}^{-1}\left(\dfrac{1}{\sqrt{2}}\right)$ 2. $\operatorname{Sin}^{-1}\left(\dfrac{\sqrt{3}}{2}\right)$ 3. $\operatorname{Cos}^{-1}\left(\dfrac{1}{2}\right)$ 4. $\operatorname{Cos}^{-1}\left(\dfrac{1}{\sqrt{2}}\right)$

5. $\operatorname{Tan}^{-1}(\sqrt{3})$ 6. $\operatorname{Tan}^{-1}\left(\dfrac{1}{\sqrt{3}}\right)$ 7. $\operatorname{Sin}^{-1}(1)$ 8. $\operatorname{Cos}^{-1}(1)$

9. $\operatorname{Sin}^{-1}\left(-\dfrac{\sqrt{3}}{2}\right)$ 10. $\operatorname{Sin}^{-1}\left(-\dfrac{1}{\sqrt{2}}\right)$ 11. $\operatorname{Cos}^{-1}\left(-\dfrac{\sqrt{3}}{2}\right)$ 12. $\operatorname{Tan}^{-1}(-1)$

13. $\operatorname{Cos}^{-1}(-1)$ 14. $\operatorname{Sin}^{-1}(-1)$ 15. $\operatorname{Sin}^{-1}(0)$ 16. $\operatorname{Tan}^{-1}(0)$

Self-test

1. Find the equation of the inverse function of $y = 3x - 6$, and draw the graphs of both functions on one set of axes.

2. State the range of principal values for $y = \text{Cos}^{-1} x$.

2._____

3. Draw the graph of $y = \text{Cos}^{-1} x$ over its range of principal values.

Find the principal value in radians in terms of π:

4. $\text{Tan}^{-1}(-\sqrt{3})$

4._____

5. $\text{Cos}^{-1}\left(-\dfrac{1}{2}\right)$

5._____

Unit

16

Cumulative Review

INTRODUCTION

In this unit you should make certain you remember all the material in all of the preceding units.

OBJECTIVE

When you have finished this unit you should be able to demonstrate that you can fulfill every objective of each preceding unit.

To prepare for this unit you should review the Self-Tests for Units 1 through 15. Do each problem of each Self-Test over again. If you cannot do a problem, or even if you have the slightest difficulty, you should:

1. Find out from the answer section the Objective for the unit to which the problem relates.
2. Review all the material in the section which has the same number as the Objective, and redo all the Exercises for the section.
3. Try the Self-Test for the unit again.

Repeat these steps until you can do each problem in each Self-Test for each of Units 1 through 15 easily and accurately.

Self-test

1. Use a special triangle or a unit point to find:
   a. tan 495°
   b. sec $\dfrac{5\pi}{2}$

   1a. _____

   1b. _____

For problems 2–4, use Table II.

2. Two angles of a triangle are 41°50′ and 55°50′, and the shortest side is 115. Find the longest side.

   2. _____

3. The base of an isosceles triangle is 12.4 inches, and the base angles are each 67°40′. Find the height of the triangle.

   3. _____

4. You are required to find the distance between two trees, one on each side of a stream. You measure along the side of the stream a distance of 20 meters and place a marker. The angle between the line from the first tree to the marker and the line of the trees is 83°50′, and the angle at the marker between the lines to the trees is 25°30′. What is the distance between the trees?

   4. _____

5. Draw one cycle of the graph of $y = 2 \sin\left(2x + \dfrac{\pi}{2}\right)$.

Prove that the equality is an identity:

6. $\sin 2x \cos x = \dfrac{2 \sin x}{1 + \tan^2 x}$

7. $\tan(2\pi - x) = -\tan x$

Solve for all values of x such that $0° \leqslant x < 360°$:

8. $\sec x - 2 \cos x = 1$

8. _____

9. $\cos 2x - \sin x = 0$

9. _____

10. Find the principal value in radians in terms of π:

a. $\text{Cos}^{-1}\left(-\dfrac{\sqrt{3}}{2}\right)$

b. $\text{Sin}^{-1}(-1)$

10a. _____

10b. _____

Unit
17
Logarithms

INTRODUCTION

In this unit you will learn about a different kind of function called the logarithmic function. The logarithmic function has many applications. Logarithms were an important computational tool from the time of their invention by John Napier in the early 1600s until the very recent general use of handheld calculators. Now, the logarithmic function is used to solve many problems which cannot be solved using ordinary algebraic operations.

OBJECTIVES

When you have finished this unit you should be able to:

1. Use the definition of a logarithm to write an exponential statement in logarithmic form, or a logarithmic statement in exponential form.
2. Find the value of a logarithmic expression, where the value can be found by inspection from the exponential form.
3. Draw the graph of a logarithmic function.
4. Use properties of logarithms to write a logarithmic expression as sums, differences, and multiples of logarithms.
5. Use a table of common logarithms to find the common logarithm of a number, or the common antilogarithm of a number.

| Section 17.1 | Definition of the Logarithm |

The **exponential function** is a function of the form

$$y = b^x,$$

where $b > 0$ and $b \neq 1$. The domain of the exponential function is the set of all real numbers, and its range is $y > 0$.

Recall from Section 15.1 that the inverse relation of a function is obtained by reversing the first and second coordinates of the ordered pairs of the function. We find the equation of the inverse of a function by interchanging x and y in the equation of the function. The domain of the function is the range of the inverse, and the range of the function is the domain of the inverse.

To construct the inverse of the exponential function

$$y = b^x,$$

we write

$$x = b^y,$$

where $b > 0$ and $b \neq 1$. The domain of $x = b^y$ is $x > 0$, and its range is the set of all real numbers. Moreover, $x = b^y$ has just one y-value for each x-value in its domain, so it is a function. The function $x = b^y$ is the inverse function of $y = b^x$.

However, there is no algebraic method by which the equation $x = b^y$ can be solved for y. Therefore, we define a function which is the solution for y of the equation $x = b^y$.

Definition: The **logarithmic function** $y = \log_b x$, where $b > 0$ and $b \neq 1$, is the solution for y of the exponential function $x = b^y$.

We say $y = \log_b x$ is the **logarithm** (or **log**) **to the base b of x**.

EXAMPLE 17.1.   Solve for y:

      a. $x = 2^y$ b. $x = 3^y$ c. $x = 10^y$

Solutions.   We use the definition of a logarithm: $y = \log_b x$ is the solution of $x = b^y$.

      a.   Since

$$x = 2^y,$$

we have $b = 2$, and

$$y = \log_2 x.$$

      b.   Since

$$x = 3^x,$$

we have $b = 3$, and

$$y = \log_3 x.$$

      c.   Since

$$x = 10^y,$$

we have $b = 10$, and

$$y = \log_{10} x.$$

The form

$$y = \log_b x$$

is called the **logarithmic form** of $x = b^y$. The form

$$x = b^y$$

is called the **exponential form** of $y = \log_b x$.

EXAMPLE 17.2.   Write in logarithmic form:

a. $10^2 = 100$ b. $2^{-3} = \frac{1}{8}$ c. $9^{\frac{1}{2}} = 3$

Solutions.   We use the definition $y = \log_b x$ is the logarithmic form of $x = b^y$.

a. For

$$10^2 = 100,$$

we have

$$b^y = x,$$

where $b = 10$, $y = 2$, and $x = 100$. Therefore, the logarithmic form

$$y = \log_b x$$

is

$$2 = \log_{10} 100$$

or

$$\log_{10} 100 = 2.$$

b. For

$$2^{-3} = \frac{1}{8},$$

we have

$$b^y = x,$$

where $b = 2$, $y = -3$, and $x = \frac{1}{8}$. Therefore, the logarithmic form

$$y = \log_b x$$

is

$$-3 = \log_2 \frac{1}{8}$$

or

$$\log_2 \frac{1}{8} = -3.$$

c. For

$$9^{\frac{1}{2}} = 3,$$

we have

$$b^y = x,$$

where $b = 9$, $y = \frac{1}{2}$, and $x = 3$. Therefore, the logarithmic form

$$y = \log_b x$$

is

$$\frac{1}{2} = \log_9 3$$

or

$$\log_9 3 = \frac{1}{2}.$$

EXAMPLE 17.3.   Write in exponential form:

      a. $\log_3 81 = 4$ b. $\log_{64} 4 = \frac{1}{3}$ c. $\log_{10} .001 = -3$

Solutions.   We use the definition $x = b^y$ is the exponential form of $y = \log_b x$.

      a. For

$$\log_3 81 = 4,$$

we have

$$\log_b x = y,$$

where $b = 3$, $x = 81$, and $y = 4$. Therefore, the exponential form

$$x = b^y$$

is

$$81 = 3^4$$

or

$$3^4 = 81.$$

      b. For

$$\log_{64} 4 = \frac{1}{3},$$

we have

$$\log_b x = y,$$

where $b = 64$, $x = 4$, and $y = \frac{1}{3}$. Therefore, the exponential form

$$x = b^y$$

is

$$4 = 64^{\frac{1}{3}}$$

or

$$64^{\frac{1}{3}} = 4.$$

c. For

$$\log_{10} .001 = -3,$$

we have

$$\log_b x = y,$$

where $b = 10$, $x = .001$, and $y = -3$. Therefore, the exponential form

$$x = b^y$$

is

$$.001 = 10^{-3}$$

or

$$10^{-3} = .001.$$

Exercise 17.1

Write in logarithmic form:

1. $4^3 = 64$ 2. $10^{-2} = .01$ 3. $64^{\frac{1}{2}} = 8$ 4. $64^{\frac{1}{3}} = 4$

5. $4^{-3} = \dfrac{1}{64}$ 6. $4^{\frac{1}{2}} = 2$ 7. $10^4 = 10{,}000$ 8. $10^{-4} = .0001$

Write in exponential form:

9. $\log_2 64 = 6$ 10. $\log_{27} 3 = \dfrac{1}{3}$ 11. $\log_{10} 100 = 2$ 12. $\log_{10} 1 = 0$

13. $\log_{36} 6 = \dfrac{1}{2}$ 14. $\log_2 \dfrac{1}{64} = -6$ 15. $\log_{10} .1 = -1$ 16. $\log_{10} .0001 = -4.$

Section 17.2   Values of Logarithms

Values of certain logarithmic expressions can be found using the definition of the logarithm. These expressions have exponential forms in which the exponent can be found by inspection.

EXAMPLE 17.4.   Find the value of $\log_2 16$.

Solution.   We write

$$\log_2 16 = y,$$

and then write the exponential form

$$2^y = 16.$$

By inspection, it is clear that $y = 4$, and therefore,

$$\log_2 16 = 4.$$

In some cases, we must do a little more work with the exponential form to find the value of a logarithmic expression.

EXAMPLE 17.5.   Find the value of $\log_3 \frac{1}{9}$.

Solution.   We write

$$\log_3 \frac{1}{9} = y,$$

and so the exponential form is

$$3^y = \frac{1}{9}$$

$$= \frac{1}{3^2}$$

Recall from algebra that $\dfrac{1}{a^n} = a^{-n}$. Then,

$$3^y = 3^{-2}.$$

Therefore, $y = -2$, and

$$\log_3 \frac{1}{9} = -2.$$

EXAMPLE 17.6.   Find the value of $\log_9 3$.

Solution.   We write

$$\log_9 3 = y,$$

and so the exponential form is

$$9^y = 3$$

$$= \sqrt{9}$$

Recall from algebra that, for any integer $n > 2$, $\sqrt[n]{a} = a^{\frac{1}{n}}$. Then,

$$9^y = 9^{\frac{1}{2}}.$$

Therefore, $y = \frac{1}{2}$, and

$$\log_9 3 = \frac{1}{2}.$$

Observe that, for each example, the goal is to produce a power on the right-hand side for which the base is the same as the base of the exponential form. This base is also the base of the logarithm.

Logarithms to the base 10 are of particular importance, and are called **common logarithms**. We can use the definition of the logarithm to find the value of the common logarithm of any number which can be written as an integral power of 10.

EXAMPLE 17.7.   Find the value of $\log_{10} 100$.

Solution.   If

$$\log_{10} 100 = y,$$

the exponential form is

$$10^y = 100,$$

so $y = 2$. Therefore,

$$\log_{10} 100 = 2.$$

EXAMPLE 17.8.   Find the value of $\log_{10} .01$.

Solution.   If

$$\log_{10} .01 = y,$$

the exponential form is

$$10^y = .01$$

$$= \frac{1}{100}$$

$$= \frac{1}{10^2}$$

$$= 10^{-2}$$

Thus $y = -2$, and

$$\log_{10} .01 = -2.$$

Exercise 17.2

Find the value of:

1. $\log_2 8$

2. $\log_5 125$

3. $\log_4 16$

4. $\log_4 \dfrac{1}{16}$

5. $\log_{16} 4$

6. $\log_{16} 2$

7. $\log_3 27$

8. $\log_{27} 3$

9. $\log_3 \dfrac{1}{27}$

10. $\log_2 \dfrac{1}{16}$

11. $\log_5 \dfrac{1}{125}$

12. $\log_{125} 5$

13. $\log_{10} 1000$

14. $\log_{10} 10,000$

15. $\log_{10} 100,000$

16. $\log_{10} .001$

17. $\log_{10} .0001$

18. $\log_{10} .00001$

19. $\log_{10} 10$

20. $\log_{10} 1$

Section 17.3 — Graphs of Logarithmic Functions

To draw the graph of a logarithmic function, we use the exponential form. We may compare the exponential form with the exponential function for which it is the inverse function.

EXAMPLE 17.9.   Draw the graph of $y = \log_2 x$.

Solution.   The exponential form of $y = \log_2 x$ is $x = 2^y$. Therefore, it is the inverse function of $y = 2^x$. Some points on the graph of $y = 2^x$ are:

$$2^0 = 1 \text{ gives } (0, 1).$$

$$2^1 = 2 \text{ gives } (1, 2).$$

$$2^2 = 4 \text{ gives } (2, 4).$$

$$2^3 = 8 \text{ gives } (3, 8).$$

$$2^{-1} = \frac{1}{2} \text{ gives } \left(-1, \frac{1}{2}\right).$$

$$2^{-2} = \frac{1}{4} \text{ gives } \left(-2, \frac{1}{4}\right).$$

$$2^{-3} = \frac{1}{8} \text{ gives } \left(-3, \frac{1}{8}\right).$$

The graph is

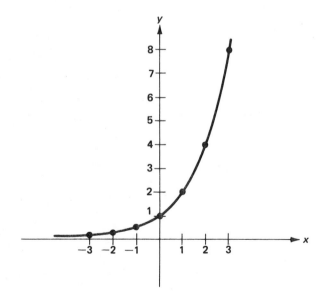

Observe that the domain is all real numbers, but the range is $y > 0$. Also, observe that the graph approaches the negative x-axis asymptotically. We now reverse the coordinates of the ordered pairs of $y = 2^x$ to draw the graph of its inverse $x = 2^y$, or $y = \log_2 x$:

$$(1, 0) \text{ gives } \log_2 1 = 0.$$

$$(2, 1) \text{ gives } \log_2 2 = 1.$$

$$(4, 2) \text{ gives } \log_2 4 = 2.$$

$$(8, 3) \text{ gives } \log_2 8 = 3.$$

$$\left(\frac{1}{2}, -1\right) \text{ gives } \log_2 \frac{1}{2} = -1.$$

$$\left(\frac{1}{4}, -2\right) \text{ gives } \log_2 \frac{1}{4} = -2.$$

$$\left(\frac{1}{8}, -3\right) \text{ gives } \log_2 \frac{1}{8} = -3.$$

We show the graphs of $y = 2^x$ and $y = \log_2 x$ on one set of axes:

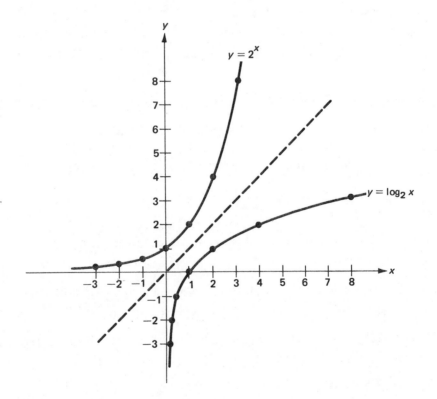

The graphs of $y = 2^x$ and its inverse $y = \log_2 x$ are symmetric with respect to the line $y = x$. Also, since x and y reverse, the graph of $y = \log_2 x$ approaches the negative y-axis asymptotically. The domain of $y = \log_2 x$ is $x > 0$ and its range is all real numbers. It is important to remember that, since the domain of a logarithmic function is $x > 0$, $\log_b A$ is defined only if $A > 0$.

Exercise 17.3

Draw the graph:

1. $y = \log_3 x$

2. $y = \log_4 x$

3. $y = \log_5 x$

4. $y = \log_{10} x$ (Use a scale of 100, 200, 300, . . . , on the x-axis.)

| Section
17.4 | Properties of Logarithms |

The definition of the logarithm can be used to prove three basic properties of logarithms. The proofs of these properties are given at the end of this unit.

> **Properties of Logarithms:**   For $A > 0$, $B > 0$, $b > 0$ and $b \neq 1$, and any real number n:
> 1. $\log_b AB = \log_b A + \log_b B$
> 2. $\log_b \dfrac{A}{B} = \log_b A - \log_b B$
> 3. $\log_b A^n = n \log_b A$

We use Properties 1, 2, and 3 to simplify logarithmic expressions.

EXAMPLE 17.10.   Write $\log_b (xy^2)$ as sums and multiples of logarithmic expressions.

Solution.   Using Property 1,

$$\log_b (xy^2) = \log_b x + \log_b y^2.$$

Then, using Property 3,

$$\log_b x + \log_b y^2 = \log_b x + 2 \log_b y.$$

EXAMPLE 17.11.   Write $\log_b \left(\dfrac{m^3}{n^2} \right)$ as differences and multiples of logarithmic expressions.

Solution.   Using Property 2,

$$\log_b \left(\frac{m^3}{n^2} \right) = \log_b m^3 - \log_b n^2.$$

Then, using Property 3,

$$\log_b m^3 - \log_b n^2 = 3 \log_b m - 2 \log_b n.$$

EXAMPLE 17.12.   Write $\log_b \left(\dfrac{xy}{z} \right)^3$ as sums, differences, and multiples of logarithmic expressions.

Solution.   Since the exponent applies to the entire expression inside the logarithm, we use Property 3 first:

$$\log_b \left(\frac{xy}{z} \right)^3 = 3 \log_b \left(\frac{xy}{z} \right).$$

Then, using Property 2,

$$3 \log_b \left(\frac{xy}{z} \right) = 3(\log_b xy - \log_b z).$$

Finally, using Property 1,

$$3(\log_b xy - \log_b z) = 3(\log_b x + \log_b y - \log_b z)$$

$$= 3 \log_b x + 3 \log_b y - 3 \log_b z.$$

If there is a radical expression inside a logarithm, we write the radical as a rational exponent. Then, we apply Property 3 to the rational exponent.

EXAMPLE 17.13.   Write $\log_b \sqrt[3]{\dfrac{p}{qr}}$ as sums, differences, and multiples of logarithmic expressions.

Solution.   We write

$$\log_b \sqrt[3]{\frac{p}{qr}}$$

as

$$\log_b \left(\frac{p}{qr} \right)^{\frac{1}{3}}$$

Then,

$$\log_b \left(\frac{p}{qr} \right)^{\frac{1}{3}} = \frac{1}{3} \log_b \left(\frac{p}{qr} \right)$$

$$= \frac{1}{3} (\log_b p - \log_b qr)$$

$$= \frac{1}{3} \left[\log_b p - (\log_b q + \log_b r) \right]$$

$$= \frac{1}{3} (\log_b p - \log_b q - \log_b r)$$

$$= \frac{1}{3} \log_b p - \frac{1}{3} \log_b q - \frac{1}{3} \log_b r.$$

EXAMPLE 17.14.   Write $\log_b \left(\dfrac{p}{\sqrt[3]{qr}} \right)$ as sums, differences, and multiples of logarithmic expressions.

Solution.   We write

$$\log_b \left(\frac{p}{\sqrt[3]{qr}} \right)$$

as

$$\log_b \left(\frac{p}{(qr)^{\frac{1}{3}}} \right).$$

Then,

$$\log_b \left(\frac{p}{(qr)^{\frac{1}{3}}} \right) = \log_b p - \log_b (qr)^{\frac{1}{3}}$$

$$= \log_b p - \frac{1}{3}\log_b (qr)$$

$$= \log_b p - \frac{1}{3}(\log_b q + \log_b r)$$

$$= \log_b p - \frac{1}{3}\log_b q - \frac{1}{3}\log_b r.$$

Observe that the coefficient $\frac{1}{3}$ does not apply to $\log_b p$ because p was not included in the original radical expression.

Exercise 17.4

Write as sums, differences, and multiples of logarithmic expressions:

1. $\log_b (xyz)$ 2. $\log_b (x^2 y^2)$ 3. $\log_b \left(\frac{xy}{z} \right)$ 4. $\log_b \left(\frac{x}{yz} \right)$

5. $\log_b (k^2 mn^3)$ 6. $\log_b \left(\frac{k^2}{mn^3} \right)$ 7. $\log_b \left(\frac{\sqrt{s}}{t^3} \right)$ 8. $\log_b \left(\frac{u^2}{\sqrt{vw}} \right)$

9. $\log_b \sqrt[3]{\frac{pq^2}{r}}$ 10. $\log_b \sqrt[4]{\frac{p^2}{qr^3}}$

**Section
17.5**   Common Logarithms

We mentioned in Section 17.2 that logarithms to the base 10, called common logarithms, are of particular importance. It is the common logarithm which was devised for use as a computational aid. With the widespread use of handheld calculators, common logarithms are decreasing in importance as a calculating tool. However, they still have importance in solving applied problems.

It is conventional to write a common logarithm as log x, leaving out the base 10. Thus, when you see log x, you may assume the base is 10. If you are using a scientific calculator, the function name "log" finds the common logarithm of a number.

We know from the definition of a logarithm how to find common logarithms of numbers which are integral powers of 10. Recall from the examples and exercises in Section 17.2:

$$\log 100{,}000 = \log 10^5 \ = 5$$
$$\log 10{,}000 = \log 10^4 \ = 4$$
$$\log 1000 = \log 10^3 \ = 3$$
$$\log 100 = \log 10^2 \ = 2$$
$$\log 10 = \log 10^1 \ = 1$$
$$\log 1 = \log 10^0 \ = 0$$
$$\log .1 = \log 10^{-1} = -1$$
$$\log .01 = \log 10^{-2} = -2$$
$$\log .001 = \log 10^{-3} = -3$$
$$\log .0001 = \log 10^{-4} = -4$$
$$\log .00001 = \log 10^{-5} = -5$$

It is clear that the value of the common logarithm of an integral power of 10 is the integral exponent. In general,

$$\log 10^n \ = \ n.$$

To find common logarithms of numbers other than integral powers of 10, we must use tables of common logarithms. The values given in these tables are found by sophisticated methods beyond the scope of this book, as are the methods used by calculators to find common logarithms.

A part of a table of logarithms is reproduced here. The full table is given in Table III at the back of this book.

N	0	1	2	3	4	5	6	7	8	9
0.0	.0000	.0043	.0086	.0128	.0170					
3.5	.5441	.5453	.5465	.5478	.5490	.5502	.5514	.5527	.5539	.5551
3.6	.5563	.5575	.5587	.5599	.5611	.5623	.5635	.5647	.5658	.5670
3.7	.5682	.5694	.5705	.5717	.5729	.5740	.5752	.5763	.5775	.5786
3.8	.5798	.5809	.5821	.5832	.5843	.5855	.5866	.5877	.5888	.5899
3.9	.5911	.5922	.5933	.5944	.5955	.5966	.5977	.5988	.5999	.6010
4.0	.6021	.6031	.6042	.6053	.6064	.6075	.6085	.6096	.6107	.6117
4.1	.6128	.6138	.6149	.6160	.6170	.6180	.6191	.6201	.6212	.6222
4.2	.6232	.6243	.6253	.6263	.6274	.6284	.6294	.6304	.6314	.6325
4.3	.6335	.6345	.6355	.6365	.6375	.6385	.6395	.6405	.6415	.6425
4.4	.6435	.6444	.6454	.6464	.6474	.6484	.6493	..6503	.6513	.6522
4.5	.6532	.6542	.6551	.6561	.6571	.6580	.6590	.6599	.6609	.6618

Notice that all of the numbers in the column marked N are between 1 and 10. Thus we can find the common logarithm of any three-digit number between 1 and 10 directly from the table.

EXAMPLE 17.15.   Find log 4.37.

Solution.   We look under N until we find 4.3. Then we go over to the right to the number in the column under 7. This number is the common logarithm of 4.37. Thus log 4.37 = .6405.

It should be mentioned that, except for the common logarithms of rational powers of 10, common logarithms are nonterminating decimals. The values given in Table III are four-place approximations of those decimals. The values given by calculators are approximations to a greater number of places, depending on the calculator. However, having observed that we are dealing with approximations, we will use the ordinary equal sign throughout the rest of this unit and the next unit. In dealing with logarithms, we do not usually say "approximately" every time a logarithm is used, or use the "approximately equal to" symbol.

To find common logarithms of numbers which are not between 1 and 10, we use Property 1 of Section 17.4.

EXAMPLE 17.16.   Find log 3820.

Solution.   From Table III, we can find log 3.82 = .5821. To find log 3820, we write

$$\log 3820 = \log 3.82(10^3)$$
$$= \log 3.82 + \log 10^3$$
$$= .5821 + 3$$
$$= 3.5821.$$

Thus, log 3820 = 3.5821. The number 3 is called the **characteristic**, and is the exponent of an integral power of 10. The number .5821 is called the **mantissa**, and is found from the table.

EXAMPLE 17.17.   Find log .0414.

Solution.   From Table III, we can find log 4.14 = .6170. To find log .0414, we write

$$\log .0414 = \log 4.14(10^{-2})$$
$$= \log 4.14 + \log 10^{-2}$$
$$= .6170 + (-2)$$
$$= .6170 - 2.$$

You must *never* write a negative characteristic in front of a positive mantissa as, for example, -2.6170. The number -2.6170 is $-2 - .6170$. However, the correct number is $-2 + .6170$. When the characteristic is negative, it is best to write the logarithm in the form

$$\log .0414 = .6170 - 2.$$

We will call this form the **characteristic-mantissa form.** If you are using a scientific calculator, you will find it gives the answer in the form -1.3830. This number is the ordinary subtraction of $.6170 - 2$:

$$0.6170 - 2.0000 = -1.3830.$$

In this section, we will expect logarithms to be written in characteristic-mantissa form. However, when we come to applications in the next unit, the subtracted form will be useful.

Finally, we need to know how to find a number given its logarithm. This process is known formally as finding the **antilogarithm**, written antilog x, but you might refer to it informally as "unlogging" a number.

EXAMPLE 17.18. Find:

a. antilog .5944   b. antilog 1.5786   c. antilog $(.5911-4)$

Solutions.   a. We find .5944 to the right of 3.9 and under 3, thus antilog .5944 = 3.93.

b. The mantissa is .5786. We find .5786 to the right of 3.7 and under 9, thus antilog .5786 = 3.79. Since the characteristic is 1,

$$\text{antilog } 1.5786 = 3.79(10^1)$$
$$= 37.9.$$

c. The mantissa is .5911. We find .5911 to the right of 3.9 and under 0, thus antilog .5911 = 3.90. Since the characteristic is -4,

$$\text{antilog } (.5911 - 4) = 3.90(10^{-4})$$
$$= .00039.$$

Exercise 17.5

Use Table III to find:

1. log 2.19
2. log 8.30
3. log 84.6
4. log .323
5. log .0055
6. log 46,700
7. log 9300
8. log 4,230,000
9. log .00645
10. log .0000399
11. antilog .5366
12. antilog .2041
13. antilog $(.8312 - 1)$
14. antilog 2.7364
15. antilog 4.3729
16. antilog 5.7709
17. antilog $(.7152 - 3)$
18. antilog $(.9494 - 4)$
19. antilog $(.9175 - 5)$
20. antilog 9.9832

Proofs of the Properties of Logarithms

PROPERTY 1. $\log_b AB = \log_b A + \log_b B$

Proof.   Let $\log_b A = x$ and $\log_b B = y$. Then, by the definition of a logarithm,

$$A = b^x \text{ and } B = b^y$$
$$AB = b^x b^y$$
$$AB = b^{x+y}.$$

Again, by the definition of a logarithm,

$$\log_b AB = x + y.$$

Substituting back for x and y,

$$\log_b AB = \log_b A + \log_b B.$$

PROPERTY 2. $\log_b \dfrac{A}{B} = \log_b A - \log_b B$

Proof.   Proceeding as in Property 1,

$$\frac{A}{B} = \frac{b^x}{b^y}$$

$$\frac{A}{B} = b^{x-y}$$

$$\log_b \frac{A}{B} = x - y$$

$$\log_b \frac{A}{b} = \log_b A - \log_b B.$$

PROPERTY 3. $\log_b A^n = n \log_b A$

Proof.   Let $\log_b A = x$. Then, by definition of a logarithm,

$$A = b^x$$

$$A^n = (b^x)^n$$

$$A^n = b^{nx}.$$

Again, by the definition of a logarithm,

$$\log_b A^n = nx.$$

Substituting back for x,

$$\log_b A^n = n \log_b A.$$

Self-test

1. Write $\log_b \left(\dfrac{x^2\sqrt{y}}{z} \right)$ as sums, differences, and multiples of logarithmic expressions.

1. _____

2. Find the value of :

   a. $\log_3 \dfrac{1}{81}$

   b. $\log_{81} 3$

2a. _____

2b. _____

3. Write $3^2 = 9$ in logarithmic form.

3. _____

4. Use Table III to find:

   a. log .634

   b. antilog 2.7033

4a. _____

4b. _____

5. Draw the graph of $y = \log_6 x$.

Unit 18 Exponential Equations

INTRODUCTION

An exponential equation is an equation with a variable in the exponent. In this unit you will learn how to solve exponential equations, and how to use exponential equations to solve some of the many types of problems they apply to. Also, you will have a brief introduction to the second important type of logarithm, the natural logarithm, and one of its applications.

OBJECTIVES

When you have finished this unit you should be able to:

1. Use common logarithms to solve exponential equations.
2. Given an appropriate formula, use common logarithms to solve applied problems involving exponential equations.
3. Given an appropriate formula, use natural logarithms to solve applied problems involving exponential equations with the base e.

Section 18.1   Solving Exponential Equations

An **exponential equation** is an equation with the variable in an exponent. Exponential equations cannot be solved by ordinary algebraic methods. The problem is to get the variable out of the exponent. This can be accomplished by using the properties of logarithms, in particular, Property 3. It must be assumed that we can take the logarithm of each side of an equation, just as we can add to, or multiply, or take the square root of each side.

EXAMPLE 18.1. Solve $5^x = 16$.

Solution. We take the common logarithm of each side of the equation:

$$\log 5^x = \log 16.$$

Applying Property 3 removes the variable from the exponent:

$$x \log 5 = \log 16.$$

Thus,

$$x = \frac{\log 16}{\log 5}$$

$$x = \frac{1.2041}{.6990}$$

$$x = 1.72.$$

You should keep in mind that this solution is only a decimal approximation of x. Indeed, it is not an awfully good approximation since there are two logarithms involved, themselves both approximations, and a round-off on the division. The only way to write the exact answer, however, is to write $x = \dfrac{\log 16}{\log 5}$, which is not very useful for practical applications. Therefore, we will write solutions as decimal approximations, usually to three significant digits.

EXAMPLE 18.2.   Solve $3^{x-1} = 8$.

Solution.   Following the method of the preceding example, we take the common logarithm of each side, apply Property 3, and then approximate x:

$$\log 3^{x-1} = \log 8$$

$$(x - 1)\log 3 = \log 8$$

$$x - 1 = \frac{\log 8}{\log 3}$$

$$x = \frac{\log 8}{\log 3} + 1$$

$$x = \frac{.9031}{.4771} + 1$$

$$x = 2.89.$$

Exercise 18.1

Solve:

1. $7^x = 24$ 2. $15^x = 200$ 3. $4^{x-1} = 11$ 4. $6^{x+1} = 21$

5. $3^{2x} = 15$ 6. $4^{-3x} = 7$ 7. $2 = 100^{.2x}$ 8. $.6 = 2^{-.4x}$

**Section
18.2**   # Applications

One application of common logarithms is to investment of an amount of money with interest compounded over a period of time. The original amount of money invested is called the **principal**. We say the interest is **compounded annually** if it is calculated once a year, **semiannually** if twice a year, **quarterly** if four times a year. How often the interest is compounded is important because interest is calculated on the total amount present at the time, the principal plus interest previously paid, so we get "interest on the interest."

The formula for calculating the total amount present at any time is

$$A = P\left(1 + \frac{r}{n}\right)^{nt},$$

where A is the total amount, P is the principal, r is the interest rate written in decimal form, n is the number of times a year the interest is calculated, and t is the number of years during which the interest accumulates.

EXAMPLE 18.3.   If \$2000 is invested at a rate of 6% compounded semiannually, how long will it take to accumulate to \$2500?

Solution. $P = 2000$, $r = .06$, $n = 2$, and $A = 2500$. We must find t. Using the formula $A = P\left(1 + \frac{r}{n}\right)^{nt}$,

$$2500 = 2000\left(1 + \frac{.06}{2}\right)^{2t}$$

$$2500 = 2000(1.03)^{2t}$$

$$\frac{2500}{2000} = 1.03^{2t}$$

$$1.25 = 1.03^{2t}.$$

Thus, we have an exponential equation. Taking the common logarithm of each side,

$$\log 1.25 = \log 1.03^{2t}$$

$$\log 1.25 = 2t \log 1.03$$

$$t = \frac{\log 1.25}{2 \log 1.03}$$

$$t = 3.77, \text{ or about 4 years.}$$

We have rounded the answer to 4 years because, since the interest is compounded semiannually, it will be calculated at $3\frac{1}{2}$ years, when the accumulated amount will be somewhat less than \$2500, and at 4 years, when the accumulated amount will be somewhat more than \$2500.

EXAMPLE 18.4.   If you invest \$1000 at $7\frac{1}{2}$% compounded quarterly, how much will you have after 10 years?

Solution. $P = 1000$, $r = .075$, $n = 4$, and $t = 10$. We must find A:

$$A = 1000\left(1 + \frac{.075}{4}\right)^{4(10)}$$

$$A = 1000(1.01875)^{40}.$$

This equation is not an exponential equation, since the variable A is not in an exponent. However, common logarithms provide an efficient way to find the value of the right-hand side.

Taking the common logarithm of each side, and carefully applying the properties of logarithms,

$$\log A = \log 1000(1.01875)^{40}$$

$$\log A = \log 1000 + \log 1.01875^{40}$$

$$\log A = \log 1000 + 40 \log 1.01875.$$

To find log 1.01875, you may use a scientific calculator or approximate using Table III. To use Table III, recall the method of linear interpolation from Section 1.5. Using this method, we find the nearest table values, and then the differences as shown in the diagram:

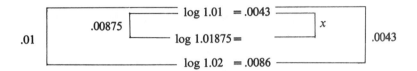

The proportion is

$$\frac{.00875}{.01} = \frac{x}{.0043}$$

$$x = \frac{(.0043)(.00875)}{.01}$$

$$= .00376.$$

Therefore,

$$\log 1.01875 = .0043 + .00376$$

$$= .00806.$$

Then,

$$\log A = \log 1000 + 40 \log 1.01875$$

$$= 3 + 40(.00806)$$

$$= 3.3224.$$

Observe that this number is clearly not the answer, since we should have significantly more than the principal of $1000. Indeed, what we have is the logarithm of the answer. To find A, we must "unlog"; that is, find the antilogarithm.

$$\begin{array}{c} .01 \left[\begin{array}{l} x \left[\begin{array}{l} \log 2.10 = .3222 \\ .3224 \end{array} \right. .0002 \\ \log 2.11 = .3243 \end{array} \right] .0021 \end{array}$$

The proportion is

$$\frac{x}{.01} = \frac{.0002}{.0021}$$

$$x = \frac{(.01)(.0002)}{.0021}$$

$$= .0009.$$

Therefore,

$$\text{antilog } .3224 = 2.10 + .0009$$

$$= 2.1009,$$

and

$$A = 2.1009(10^3)$$

$$= 2100.9$$

or

$$\$2100.$$

Using a scientific calculator, use the "log" key to find log $1.01875 = .00806762$. Multiply by 40 and add 3 to find log $A = 3.3227049$. To find the antilog, press "inv" and "log" to obtain 2102.3493, or \$2100.

Most banks now compound interest continuously, as if it were being calculated not once, or twice, or four times a year, but every instant. The formula for this calculation is

$$A = Pe^{rt},$$

where A, P, r, and t are as before, and e is a special constant. Like π, e is a number that is found many places in mathematics and science. Also like π, e is an irrational number, which may be approximated to as many places as we wish by a decimal. A common approximation for e is 2.718. If you are using Table III you may find log 2.718 using linear interpolation:

The proportion is

$$\frac{.008}{.01} = \frac{x}{.0016}$$

$$x = \frac{(.0016)(.008)}{.01}$$

$$= .00128.$$

Therefore,

$$\log 2.718 = .4330 + .00128$$

$$= .43428.$$

Remember that linear interpolation gives an approximation. The number given by a scientific calculator is .43425 to five decimal places. It is common to use $\log 2.718 = .4343$, which is either number rounded to four decimal places.

EXAMPLE 18.5. Suppose you put \$500 in a savings account on which the interest is $5\frac{1}{4}\%$ compounded continuously. How much will you have at the end of a year?

Solution. $P = 500$, $r = .0525$, and $t = 1$. We must find A:

$$A = 500(2.718)^{.0525(1)}$$

$$A = 500(2.718)^{.0525}$$

$$\log A = \log 500(2.718)^{.0525}$$

$$\log A = \log 500 + \log 2.718^{.0525}$$

$$\log A = \log 500 + .0525 \log 2.718$$

$$\log A = 2.7218$$

$$A = \text{antilog } 2.7218$$

$$A = \$527.$$

Another application in which the constant e appears is growth and decay problems. Actually, the preceding example is a growth problem, specifically, growth of money. Often growth problems involve the growth of a population such as a bacteria culture. Decay problems often involve disintegration of radioactive materials. For both, the formula is

$$N = N_0 e^{kt},$$

where N is the amount after time t, N_0 is the initial amount, and k is a given constant. For growth problems k is positive, and for decay problems k is negative.

EXAMPLE 18.6. A radioactive material decays according to the formula $N = N_0 e^{-.2t}$, where t is in seconds. If there are 600 grams of the material initially, how long will it take for the material to be reduced to 100 grams?

Solution. $N = 100$, $N_0 = 600$, and we must find t. Since $N = N_0 e^{-.2t}$,

$$100 = 600(2.718)^{-.2t}$$

$$\frac{100}{600} = 2.718^{-.2t}$$

$$\frac{1}{6} = 2.718^{-.2t}.$$

To avoid unnecessary rounding approximations, we will leave $\frac{1}{6}$ in fraction form. Then, solving the exponential equation,

$$\log \frac{1}{6} = \log 2.718^{-.2t}$$

$$\log 1 - \log 6 = -.2t \log 2.718$$

$$t = \frac{\log 1 - \log 6}{-.2 \log 2.718}$$

$$t = \frac{0 - .7782}{-.2(.4343)}$$

$$t = \frac{.7782}{.2(.4343)}$$

$$t = 8.96 \text{ seconds.}$$

Observe that, since t is in an exponent, we do not have $\log t$ and we do not find an antilog.

Exercise 18.2

Use $A = P\left(1 + \frac{r}{n}\right)^{nt}$:

1. If you invest \$3000 at 9% compounded annually, how much will you have after 5 years?

2. If you invest \$600 at 4% compounded quarterly, how much will you have after 2 years?

3. How long will it take for \$5000 invested at 6% compounded semiannually to grow to \$10,000?

4. If \$1500 is invested at 8% compounded quarterly, how long will it take to reach \$2000?

Use $A = Pe^{rt}$ for problems 5 and 6:

5. How long will it take for \$1000 invested at 7% compounded continuously to reach \$1500?

6. If you invest \$10,000 at $7\frac{1}{2}$% compounded continuously, how much will you have after 10 years?

7. A bacteria culture grows according to the formula $N = N_0 e^{.5t}$, where t is in days. If there are 100 bacteria initially, how many will there be after 7 days?

8. A bacteria culture grows according to the formula $N = N_0 e^{.02t}$, where t is in hours. If the initial amount is 3000 bacteria, how long will it take for the population to double?

9. A radioactive material decays according to the formula $N = N_0 e^{-.2t}$, where t is in seconds. After 10 seconds, how much is left of an initial amount of 50 grams?

10. A radioactive material decays according to the formula $N = N_0 e^{-.004t}$, where t is in hours. How long will it take an initial amount of 1000 grams to reduce to 100 grams?

11.* Dating of ancient remains can be done using a radioactive isotope of carbon called carbon 14. If carbon 14 decays according to the formula $N = N_0 e^{-.0001245t}$, where t is in years, what is its half-life; that is, how long will it take an initial amount of N_0 to reduce to $\frac{1}{2} N_0$?

12.* Suppose it is known that the initial amount of carbon 14 in a piece of wood is N_0, and now only $\frac{1}{5} N_0$ remains. How old is the piece of wood?

*Problems 11 and 12 are adapted from William E. Boyce and Richard C. DiPrima, *Elementary Differential Equations*, 2nd Edition, John Wiley & Sons, Inc., 1969, p. 58.

Section 18.3   Natural Logarithms

In the examples and exercises in Section 18.2, we saw several exponential equations involving the constant e. This constant occurs so often in such equations that it is defined as the base of another type of logarithm. A **natural logarithm** is a logarithm to the base e. We use $\ln x$ to indicate the natural logarithm of x. When you see $\ln x$, you may assume $\log_e x$ is meant. If you are using a scientific calculator, the function name "$\ln x$" finds the natural logarithm of a number.

First, we observe that $\ln e = 1$. Since $\ln e = \log_e e$, if $y = \log_e e$, then by the definition of a logarithm, $e^y = e$, and so $y = 1$.

To find other values of natural logarithms we must use tables. Table IV at the back of this book is a table of natural logarithms for three-digit numbers between 1 and 10.

EXAMPLE 18.7.   Find $\ln 3.92$.

Solution.   Referring to Table IV, we look to the right of 3.9 and under 2. We find the number 3661. However, the first number in this group has a whole number 1 in front. Thus $\ln 3.92 = 1.3661$.

If you look at the numbers next to 2.7, and also the last number next to 7.3, you will see asterisks in front of them. The asterisks mean we should use the whole numbers, 1 or 2, for the group which follows with these numbers. If there is no indication of a whole number with the first number of a group, the number is written as a decimal less than 1.

EXAMPLE 18.8.   Find:
   a. $\ln 1.73$ b. $\ln 2.77$

Solutions.

   a.   Since the number to the right of 1.7 and under 3 is 5481, with no whole number in front of the group, $\ln 1.73 = .5481$.

   b.   Since the number to the right of 2.7 and under 7 is 0188, and there is an asterisk in front of the number, $\ln 2.77 = 1.0188$.

We cannot use characteristics which are powers of 10 to find natural logarithms of numbers not between 1 and 10. Powers of 10 go with common logarithms, which are to the base 10. However, we can use a power of 10 times $\ln 10$, given that $\ln 10 = 2.3026$.

EXAMPLE 18.9.   Find $\ln 578$.

Solution.
$$\ln 578 = \ln 5.78(10^2)$$
$$= \ln 5.78 + \ln 10^2$$
$$= \ln 5.78 + 2 \ln 10$$
$$= 1.7544 + 2(2.3026)$$
$$= 6.3596.$$

If you have a calculator with the function "$\ln x$," you do not have to go through the process in the preceding example.

Exponential equations involving e are often easier to solve using natural logarithms than using common logarithms.

EXAMPLE 18.10.   A radioactive material decays according to the formula $N = N_0 e^{-.5t}$, where t is in seconds. If there are 800 grams of the material initially, how long will it take for the material to reduce to 200 grams?

Solution. $N = 200$, $N_0 = 800$, and we must find t:

$$200 = 800e^{-.5t}$$

$$\frac{200}{800} = e^{-.5t}$$

$$.25 = e^{-.5t}$$

We use natural logarithms to solve the exponential equation, recalling that $\ln e = 1$:

$$\ln .25 = \ln e^{-.5t}$$

$$\ln .25 = -.5t \ln e$$

$$\ln .25 = -.5t$$

$$t = \frac{\ln .25}{-.5}$$

$$t = 2.77 \text{ seconds.}$$

If you are using Table IV, you can find $\ln .25$ by calculating

$$\ln .25 = \ln 2.5(10^{-1})$$

$$= \ln 2.5 - \ln 10$$

$$= .9163 - 2.3026$$

$$= -1.3863.$$

Then, divide by $-.5$ to find t.

To find t using a scientific calculator, enter .25 and press the "ln x" key. The result is the negative number -1.3862944. Press "$+/-$" to change sign, and divide by .5 to obtain 2.7725887.

EXAMPLE 18.11.   Suppose you invest $5000 in a savings account at 6% compounded continuously. How much will you have in the account at the end of 3 years?

Solution.   The formula, from Section 18.2, is $A = Pe^{rt}$. We have $P = 5000$, $r = .06$, $t = 3$, and we must find A:

$$A = 5000e^{.06(3)}$$

Using natural logarithms,

$$\ln A = \ln 5000e^{.18}$$

$$= \ln 5000 + .18 \ln e$$

$$= \ln 5000 + .18.$$

We find ln 5000 using either a scientific calculator or Table IV and the method of Example 18.9. Then,

$$\ln A = 8.5172 + .18$$
$$= 8.6972.$$

To find the natural antilogarithm of 8.6972 using a scientific calculator, press "inv" and "ln x" to obtain 5986. To use Table IV, reverse the process of Example 18.9:

$$\ln A = 8.6972$$
$$= 1.7894 + 3(2.3026).$$

Therefore,

$$A = 5.99(10^3)$$
$$= 5990.$$

The amount is \$5986, or \$5990 to three significant digits. Clearly, this problem is much easier to solve, and can be solved more accurately, using a scientific calculator. In fact, since the equation is not an exponential equation because the variable A is not in an exponent, the problem can be solved using a scientific calculator without using logarithms at all. Observe that $y = e^x$ is the inverse function of $y = \ln x$. Enter .18, and press the "inv" and "ln x" keys to find $e^{.18} = 1.1972174$. Then, multiply by 5000 to obtain 5986.0868, or \$5986.

Exercise 18.3

Solve using natural logarithms:

1. A bacteria culture grows according to the formula $N = N_0 e^{.69t}$, where t is in hours. If there are 500 bacteria initially, how long will it take for the culture to grow to 1000 bacteria?

2. A radioactive material decays according to the formula $N = N_0 e^{-.092t}$, where t is in seconds. If 10 grams are present initially, how long will it take to reduce the material to 4 grams?

3. Suppose you invest \$2000 at $5\frac{1}{2}\%$ compounded continuously. Use $A = Pe^{rt}$ to find how much you will have at the end of 1 year.

4. Suppose you invest \$1000 at $6\frac{1}{2}\%$ compounded continuously. Use $A = Pe^{rt}$ to find how long it will take for your investment to double.

5. Radium decays according to the formula $N = N_0 e^{-.000411t}$, where t is in years. If there are 6 milligrams of radium in a substance initially, how much will there be 100 years later?

6.* The formula $P = P_0 e^{-kt}$ represents the amount of pollutant P in a lake after t years, where P_0 is the initial amount of pollutant, and then only clear water enters the lake. The constant $k = \dfrac{r}{V}$, where r is the rate at which the clear water enters the lake and V is the volume of the lake. Find how long it will take to reduce the amount of pollution to 10% of its current amount, that is, to reduce P_0 to $\frac{1}{10}P_0$, for each of these Great Lakes:

Lake	V in km³	r in km³/yr
Superior	12,200	65.2
Michigan	4900	158
Erie	460	175
Ontario	1600	209

*Adapted from Boyce and DiPrima, *Elementary Differential Equations*, 2nd Edition, pp. 60-61.

Self-test

Solve using common logarithms:

1. $5^{x+1} = 20$

1. _____

2. $500 is invested at 8% compounded quarterly. Use $A = P\left(1 + \dfrac{r}{n}\right)^{nt}$ to find how much will accumulate in 5 years.

2. _____

Solve using common or natural logarithms:

3. $500 is invested at $5\frac{1}{2}\%$ compounded continuously. Use $A = Pe^{rt}$ to find how long it will take to grow to $550.

3. _____

4. A bacteria culture grows according to the formula $N = N_0 e^{.04t}$, where t is in hours. If there are 300 bacteria initially, how many will there be after 12 hours?

4. _____

5. A radioactive material decays according to the formula $N = N_0 e^{-.5t}$, where t is in seconds. How long will it take an initial amount of 20 grams to reduce to 2 grams?

5. _____

Imaginary Numbers

INTRODUCTION

In the preceding units, radicals where the radicand is negative have been ignored. Such radicals do not represent real numbers. However, it is quite possible for a radicand to be negative. For example, the quadratic formula can give the square root of a negative number just as well as a positive number. In this unit you will learn the definition of imaginary numbers involving square roots of negative numbers. You will learn how to add, subtract, multiply and divide imaginary numbers. Then, you will use square roots of negative numbers to solve quadratic equations which have imaginary number solutions.

OBJECTIVES

When you have finished this unit you should be able to:

1. Write the square root of a negative number as a pure imaginary number.
2. Add and subtract imaginary numbers.
3. Multiply pure imaginary numbers, an imaginary by a pure imaginary number, or two imaginary numbers.
4. Divide by a pure imaginary number, or an imaginary number.
5. Solve quadratic equations where the solutions are imaginary numbers.

Pure Imaginary Numbers

Consider the number $\sqrt{-1}$. Suppose $x = \sqrt{-1}$. Then $x^2 = -1$; but this is not possible because the square of a number is always zero or positive. That is, the square of a **real number** is always zero or positive. We need to define a new kind of number represented by $\sqrt{-1}$.

Definition:

$$i = \sqrt{-1}$$

and

$$i^2 = -1$$

The letter i is used to stand for $\sqrt{-1}$. Any number of the form bi, where b is a real number, is called a **pure imaginary number**. These are some examples of pure imaginary numbers:

$$2i, \quad -i, \quad \frac{1}{2}i, \quad i\sqrt{2}.$$

Observe that, in the last number, we have written i in front of b. This order is conventional when b is a square root, because then it is clear that i is not under the radical.

EXAMPLE 19.1.   Write $\sqrt{-4}$ as a pure imaginary number.

Solution.   Since

$$\sqrt{-4} = \sqrt{(-1)4}$$
$$= 2\sqrt{-1},$$

and since $i = \sqrt{-1}$, we have

$$2\sqrt{-1} = 2i.$$

EXAMPLE 19.2.   Write $\sqrt{-10}$ as a pure imaginary number.

Solution.

$$\sqrt{-10} = \sqrt{(-1)10}$$
$$= \sqrt{-1}\sqrt{10}$$
$$= i\sqrt{10}.$$

Since $\sqrt{10}$ cannot be simplified, the pure imaginary number $i\sqrt{10}$ cannot be simplified further.

EXAMPLE 19.3.   Write $\sqrt{-12}$ as a pure imaginary number.

Solution.

$$\sqrt{-12} = \sqrt{(-1)12}$$
$$= \sqrt{-1}\sqrt{12}$$
$$= i\sqrt{12}.$$

But $\sqrt{12} = 2\sqrt{3}$, therefore

$$i\sqrt{12} = 2i\sqrt{3}.$$

EXAMPLE 19.4.   Write $-\sqrt{-3}$ as a pure imaginary number.

Solution.   It is a common error to make the two negatives into a positive. However,

$$\sqrt{-3} = i\sqrt{3},$$

therefore

$$-\sqrt{-3} = -i\sqrt{3}.$$

The two negatives have no effect on one another. The negative under the square root simply makes the number a pure imaginary number.

Real numbers and pure imaginary numbers can be added to make **complex numbers**. For example, $3 + 2i$ is a complex number. It has a **real part** 3, and a **pure imaginary part** $2i$. In general, a complex number is a number of the form $a + bi$, where a is the real part and bi is the pure imaginary part.

If $b = 0$, then $bi = 0$, and we have a number $a + 0 = a$. In this case, the complex number is a real number. If $b \neq 0$, but $a = 0$, then the number is $0 + bi = bi$. In this case, the complex number is a pure imaginary number. If neither $a = 0$ nor $b = 0$, then $a + bi$ is called simply an **imaginary number.** For example,

$3 + 0i = 3$ is a real number.
$0 + 2i = 2i$ is a pure imaginary number.
$3 + 2i$ is an imaginary number.

Real numbers, pure imaginary numbers, and imaginary numbers, all are complex numbers.

EXAMPLE 19.5.   Write $12 + \sqrt{-25}$ as an imaginary number.

Solution.   First, we write $\sqrt{-25}$ as a pure imaginary number:

$$\sqrt{-25} = \sqrt{(-1)25}$$
$$= 5\sqrt{-1}$$
$$= 5i.$$

Then, we write the imaginary number

$$12 + \sqrt{-25} = 12 + 5i.$$

Observe that the negative under the square root creates an imaginary number, but in no way affects the sign between the real and pure imaginary parts of the imaginary number.

EXAMPLE 19.6.   Write $\sqrt{3} - \sqrt{-48}$ as an imaginary number.

Solution.   The pure imaginary part is

$$\sqrt{-48} = \sqrt{(-1)48}$$
$$= \sqrt{-1}\,\sqrt{48}$$
$$= i\sqrt{48}$$
$$= 4i\sqrt{3}$$

Thus the imaginary number is

$$\sqrt{3} - \sqrt{-48} = \sqrt{3} - 4i\sqrt{3}$$

The name "imaginary number" is the result of an unfortunate twist of historical fate. Imaginary numbers are no more "imaginary," in the sense of artificial, than real numbers. The symbol $2i$ is a symbol made up to stand for a numerical concept, just like the symbol -2 or the symbol $\sqrt{2}$ are made up to stand for numerical concepts.

Exercise 19.1

Write as a pure imaginary number:

1. $\sqrt{-36}$ 2. $-\sqrt{-9}$ 3. $\sqrt{-7}$ 4. $-\sqrt{-11}$

5. $\sqrt{-18}$ 6. $\sqrt{-27}$ 7. $\sqrt{-80}$ 8. $\sqrt{-98}$

9. $-\sqrt{-50}$ 10. $-\sqrt{-45}$ 11. $\sqrt{-\frac{1}{4}}$ 12. $\sqrt{-\frac{9}{25}}$

13. $-\sqrt{-\frac{16}{9}}$ 14. $-\sqrt{-\frac{1}{16}}$ 15. $\sqrt{-\frac{5}{4}}$ 16. $-\sqrt{-\frac{8}{9}}$

Write as an imaginary number:

17. $9 + \sqrt{-81}$ 18. $4 - \sqrt{-64}$ 19. $10 - \sqrt{-20}$ 20. $\sqrt{3} + \sqrt{-72}$

21. $\frac{3}{4} - \sqrt{-\frac{9}{4}}$ 22. $\frac{2}{3} + \sqrt{-\frac{2}{9}}$

Section 19.2	Addition and Subtraction

To add imaginary numbers we combine like terms. In many cases, the real parts of imaginary numbers are like terms, and the pure imaginary parts are like terms.

EXAMPLE 19.7.   Add $(4 + 2i) + (6 - 4i)$.

Solution.   First, we remove the parentheses:

$$(4 + 2i) + (6 - 4i) = 4 + 2i + 6 - 4i.$$

Then, we combine the real parts, which are like terms, and the pure imaginary parts, which are like terms:

$$4 + 2i + 6 - 4i = 10 - 2i.$$

To subtract imaginary numbers, we follow the same procedure, but we must be careful in removing the parentheses.

EXAMPLE 19.8.   Subtract $(3 - i) - (2 - 5i)$.

Solution.   We remove the parentheses, being careful with the signs of the second number:

$$(3 - i) - (2 - 5i) = 3 - i - 2 + 5i.$$

Now, we can combine the real parts and the pure imaginary parts:

$$3 - i - 2 + 5i = 1 + 4i.$$

In the examples above, the sum or difference of imaginary numbers is also an imaginary number. It is possible, in special cases, for the sum or difference to be a real number or a pure imaginary number.

EXAMPLE 19.9.   Add $(5 - 2i) + (1 + 2i)$.

Solution.   Removing parentheses, and combining like terms,

$$(5 - 2i) + (1 + 2i) = 5 - 2i + 1 + 2i$$
$$= 6 + 0i$$
$$= 6,$$

which is a real number.

EXAMPLE 19.10.   Subtract $(4 - 5i) - (4 + 3i)$.

Solution.   Carefully removing parentheses, and combining like terms,

$$(4 - 5i) - (4 + 3i) = 4 - 5i - 4 - 3i$$
$$= 0 - 8i$$
$$= -8i,$$

which is a pure imaginary number.

It is possible that the real parts, or the pure imaginary parts, or both, will not be like terms.

EXAMPLE 19.11.   Add $(\sqrt{2} + i\sqrt{3}) + (\sqrt{2} + i\sqrt{5})$.

Solution. $$(\sqrt{2} + i\sqrt{3}) + (\sqrt{2} + i\sqrt{5}) = \sqrt{2} + i\sqrt{3} + \sqrt{2} + i\sqrt{5}$$
$$= 2\sqrt{2} + i\sqrt{3} + i\sqrt{5}.$$

Since $\sqrt{3} + \sqrt{5}$ cannot be combined, $i\sqrt{3} + i\sqrt{5}$ cannot be combined, and the addition cannot be carried further. The pure imaginary part can be written as one term by factoring out the common factor i:

$$2\sqrt{2} + i\sqrt{3} + i\sqrt{5} = 2\sqrt{2} + (\sqrt{3} + \sqrt{5})i.$$

An imaginary number is often written in this form.

If the pure imaginary parts are not already written in terms of i, this should be done first.

EXAMPLE 19.12.   Subtract $(3 - \sqrt{-4}) - (2 - \sqrt{-9})$.

Solution.   Writing the numbers in imaginary number form,

$$(3 - \sqrt{-4}) - (2 - \sqrt{-9}) = (3 - 2i) - (2 - 3i).$$

Now, continuing as before, we have

$$(3 - 2i) - (2 - 3i) = 3 - 2i - 2 + 3i$$
$$= 1 + i.$$

Observe that we would not have been able to combine the pure imaginary parts were they in the form $\sqrt{-4}$ and $\sqrt{-9}$. It is important always to write imaginary and pure imaginary numbers in terms of i.

Exercise 19.2

Add or subtract as indicated:

1. $2i + 3i$

2. $4i - i$

3. $(1 + 2i) + (2 + 4i)$

4. $(5 - 2i) + (2 - i)$

5. $(8 + 6i) - (5 + 2i)$

6. $(4 - i) - (5 - 3i)$

7. $(10 - 8i) - (4 + 4i)$

8. $(2 + 3i) + (3 - 6i)$

9. $(-5 + 6i) + (3 - 2i)$

10. $(-6 - 8i) - (-1 + 2i)$

11. $(6 + 8i) - (6 + 4i)$

12. $(3 + i) + (-3 - 2i)$

13. $(7 + 5i) - (3 + 5i)$

14. $(4 + 5i) + (4 - 5i)$

15. $(2 + i) - (3 - 4i) + (6 - 8i)$

16. $(10 - 5i) - (3 + 4i) - (7 - 2i)$

17. $(4 - \sqrt{-25}) + (4 - \sqrt{-81})$

18. $(6 - \sqrt{-36}) - (-3 - \sqrt{-49})$

19. $(\sqrt{2} + \sqrt{-8}) + (\sqrt{2} + \sqrt{-18})$

20. $(\sqrt{12} + \sqrt{-20}) + (\sqrt{3} - \sqrt{-12})$

| Section 19.3 | Multiplication |

To multiply imaginary numbers, we recall that $i^2 = -1$. Thus, for example,

$$(3i)(4i) = 12i^2 = 12(-1) = -12.$$

It is very important to write pure imaginary numbers in terms of i before multiplying. If a and b are both negative, then $\sqrt{a}\,\sqrt{b}$ is not equal to $\sqrt{ab}$.

EXAMPLE 19.13.   Multiply $\sqrt{-5}\,\sqrt{-5}$.

Solution.   It is a common error to write $\sqrt{(-5)(-5)}$, and then $\sqrt{25} = 5$. However, since both radicands are negative, we must write the numbers as pure imaginary numbers in terms of i:

$$\sqrt{-5}\,\sqrt{-5} = (i\sqrt{5})(i\sqrt{5}).$$

Now, we have

$$(i\sqrt{5})(i\sqrt{5}) = i^2(\sqrt{5}\,\sqrt{5})$$

$$= (-1)(5)$$

$$= -5.$$

The product is -5, not 5, because $i^2 = -1$.

To multiply an imaginary number by a pure imaginary, we follow the usual procedure for multiplying by a single term, remembering that $i^2 = -1$.

EXAMPLE 19.14. Multiply $3i(2 + 6i)$.

Solution. Multiplying each term of the imaginary number by $3i$,

$$3i(2 + 6i) = 3i(2) + (3i)(6i)$$
$$= 6i + 18i^2.$$

Now, since $i^2 = -1$, we write

$$6i + 18i^2 = 6i + 18(-1)$$
$$= 6i - 18.$$

EXAMPLE 19.15. Multiply $2i(5 - 4i)$.

Solution. Multiplying each term by $2i$, and using $i^2 = -1$,

$$2i(5 - 4i) = 2i(5) - (2i)(4i)$$
$$= 10i - 8i^2$$
$$= 10i - 8(-1)$$
$$= 10i + 8.$$

When the real part of an imaginary number is positive, we usually write the real part first:

$$10i + 8 = 8 + 10i.$$

We multiply two imaginary numbers in the same way we multiply two binomials. Again, it is important to use $i^2 = -1$.

EXAMPLE 19.16. Multiply $(5 - 2i)(3 - 6i)$.

Solution. First, we multiply the different parts of the imaginary numbers as if they were binomials:

$$(5 - 2i)(3 - 6i) = 15 - 30i - 6i + 12i^2.$$

We may combine like terms:

$$15 - 30i - 6i + 12i^2 = 15 - 36i + 12i^2.$$

A common error is to stop at this point. However, using $i^2 = -1$, we can combine more terms:

$$15 - 36i + 12i^2 = 15 - 36i + 12(-1)$$
$$= 15 - 36i - 12$$
$$= 3 - 36i.$$

The product of two imaginary numbers always can be reduced by using $i^2 = -1$. However, as with addition and subtraction, we may not always have like terms.

EXAMPLE 19.17.   Multiply $(\sqrt{3} + i\sqrt{2})(\sqrt{3} + i\sqrt{5})$.

Solution.   Multiplying the different parts, and using $i^2 = -1$, we have

$$
\begin{aligned}
(\sqrt{3} + i\sqrt{2})(\sqrt{3} + i\sqrt{5}) &= 3 + i\sqrt{15} + i\sqrt{6} + i^2\sqrt{10} \\
&= 3 + i\sqrt{15} + i\sqrt{6} + (-1)\sqrt{10} \\
&= 3 + i\sqrt{15} + i\sqrt{6} - \sqrt{10} \\
&= 3 - \sqrt{10} + (\sqrt{15} + \sqrt{6})i.
\end{aligned}
$$

Although the terms cannot be combined, the resulting number is an imaginary number. The real part is $3 - \sqrt{10}$, and the pure imaginary part is $(\sqrt{15} + \sqrt{6})i$.

It is possible for the product of two imaginary numbers to be a real number or a pure imaginary number.

EXAMPLE 19.18.   Multiply $(2 + 3i)(2 - 3i)$.

Solution.
$$
\begin{aligned}
(2 + 3i)(2 - 3i) &= 4 - 6i + 6i - 9i^2 \\
&= 4 - 9i^2 \\
&= 4 - 9(-1) \\
&= 4 + 9 \\
&= 13.
\end{aligned}
$$

Exercise 19.3

Multiply:

1. $(7i)(2i)$
2. $(4i)i$
3. $\sqrt{-4}\,\sqrt{-4}$
4. $\sqrt{-6}\,\sqrt{-6}$
5. $\sqrt{-2}\,\sqrt{-8}$
6. $\sqrt{-3}\,\sqrt{-6}$
7. $8i(3 + 2i)$
8. $2i(6 - 4i)$
9. $-i(4 + i)$
10. $-5i(5 - 2i)$
11. $(2 + 3i)(4 + i)$
12. $(3 + 5i)(2 + 3i)$
13. $(6 - 2i)(3 + 4i)$
14. $(2 - 5i)(6 - 3i)$
15. $(5 + 4i)(5 - 4i)$
16. $(2 - i)(2 + i)$
17. $(3 + 2i)(2 + 3i)$
18. $(4 - 2i)(2 - 4i)$

19. $(2 + i\sqrt{2})(3 + i\sqrt{2})$

20. $(\sqrt{3} + 2i)(2\sqrt{3} - i)$

21. $(\sqrt{3} - i\sqrt{5})(\sqrt{3} + i)$

22. $(\sqrt{2} + i\sqrt{6})(\sqrt{2} - i\sqrt{6})$

23   $(2 + \sqrt{-3})(3 + \sqrt{-5})$

24. $(\sqrt{3} + \sqrt{-6})(\sqrt{6} + \sqrt{-3})$

Section
19.4

Division

Division of imaginary numbers is usually written in the form of a fraction. The division is accomplished by making the denominator of the fraction a real number. If the denominator is a pure imaginary number, we need only multiply the numerator and the denominator by i.

EXAMPLE 19.19.   Divide $\dfrac{5}{3i}$.

Solution.   Multiplying the numerator and the denominator by i,

$$\frac{5}{3i} = \frac{5i}{3i^2}.$$

Recalling that $i^2 = -1$,

$$\frac{5i}{3i^2} = \frac{5i}{3(-1)}$$

$$= \frac{5i}{-3}$$

$$= -\frac{5i}{3}.$$

The denominator is a real number. The entire quotient is a pure imaginary number which may be written

$$-\frac{5i}{3}$$

or

$$-\frac{5}{3}i.$$

EXAMPLE 19.20.   Divide $\dfrac{3 + 10i}{2i}$.

Solution.   Again, we begin by multiplying the numerator and the denominator by i:

$$\frac{3 + 10i}{2i} = \frac{(3 + 10i)i}{2i^2}.$$

Now, we must multiply the expression in the numerator:

$$\frac{(3 + 10i)i}{2i^2} = \frac{3i + 10i^2}{2i^2}.$$

Using $i^2 = -1$ in both the numerator and the denominator,

$$\frac{3i + 10i^2}{2i^2} = \frac{3i + 10(-1)}{2(-1)}$$

$$= \frac{3i - 10}{-2}$$

$$= -\frac{3i - 10}{2}.$$

Since $-(3i - 10) = 10 - 3i$, a good form for the resulting imaginary number is

$$\frac{10 - 3i}{2}.$$

Another common form is

$$5 - \frac{3}{2}i,$$

which shows the real part, 5, and the pure imaginary part, $-\frac{3}{2}i$.

EXAMPLE 19.21.   Divide $\dfrac{6 - 4i}{5i}$.

Solution.

$$\frac{6 - 4i}{5i} = \frac{(6 - 4i)i}{5i^2}$$

$$= \frac{6i - 4i^2}{5i^2}$$

$$= \frac{6i - 4(-1)}{5(-1)}$$

$$= \frac{6i + 4}{-5}$$

$$= -\frac{6i + 4}{5}.$$

This imaginary number can be written in the form

$$-\frac{4 + 6i}{5}$$

or

$$\frac{-4 - 6i}{5}$$

or

$$-\frac{4}{5} - \frac{6}{5}i.$$

If a division is written as a fraction and the denominator is an imaginary number, we multiply the numerator and the denominator of the fraction by the **conjugate** of the denominator. The conjugate of an imaginary number $a + bi$ is $a - bi$, and vice versa. The product of two conjugate imaginary numbers always is a real number:

$$(a + bi)(a - bi) = a^2 - abi + abi - b^2i^2$$

$$= a^2 - b^2i^2$$

$$= a^2 - b^2(-1)$$

$$= a^2 + b^2.$$

Since a and b are real numbers, and no i appears, $a^2 + b^2$ is a real number.

EXAMPLE 19.22.   Divide $\dfrac{2 + 3i}{2 + i}$.

Solution.   The conjugate of $2 + i$ is $2 - i$, so we multiply the numerator and the denominator by $2 - i$:

$$\frac{2 + 3i}{2 + i} = \frac{(2 + 3i)(2 - i)}{(2 + i)(2 - i)}.$$

Multiplying the imaginary numbers in the numerator and denominator, we have

$$\frac{(2 + 3i)(2 - i)}{(2 + i)(2 - i)} = \frac{4 - 2i + 6i - 3i^2}{4 - 2i + 2i - i^2}$$

$$= \frac{4 + 4i - 3i^2}{4 - i^2}.$$

Then, using $i^2 = -1$,

$$\frac{4 + 4i - 3i^2}{4 - i^2} = \frac{4 + 4i - 3(-1)}{4 - (-1)}$$

$$= \frac{4 + 4i + 3}{4 + 1}$$

$$= \frac{7 + 4i}{5}.$$

The denominator is a real number, and we may write the resulting imaginary number in the form

$$\frac{7 + 4i}{5}$$

or

$$\frac{7}{5} + \frac{4}{5}i.$$

EXAMPLE 19.23.   Divide $\dfrac{3 - 4i}{4 - 2i}$.

Solution. We multiply the numerator and the denominator by $4 + 2i$, the conjugate of the denominator:

$$\frac{3 - 4i}{4 - 2i} = \frac{(3 - 4i)(4 + 2i)}{(4 - 2i)(4 + 2i)}.$$

Multiplying the imaginary numbers, combining like terms, and using $i^2 = -1$, we have

$$\frac{(3 - 4i)(4 + 2i)}{(4 - 2i)(4 + 2i)} = \frac{12 + 6i - 16i - 8i^2}{16 + 8i - 8i - 4i^2}$$

$$= \frac{12 - 10i - 8i^2}{16 - 4i^2}$$

$$= \frac{12 - 10i - 8(-1)}{16 - 4(-1)}$$

$$= \frac{12 - 10i + 8}{16 + 4}$$

$$= \frac{20 - 10i}{20}.$$

We reduce the number by dividing each term by 10:

$$\frac{20 - 10i}{20} = \frac{2 - i}{2}.$$

This number may be written

$$\frac{2 - i}{2}$$

or

$$1 - \frac{1}{2}i.$$

Exercise 19.4

Divide:

1. $\dfrac{5}{6i}$ 2. $\dfrac{-2}{3i}$ 3. $\dfrac{6}{8i}$ 4. $\dfrac{25}{5i}$ 5. $\dfrac{3}{\sqrt{-9}}$

6. $\dfrac{\sqrt{2}}{\sqrt{-16}}$ 7. $\dfrac{5 + 8i}{2i}$ 8. $\dfrac{3 - 4i}{9i}$ 9. $\dfrac{-2 + 8i}{5i}$ 10. $\dfrac{1 + i}{i}$

11. $\dfrac{5i}{2 + 4i}$ 12. $\dfrac{-5i}{2 - i}$ 13. $\dfrac{6 + 3i}{2 - i}$ 14. $\dfrac{4 + 5i}{2 - 3i}$ 15. $\dfrac{2 - 3i}{1 - 4i}$

16. $\dfrac{3 - i}{4 + 3i}$ 17. $\dfrac{5 - 10i}{6 - 2i}$ 18. $\dfrac{7 - 5i}{3 + 3i}$ 19. $\dfrac{3 + i}{3 - i}$ 20. $\dfrac{1 - 2i}{1 + 2i}$

21. $\dfrac{4 + 3i}{3 - 4i}$ 22. $\dfrac{1 - 7i}{7 + i}$ 23. $\dfrac{2\sqrt{5} + i}{1 + i\sqrt{5}}$ 24. $\dfrac{2\sqrt{2} + i}{\sqrt{2} - i}$

Imaginary Solutions

In algebra, we solve equations such as $x^2 - 4 = 0$ by writing

$$x^2 - 4 = 0$$
$$x^2 = 4$$
$$x = \pm\sqrt{4}$$
$$x = \pm 2.$$

If we encounter an equation such as $x^2 + 4 = 0$, however, we have no real number solution. Using pure imaginary numbers, now we can find solutions to such equations.

EXAMPLE 19.24. Solve $x^2 + 4 = 0$.

Solution. Following the method above,

$$x^2 + 4 = 0$$
$$x^2 = -4$$
$$x = \pm\sqrt{-4}.$$

Since we can write $\sqrt{-4}$ as the pure imaginary number $2i$, the solutions are

$$x = \pm\sqrt{-4}$$
$$x = \pm 2i.$$

The solutions are $2i$ and $-2i$.

Recall that the quadratic equation in standard form,

$$ax^2 + bx + c = 0,$$

can be solved by the quadratic formula,

$$x = \frac{-b \pm \sqrt{b^2 - 4ac}}{2a}$$

Using the quadratic formula, there are equations where the number under the radicand is negative. The solutions to such equations are imaginary numbers.

EXAMPLE 19.25. Solve $2x^2 - 3x + 2 = 0$.

Solution. In the quadratic formula, $a = 2$, $b = -3$, and $c = 2$:

$$x = \frac{-b \pm \sqrt{b^2 - 4ac}}{2a}$$

$$= \frac{-(-3) \pm \sqrt{(-3)^2 - 4(2)(2)}}{2(2)}$$

$$= \frac{3 \pm \sqrt{9 - 16}}{4}$$

$$= \frac{3 \pm \sqrt{-7}}{4}$$

$$= \frac{3 \pm i\sqrt{7}}{4}.$$

The solutions are the imaginary numbers $\dfrac{3 + i\sqrt{7}}{4}$ and $\dfrac{3 - i\sqrt{7}}{4}$. Imaginary solutions to quadratic equations always will be conjugates of one another.

EXAMPLE 19.26.   Solve $x^2 + 2x + 4 = 0$.

Solution.   In the quadratic formula, $a = 1$, $b = 2$, and $c = 4$:

$$x = \frac{-b \pm \sqrt{b^2 - 4ac}}{2a}$$

$$= \frac{-2 \pm \sqrt{2^2 - 4(1)(4)}}{2(1)}$$

$$= \frac{-2 \pm \sqrt{4 - 16}}{2}$$

$$= \frac{-2 \pm \sqrt{-12}}{2}$$

$$= \frac{-2 \pm i\sqrt{12}}{2}.$$

Of course, we reduce imaginary number solutions whenever possible:

$$\frac{-2 \pm i\sqrt{12}}{2} = \frac{-2 \pm 2i\sqrt{3}}{2}$$

$$= -1 \pm i\sqrt{3}.$$

The solutions are $-1 + i\sqrt{3}$ and $-1 - i\sqrt{3}$.

We should realize that imaginary number solutions are no less "real" than real number solutions. Imaginary numbers have applications ranging from theory of electricity to a geometry that might turn out to describe the shape of the universe. However, we cannot plot imaginary numbers on the number line.

Exercise 19.5

Solve:

1. $x^2 + 1 = 0$

2. $x^2 + 25 = 0$

3. $x^2 + 45 = 0$

4. $4x^2 + 48 = 0$

5. $x^2 + x + 1 = 0$

6. $2x^2 - 5x + 5 = 0$

7. $x^2 - 2x + 2 = 0$

8. $x^2 + 2x + 26 = 0$

9. $9 = 2x - 2x^2$

10. $3x - 4 = 3x^2$

Self-test

1. Write $-\sqrt{-90}$ as a pure imaginary number.

1. _____

Perform the operation as indicated:

2. $(5 - i)(4 - 6i)$

2. _____

3. $(5 + 2i) - (3 - 4i)$

3. _____

4. $\dfrac{-5i}{4 - 3i}$

4. _____

5. Solve $x^2 + 2x + 15 = 0$.

5. _____

Unit 20 Polar Form

INTRODUCTION

In this unit you will learn how to interpret a complex number as a vector, where the complex number may be real, pure imaginary, or imaginary. The complex number can then be written in a form called its polar form. Once they are written in polar form, there are very easy methods for multiplying and dividing complex numbers. Then, you will learn a theorem called De Moivre's theorem. This theorem extends the methods for multiplying and dividing complex numbers written in polar form to finding roots and powers of complex numbers.

OBJECTIVES

When you have finished this unit you should be able to:

1. Change the form of a complex number, real, pure imaginary, or imaginary, from rectangular form to polar form and from polar form to rectangular form.
2. Multiply or divide complex numbers written in polar form.
3. Use De Moivre's theorem to find powers or roots of complex numbers.

Section 20.1

Complex Numbers in Polar Form

Recall from Section 7.3 that a vector can be expressed as the resultant of its horizontal and vertical components. In the Cartesian coordinate system, suppose a vector has its base at the origin and lies in the first quadrant. Then, its horizontal component is the x-coordinate of the point at its tip, and its vertical component is the y-coordinate of the point at its tip:

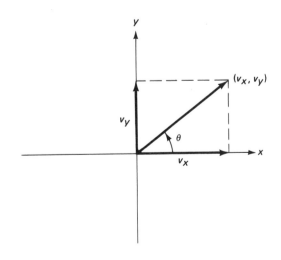

317

We also recall that

$$\cos \theta = \frac{v_x}{r}$$

or

$$v_x = r \cos \theta,$$

and

$$\sin \theta = \frac{v_y}{r}$$

or

$$v_y = r \sin \theta.$$

Therefore, the coordinates of the point at the tip of the vector are (v_x, v_y), or $(r \cos \theta, r \sin \theta)$.

Now, we consider any complex number $a + bi$, where a and b are any real numbers. A complex number written in the form $a + bi$ is said to be in **rectangular form**. If the complex number is imaginary, that is, if $b \neq 0$, the complex number cannot be plotted as a point on the number line. However, any complex number written in rectangular form may be represented as a vector in the Cartesian coordinate system (also called the **rectangular** coordinate system), with the base of the vector at the origin and its tip at the point (a, b).

EXAMPLE 20.1. Draw the vector which represents the complex number $2 - 3i$.

Solution. The complex number $2 - 3i$ is represented by the vector with its base at the origin and its tip at the point $(2, -3)$:

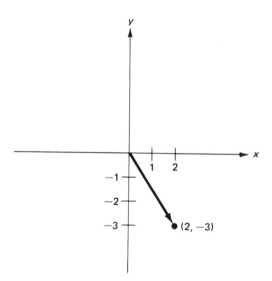

EXAMPLE 20.2. Draw the vector which represents the complex number -5.

Solution. The real number -5 is interpreted as the complex number $-5 + 0i$. Therefore, it is represented by the vector with its base at the origin and its tip at the point $(-5, 0)$:

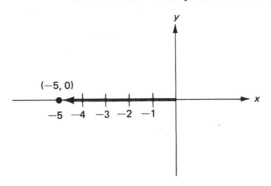

Observe that the vector which represents a real number lies along the x-axis.

EXAMPLE 20.3.   Draw the vector which represents the complex number $4i$.

Solution.   The pure imaginary number $4i$ is interpreted as the complex number $0 + 4i$. Therefore, it is represented by the vector with its base at the origin and its tip at the point $(0, 4)$:

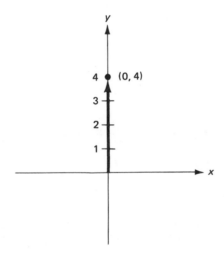

Observe that the vector which represents a pure imaginary number lies along the y-axis.

Now, suppose $a + bi$ is a complex number represented by a vector in the first quadrant, where r is the length of the vector and θ is the angle between the positive x-axis and the vector:

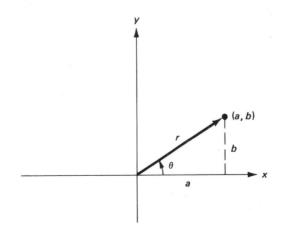

Then,

$$a = r \cos \theta$$

and

$$b = r \sin \theta.$$

Also,

$$r = \sqrt{a^2 + b^2} \ .$$

These formulas can be extended to complex numbers represented by vectors in any quadrant, or on an axis, where r is the length of the vector and θ is an angle in standard position with the vector on its terminal side. Observe that we take r to be nonnegative.

Consider any complex number in the rectangular form $a + bi$. Since $a = r \cos \theta$ and $b = r \sin \theta$, we have

$$a + bi = r \cos \theta + (r \sin \theta)i$$

$$= r(\cos \theta + i \sin \theta).$$

The form $r(\cos \theta + i \sin \theta)$ is the **polar form** or **trigonometric form** of the complex number. The polar form is often abbreviated

$$r \operatorname{cis} \theta.$$

The number r is the **absolute value** of the complex number, and is nonnegative. The angle θ is called the **argument** of the complex number.

EXAMPLE 20.4.   Write $-1 + i\sqrt{3}$ in polar form.

Solution.   The absolute value is

$$r = \sqrt{a^2 + b^2}$$

$$= \sqrt{1 + 3}$$

$$= \sqrt{4}$$

$$= 2.$$

The argument is an angle θ such that

$$r \cos \theta = a \text{ and } r \sin \theta = b.$$

Thus,

$$2 \cos \theta = -1 \text{ and } 2 \sin \theta = \sqrt{3}$$

or

$$\cos \theta = -\frac{1}{2} \text{ and } \sin \theta = \frac{\sqrt{3}}{2}.$$

The angle θ is in the second quadrant and its reference angle is 60°. Therefore,

$$\theta = 120°.$$

The polar form is

$$2(\cos 120° + i \sin 120°)$$

or

$$2 \text{ cis } 120°.$$

EXAMPLE 20.5.   Write $1 - i$ in polar form.

Solution.   The absolute value is

$$r = \sqrt{1 + 1}$$
$$= \sqrt{2}.$$

Since

$$\sqrt{2} \cos \theta = 1 \text{ and } \sqrt{2} \sin \theta = -1$$

or

$$\cos \theta = \frac{1}{\sqrt{2}} \text{ and } \sin \theta = -\frac{1}{\sqrt{2}},$$

θ is in the fourth quadrant and the reference angle is 45°. Therefore, the argument is

$$\theta = 315°.$$

The polar form is

$$\sqrt{2} \, (\cos 315° + i \sin 315°)$$

or

$$\sqrt{2} \text{ cis } 315°.$$

EXAMPLE 20.6.   Write 2 in polar form.

Solution.   The absolute value is 2. Also,

$$2 \cos \theta = 2$$

or

$$\cos \theta = 1.$$

Therefore,

$$\theta = 0°.$$

The polar form is

$$2(\cos 0° + i \sin 0°)$$

or

$$2 \operatorname{cis} 0°.$$

Recall that real numbers are represented by vectors lying on the x-axis. The argument of any positive real number is $0°$, and the argument of any negative real number is $180°$.

EXAMPLE 20.7.   Write $-2i$ in polar form.

Solution.   The absolute value is again 2. Also,

$$2 \sin \theta = -2$$

or

$$\sin \theta = -1.$$

Therefore,

$$\theta = 270°.$$

The polar form is

$$2(\cos 270° + i \sin 270°)$$

or

$$2 \operatorname{cis} 270°.$$

Recall that pure imaginary numbers are represented by vectors lying on the y-axis. The argument of any positive pure imaginary number is $90°$, and the argument of any negative pure imaginary number is $270°$.

Given the polar form of any complex number, we can write the complex number in rectangular form by substituting values for $\sin \theta$ and $\cos \theta$.

EXAMPLE 20.8.   Write $2 \operatorname{cis} 225°$ in rectangular form.

Solution.

$$2 \operatorname{cis} 225° = 2(\cos 225° + i \sin 225°)$$

$$= 2\left[-\frac{1}{\sqrt{2}} + i\left(-\frac{1}{\sqrt{2}} \right) \right]$$

$$= -\frac{2}{\sqrt{2}} - \frac{2i}{\sqrt{2}}.$$

Using the rationalized form $\dfrac{\sqrt{2}}{2}$ for $\dfrac{1}{\sqrt{2}}$, we have the simpler rectangular form

$$2\left[-\frac{\sqrt{2}}{2} + i\left(-\frac{\sqrt{2}}{2} \right) \right] = -\sqrt{2} - i\sqrt{2}.$$

EXAMPLE 20.9.   Write $\sqrt{3}$ cis 90° in rectangular form.

Solution.

$$\sqrt{3} \text{ cis } 90° = \sqrt{3}\,(\cos 90° + i\sin 90°)$$

$$= \sqrt{3}\,\big[0 + i(1)\big]$$

$$= i\sqrt{3}\,.$$

Since the argument is 90°, the complex number is a pure imaginary number.

EXAMPLE 20.10.   Write 4 cis 0° in rectangular form.

Solution.

$$4 \text{ cis } 0° = 4(\cos 0° + i\sin 0°)$$

$$= 4\big[1 + i(0)\big]$$

$$= 4.$$

Since the argument is 0°, the complex number is a real number.

If the absolute value r of a complex number is zero, then the number is zero. Observe that

$$0 \text{ cis }\theta = 0$$

for any value of the argument θ.

Exercise 20.1

Draw the vector which represents the complex number:

1. $1 + 2i$ 2. $3 - 4i$ 3. $-4 - i$ 4. $-3 + 2i$

5. 3 6. -2 7. i 8. $-2i$

Write the complex number in polar form:

9. $\sqrt{3} + i$ 10. $1 + i$ 11. $-1 - i$ 12. $-\sqrt{3} + i$

13. $\sqrt{2} + i\sqrt{2}$ 14. $2 - 2i$ 15. $-\sqrt{3} + 3i$ 16. $-3 - i\sqrt{3}$

17. 1 18. -2 19. $-3i$ 20. i

Write the complex number in rectangular form:

21. $2 \text{ cis } 60°$ 22. $\sqrt{2} \text{ cis } 45°$ 23. $\sqrt{2} \text{ cis } 135°$ 24. $2 \text{ cis } 330°$

25. $2 \text{ cis } 315°$ 26. $2\sqrt{2} \text{ cis } 135°$ 27. $6 \text{ cis } 150°$ 28. $2\sqrt{3} \text{ cis } 240°$

29. $2 \operatorname{cis} 90°$ 30. $3 \operatorname{cis} 180°$ 31. $\operatorname{cis} 270°$ 32. $\operatorname{cis} 0°$

33. $0 \operatorname{cis} 0°$ 34. $0 \operatorname{cis} 90°$

Section 20.2

Multiplication and Division

We can use identities from Unit 14 to derive an easy way to multiply or divide two complex numbers when the numbers are written in polar form. Suppose $r_1 \operatorname{cis} \theta_1$ and $r_2 \operatorname{cis} \theta_2$ are two complex numbers. Then,

$$(r_1 \operatorname{cis} \theta_1)(r_2 \operatorname{cis} \theta_2) = (r_1 r_2)(\operatorname{cis} \theta_1 \operatorname{cis} \theta_2).$$

Therefore, the absolute value of the product is the product of the absolute values.

Now, we recall that $\operatorname{cis} \theta$ is an abbreviation for $\cos \theta + i \sin \theta$. Then,

$$(r_1 r_2)(\operatorname{cis} \theta_1 \operatorname{cis} \theta_2) = (r_1 r_2)\big[(\cos \theta_1 + i \sin \theta_1)(\cos \theta_2 + i \sin \theta_2)\big]$$

$$= (r_1 r_2)\big(\cos \theta_1 \cos \theta_2 + i \cos \theta_1 \sin \theta_2 + i \sin \theta_1 \cos \theta_2 + i^2 \sin \theta_1 \sin \theta_2\big)$$

$$= (r_1 r_2)\big[(\cos \theta_1 \cos \theta_2 - \sin \theta_1 \sin \theta_2) + i(\sin \theta_1 \cos \theta_2 + \cos \theta_1 \sin \theta_2)\big].$$

Using the identities for the sum of two angles from Section 14.1,

$$(r_1 r_2)(\operatorname{cis} \theta_1 \operatorname{cis} \theta_2) = (r_1 r_2)\big[\cos(\theta_1 + \theta_2) + i \sin(\theta_1 + \theta_2)\big]$$

$$= (r_1 r_2) \operatorname{cis}(\theta_1 + \theta_2).$$

Thus the argument of the product is the *sum* of the arguments.

EXAMPLE 20.11. Find the product of $3 \operatorname{cis} 90°$ and $4 \operatorname{cis} 180°$.

Solution. Using the formula derived above,

$$(3 \operatorname{cis} 90°)(4 \operatorname{cis} 180°) = (3 \cdot 4)\big[\operatorname{cis}(90° + 180°)\big]$$

$$= 12 \operatorname{cis} 270°.$$

Observe that $3 \operatorname{cis} 90° = 3i$ and $4 \operatorname{cis} 180° = -4$, and so the product is indeed $12 \operatorname{cis} 270° = -12i$.

When the sum of the arguments is $360°$ or more, we may use the least positive coterminal angle so that the argument of the product is an angle θ such that $0° \leqslant \theta < 360°$.

EXAMPLE 20.12. Find the product of $\sqrt{2} \operatorname{cis} 315°$ and $\sqrt{2} \operatorname{cis} 270°$.

Solution.

$$(\sqrt{2} \operatorname{cis} 315°)(\sqrt{2} \operatorname{cis} 270°) = (\sqrt{2}\ \sqrt{2}\ )\big[\operatorname{cis}(315° + 270°)\big]$$

$$= 2 \operatorname{cis} 585°$$

$$= 2 \operatorname{cis} 225°.$$

Here we observe that $\sqrt{2} \operatorname{cis} 315° = 1 - i$ and $\sqrt{2} \operatorname{cis} 270° = -i\sqrt{2}$. The product is

$$
\begin{aligned}
(1 - i)(-i\sqrt{2}) &= -i\sqrt{2} + i^2\sqrt{2} \\
&= -i\sqrt{2} - \sqrt{2} \\
&= -\sqrt{2} - i\sqrt{2}.
\end{aligned}
$$

You should check that this result is the same as $2 \operatorname{cis} 225°$.

When complex numbers are written in rectangular form, it is often convenient to write them in polar form and multiply using the polar forms.

EXAMPLE 20.13.   Find the product of $1 - i\sqrt{3}$ and $-2 + 2i\sqrt{3}$.

Solution.   Using the method of the preceding section,

$$
1 - i\sqrt{3} = 2 \operatorname{cis} 300°
$$

and

$$
-2 + 2i\sqrt{3} = 4 \operatorname{cis} 120°.
$$

Then,

$$
\begin{aligned}
(1 - i\sqrt{3})(-2 + 2i\sqrt{3}) &= (2 \operatorname{cis} 300°)(4 \operatorname{cis} 120°) \\
&= 8 \operatorname{cis} 420° \\
&= 8 \operatorname{cis} 60°.
\end{aligned}
$$

You should check that this result is the same as the result obtained by multiplying the complex numbers in rectangular form.

We use the polar forms of complex numbers in a similar way to find the quotient of two complex numbers. Suppose again that the two complex numbers in polar form are $r_1 \operatorname{cis} \theta_1$ and $r_2 \operatorname{cis} \theta_2$. Then,

$$
\frac{r_1 \operatorname{cis} \theta_1}{r_2 \operatorname{cis} \theta_2} = \frac{r_1}{r_2}\left(\frac{\cos \theta_1 + i \sin \theta_1}{\cos \theta_2 + i \sin \theta_2}\right).
$$

Next, we use the conjugate of the denominator to perform the division:

$$
\frac{r_1 \operatorname{cis} \theta_1}{r_2 \operatorname{cis} \theta_2} = \frac{r_1}{r_2}\left[\frac{(\cos \theta_1 + i \sin \theta_1)(\cos \theta_2 - i \sin \theta_2)}{(\cos \theta_2 + i \sin \theta_2)(\cos \theta_2 - i \sin \theta_2)}\right]
$$

$$
= \frac{r_1}{r_2}\left[\frac{\cos \theta_1 \cos \theta_2 - i \cos \theta_1 \sin \theta_2 + i \sin \theta_1 \cos \theta_2 - i^2 \sin \theta_1 \sin \theta_2}{\cos \theta_2 \cos \theta_2 - i \sin \theta_2 \cos \theta_2 + i \sin \theta_2 \cos \theta_2 - i^2 \sin \theta_2 \sin \theta_2}\right]
$$

$$
= \frac{r_1}{r_2}\left[\frac{(\cos \theta_1 \cos \theta_2 + \sin \theta_1 \sin \theta_2) + i(\sin \theta_1 \cos \theta_2 - \cos \theta_1 \sin \theta_2)}{\cos^2 \theta_2 + \sin^2 \theta_2}\right]
$$

$$
= \frac{r_1}{r_2}\left[\frac{\cos(\theta_1 - \theta_2) + i \sin(\theta_1 - \theta_2)}{1}\right]
$$

$$
= \frac{r_1}{r_2}\left[\cos(\theta_1 - \theta_2) + i \sin(\theta_1 - \theta_2)\right]
$$

$$
= \frac{r_1}{r_2} \operatorname{cis}(\theta_1 - \theta_2).
$$

Thus the absolute value of the quotient is the quotient of the absolute values, and the argument of the quotient is the *difference* of the absolute values.

EXAMPLE 20.14.   Find the quotient of $\sqrt{2}$ cis 135° and 2 cis 90°.

Solution.   Using the formula derived above,

$$\frac{\sqrt{2} \text{ cis } 135°}{2 \text{ cis } 90°} = \frac{\sqrt{2}}{2} \text{cis}(135° - 90°)$$

$$= \frac{\sqrt{2}}{2} \text{ cis } 45°.$$

Observe that $\sqrt{2}$ cis 135° $= -1 + i$ and 2 cis 90° $= i$. The quotient is

$$\frac{-1 + i}{i} = \frac{(-1 + i)i}{i \cdot i}$$

$$= \frac{-i + i^2}{i^2}$$

$$= \frac{-i - 1}{-1}$$

$$= i + 1$$

$$= 1 + i,$$

which in polar form is

$$\frac{1}{\sqrt{2}} \text{ cis } 45°$$

or

$$\frac{\sqrt{2}}{2} \text{ cis } 45°.$$

EXAMPLE 20.15.   Find the quotient of $2 + 2i\sqrt{3}$ and $1 - i\sqrt{3}$.

Solution.   We find the polar forms of the numbers:

$$2 + 2i\sqrt{3} = 4 \text{ cis } 60°$$

and

$$1 - i\sqrt{3} = 2 \text{ cis } 300°.$$

Then,

$$\frac{2 + 2i\sqrt{3}}{1 - i\sqrt{3}} = \frac{4 \text{ cis } 60°}{2 \text{ cis } 300°}$$

$$= \frac{4}{2} \text{cis}(60° - 300°)$$

$$= 2 \text{ cis}(-240°).$$

Using the least positive coterminal angle,

$$2\operatorname{cis}(-240°) = 2\operatorname{cis}120°.$$

Exercise 20.2

Find the product:

1. $(\operatorname{cis}0°)(\operatorname{cis}90°)$

2. $(4\operatorname{cis}90°)(2\operatorname{cis}270°)$

3. $(\sqrt{2}\operatorname{cis}135°)(\sqrt{2}\operatorname{cis}90°)$

4. $(2\operatorname{cis}210°)(2\sqrt{3}\operatorname{cis}180°)$

5. $(2\operatorname{cis}120°)(4\operatorname{cis}300°)$

6. $(\sqrt{2}\operatorname{cis}225°)(2\sqrt{2}\operatorname{cis}315°)$

Write in polar form and find the product:

7. $(2)(2i)$

8. $(-3)(-4i)$

9. $(i)(1 + i)$

10. $(-i)(-1 + i\sqrt{3})$

11. $(1 - i)(2 - 2i)$

12. $(2\sqrt{3} + 2i)(-\sqrt{3} + i)$

13. $(\sqrt{3} + i)(-1 + i)$

14. $(1 - i)(-1 + i\sqrt{3})$

Find the quotient:

15. $\dfrac{\operatorname{cis}270°}{\operatorname{cis}180°}$

16. $\dfrac{6\operatorname{cis}270°}{3\operatorname{cis}90°}$

17. $\dfrac{2\sqrt{2}\operatorname{cis}315°}{2\operatorname{cis}270°}$

18. $\dfrac{4\operatorname{cis}240°}{2\operatorname{cis}180°}$

19. $\dfrac{2\operatorname{cis}60°}{6\operatorname{cis}120°}$

20. $\dfrac{\sqrt{2}\operatorname{cis}135°}{\sqrt{2}\operatorname{cis}315°}$

Write in polar form and find the quotient:

21. $\dfrac{4}{2i}$

22. $\dfrac{-2}{-4i}$

23. $\dfrac{1 + i}{i}$

24. $\dfrac{1 - i\sqrt{3}}{-i}$

25. $\dfrac{3}{2 - 2i\sqrt{3}}$

26. $\dfrac{i\sqrt{2}}{2 + 2i}$

27. $\dfrac{\sqrt{2} + i\sqrt{2}}{1 - i}$

28. $\dfrac{-1 + i\sqrt{3}}{\sqrt{3} - i}$

29. $\dfrac{-\sqrt{3} + i}{1 - i}$

30. $\dfrac{1 + i\sqrt{3}}{1 + i}$

Section 20.3

De Moivre's Theorem

We can extend the method for finding products to a method for finding powers and roots of complex numbers. This method is **De Moivre's theorem,** named for the French mathematician Abraham De Moivre (1667–1754).

> **De Moivre's theorem:**   For any complex number in the polar form $r \operatorname{cis} \theta$, the nth power is given by $(r \operatorname{cis} \theta)^n = r^n \operatorname{cis} n\theta$.

Observe that, using the product rule to square a complex number,

$$(r \operatorname{cis} \theta)^2 = r \cdot r \operatorname{cis}(\theta + \theta)$$

$$= r^2 \operatorname{cis} 2\theta.$$

To cube the complex number,

$$(r \operatorname{cis} \theta)^3 = (r \operatorname{cis} \theta)^2 (r \operatorname{cis} \theta)$$

$$= (r^2 \operatorname{cis} 2\theta)(r \operatorname{cis} \theta)$$

$$= r^2 \cdot r \operatorname{cis}(2\theta + \theta)$$

$$= r^3 \operatorname{cis} 3\theta.$$

The generalization of these computations to the nth power is De Moivre's theorem. The general proof of De Moivre's theorem uses a process beyond the scope of this book. The theorem itself, however, is easy to use.

EXAMPLE 20.16.   Find $(1 - i)^4$.

Solution.   First, we write $1 - i$ in polar form:

$$1 - i = \sqrt{2} \operatorname{cis} 315°.$$

Then, using De Moivre's theorem,

$$(1 - i)^4 = (\sqrt{2} \operatorname{cis} 315°)^4$$

$$= (\sqrt{2})^4 \operatorname{cis}(4 \cdot 315°)$$

$$= 4 \operatorname{cis} 1260°.$$

It is a simple calculation to find the least positive coterminal angle:

$$1260° - 3(360°) = 1260° - 1080°$$

$$= 180°.$$

Therefore,

$$(1 - i)^4 = 4 \operatorname{cis} 180°$$

or

$$(1 - i)^4 = -4.$$

We can also use De Moivre's theorem with an exponent $\frac{1}{n}$ to find roots of complex numbers. Recall from algebra that, for any real number a,

$$\sqrt[n]{a} = a^{\frac{1}{n}}.$$

(For real numbers, a must be positive when n is even.) Similarly, for any complex number $r \operatorname{cis} \theta$, an nth root is given by

$$(r \operatorname{cis} \theta)^{\frac{1}{n}}.$$

Using De Moivre's theorem with $\frac{1}{n}$ in place of n,

$$(r \operatorname{cis} \theta)^{\frac{1}{n}} = r^{\frac{1}{n}} \operatorname{cis}\left(\frac{1}{n} \cdot \theta\right)$$

$$= r^{\frac{1}{n}} \operatorname{cis} \frac{\theta}{n}.$$

EXAMPLE 20.17. Find the square roots of $1 + i\sqrt{3}$.

Solution.   In polar form,

$$1 + i\sqrt{3} = 2 \operatorname{cis} 60°.$$

Therefore, a square root is

$$(2 \operatorname{cis} 60°)^{\frac{1}{2}} = 2^{\frac{1}{2}} \operatorname{cis} \frac{60°}{2}$$

$$= \sqrt{2} \operatorname{cis} 30°$$

$$= \sqrt{2}\left[\frac{\sqrt{3}}{2} + i\left(\frac{1}{2}\right)\right]$$

$$= \frac{\sqrt{6} + i\sqrt{2}}{2}.$$

However, recall from algebra that real numbers have two square roots, one positive and one negative. For example, the square roots of 4 are 2 and -2. We must look for a second square root of $1 + i\sqrt{3}$. Clearly, the other square root could be written

$$-\frac{\sqrt{6} + i\sqrt{2}}{2} = -\frac{\sqrt{6}}{2} - \frac{i\sqrt{2}}{2}.$$

We find the polar form of this complex number. For the absolute value,

$$r^2 = \left(-\frac{\sqrt{6}}{2}\right)^2 + \left(-\frac{\sqrt{2}}{2}\right)^2$$

$$= \frac{6}{4} + \frac{2}{4}$$

$$= 2.$$

Therefore, the absolute value is $\sqrt{2}$. To find the argument,

$$\cos\theta = \frac{-\dfrac{\sqrt{6}}{2}}{\sqrt{2}}$$

$$= -\frac{\sqrt{6}}{2} \cdot \frac{1}{\sqrt{2}}$$

$$= -\frac{\sqrt{3}}{2}$$

and

$$\sin\theta = \frac{-\dfrac{\sqrt{2}}{2}}{\sqrt{2}}$$

$$= -\frac{\sqrt{2}}{2} \cdot \frac{1}{\sqrt{2}}$$

$$= -\frac{1}{2}.$$

Therefore, the argument is $\theta = 210°$. We can find the polar form of the second square root more simply by using an angle coterminal with the original angle. To find the second square root of $2 \operatorname{cis} 60°$, we use the angle

$$60° + 360° = 420°.$$

Then,

$$(2 \operatorname{cis} 420°)^{\frac{1}{2}} = 2^{\frac{1}{2}} \operatorname{cis} \frac{420°}{2}$$

$$= \sqrt{2} \operatorname{cis} 210°.$$

In general, any complex number has n nth roots. That is, there are two square roots, three cube roots, four fourth roots, and so on. We can find all the nth roots by using the arguments

$$\theta$$
$$\theta + 360°$$
$$\theta + 2(360°)$$
$$\theta + 3(360°)$$

and so on to

$$\theta + (n - 1)(360°).$$

EXAMPLE 20.18.   Find the fourth roots of -4.

Solution.   In polar form,

$$- 4 = 4 \operatorname{cis} 180°.$$

We use the four arguments

$$180°$$
$$180° + 360° = 540°$$
$$180° + 2(360°) = 900°$$
$$180° + 3(360°) = 1260°.$$

(Recall from Example 20.16 that $1260°$ was an argument of the fourth power of $1 - i$. Since the fourth power of $1 - i$ was -4, one fourth root of -4 should be $1 - i$.) Now,

$$(4 \operatorname{cis} 180°)^{\frac{1}{4}} = 4^{\frac{1}{4}} \operatorname{cis} \frac{180°}{4}.$$

Recall from algebra that

$$4^{\frac{1}{4}} = (2^2)^{\frac{1}{4}} = 2^{\frac{1}{2}} = \sqrt{2}.$$

Then, one fourth root is

$$\sqrt{2} \operatorname{cis} 45° = 1 + i.$$

Similarly,

$$(4 \operatorname{cis} 540°)^{\frac{1}{4}} = 4^{\frac{1}{4}} \operatorname{cis} \frac{540°}{4}$$

$$= \sqrt{2} \operatorname{cis} 135°$$

$$= -1 + i,$$

$$(4 \operatorname{cis} 900°)^{\frac{1}{4}} = 4^{\frac{1}{4}} \operatorname{cis} \frac{900°}{4}$$

$$= \sqrt{2} \operatorname{cis} 225°$$

$$= -1 - i,$$

and

$$(4 \operatorname{cis} 1260°)^{\frac{1}{4}} = 4^{\frac{1}{4}} \operatorname{cis} \frac{1260°}{4}$$

$$= \sqrt{2} \operatorname{cis} 315°$$

$$= 1 - i.$$

Finally, we observe that the vectors representing the four fourth roots spread evenly around the coordinate system:

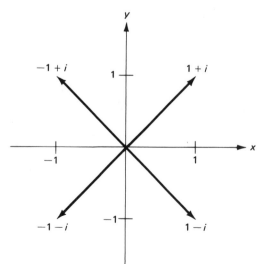

We know that for any positive integer n, an nth root of 1 is 1. However, 1 should have $n - 1$ other nth roots. These roots are called the **roots of unity**. The two square roots of unity are of course 1 and -1.

EXAMPLE 20.19.   Find and graph the three cube roots of unity.

Solution.   In polar form,

$$1 = 1 \operatorname{cis} 0°.$$

The three arguments are

$$0°$$
$$0° + 360° = 360°$$
$$0° + 2(360°) = 720°.$$

Then,

$$(1 \operatorname{cis} 0°)^{\frac{1}{3}} = 1^{\frac{1}{3}} \operatorname{cis} \frac{0°}{3}$$

$$= 1 \operatorname{cis} 0°$$

$$= 1.$$

The other two roots are

$$(1 \operatorname{cis} 360°)^{\frac{1}{3}} = 1^{\frac{1}{3}} \operatorname{cis} \frac{360°}{3}$$

$$= 1 \operatorname{cis} 120°$$

$$= -\frac{1}{2} + \frac{i\sqrt{3}}{2}$$

and

$$(1 \operatorname{cis} 720°)^{\frac{1}{3}} = 1^{\frac{1}{3}} \operatorname{cis} \frac{720°}{3}$$

$$= 1 \operatorname{cis} 240°$$

$$= -\frac{1}{2} - \frac{i\sqrt{3}}{2}.$$

The graph is

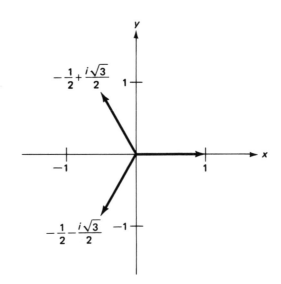

Exercise 20.3

Find the power, and write it in rectangular form:

1. $(1 + i)^4$ 2. $(-1 - i)^3$ 3. $(\sqrt{3} - i)^5$ 4. $(-1 + i\sqrt{3})^6$

5. $\left(-\dfrac{1}{\sqrt{2}} + \dfrac{i}{\sqrt{2}}\right)^6$ 6. $\left(\dfrac{1}{2} + \dfrac{i\sqrt{3}}{2}\right)^{12}$

Find the roots, write them in rectangular form, and draw their graph:

7. the square roots of $-1 + i\sqrt{3}$

8. the square roots of $\dfrac{1}{2} - \dfrac{i\sqrt{3}}{2}$

9. the cube roots of -8

10. the cube roots of $-i$

11. the fourth roots of -81

12. the fourth roots of $-\dfrac{1}{2} + \dfrac{i\sqrt{3}}{2}$

13. the fourth roots of unity

14. the sixth roots of unity

Self-test

1. Write $\sqrt{2}\ \text{cis}\ 315°$ in rectangular form.

Multiply or divide in polar form:

2. $(2\sqrt{3}\ \text{cis}\ 240°)(4\sqrt{3}\ \text{cis}\ 210°)$

3. $\dfrac{2\sqrt{2}\ \text{cis}\ 90°}{\sqrt{2}\ \text{cis}\ 135°}$

4. Find $\left(\dfrac{1}{\sqrt{2}} + \dfrac{i}{\sqrt{2}}\right)^{14}$ and write it in rectangular form.

5. Find the fourth roots of -16, write them in rectangular form, and draw the graph.

1._____

2._____

3._____

4._____

5._____
graph:

Unit
21

Polar Coordinates

INTRODUCTION

Throughout this book you have used the coordinate system called the Cartesian or rectangular coordinate system. In the rectangular coordinate system points are located by their distances from two perpendicular lines. In this unit you will learn about the polar coordinate system. In the polar coordinate system points are located by a distance and an angle. You will learn how to change coordinates between the rectangular and polar systems. Then, you will learn how to draw some basic graphs in the polar coordinate system.

OBJECTIVES

When you have finished this unit you should be able to:

1. Plot points in the polar coordinate system.
2. Change coordinates from rectangular form to polar form and from polar form to rectangular form.
3. Draw the graph of an equation given in polar coordinates.

Section 21.1 — The Polar Coordinate System

Suppose (x_0, y_0) represents a point in the Cartesian or rectangular coordinate system. Then, there is exactly one point in the rectangular coordinate system with coordinates (x_0, y_0). For example, the unique point with coordinates $(2, -3)$ is

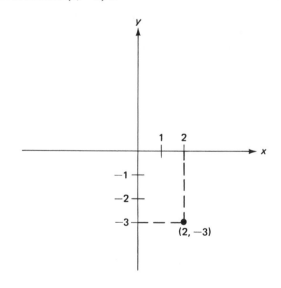

Moreover, given any point in the rectangular coordinate system, we can find a unique pair of coordinates which represents that point. That is, for each point in the rectangular coordinate system there is exactly one pair of coordinates. For example, the point P indicated is represented only by the coordinates $(2, -3)$:

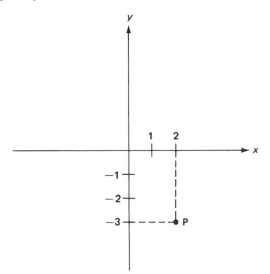

This coordinate system is called the Cartesian coordinate system for its inventor, René Descartes (1596–1650). Observe that it is also called the rectangular coordinate system because the lines to a point not on the axes, with the axes, make a rectangle.

Now, consider an ordered pair of the form (r, θ), where r is any real number and θ is an angle given in degrees or radians. The numbers r and θ are the **polar coordinates** of a point in the **polar coordinate system**. To locate a point (r_0, θ_0) in the polar coordinate system, we first construct the angle in standard position with measure θ_0. Then, if r_0 is positive, we measure the distance r_0 from the origin along the terminal side of the angle. In this way, we locate a unique point with polar coordinates (r_0, θ_0). Because distances are always taken from the origin, the origin is called the **pole** of the polar coordinate system. Angles in standard position are measured from the positive x-axis, which is called the **polar axis**.

EXAMPLE 21.1.   Plot the point $(2, 30°)$ in the polar coordinate system.

Solution.   We draw a 30° angle in standard position, and then measure 2 units from the origin along the terminal side of the angle:

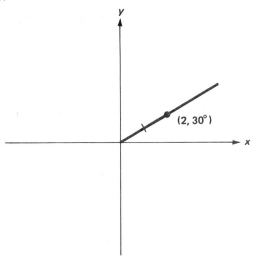

The point we have located is unique. There is no other point in the polar coordinate system with coordinates (2, 30°). However, observe that there are many other pairs of polar coordinates which represent this point. In particular, any angle coterminal with 30° may represent the point. For example, (2, 390°) and (2, −330°) both represent the same point as (2, 30°). There are infinitely many other such pairs of coordinates. In general, we will choose the pair where $0° \leqslant \theta < 360°$.

If $\theta = 0°$, we have a point on the polar axis. For example, (2, 0°) lies on the positive x-axis, 2 units from the origin. If $\theta = 180°$, we have a point on the negative x-axis. For example, (2, 180°) is on the negative x-axis, 2 units from the origin:

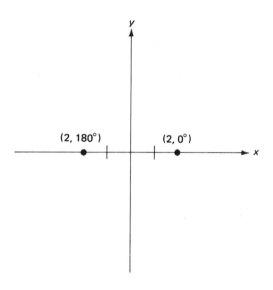

Observe that the coordinate 2 is not negative. The 180° angle causes the point to lie to the left of the origin.

Similarly, if $\theta = 90°$ we have a point on the positive y-axis, and if $\theta = 270°$ we have a point on the negative y-axis:

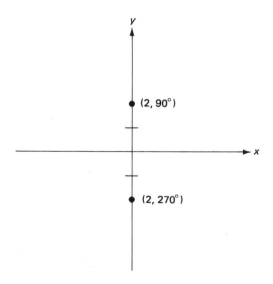

Again, observe that the coordinate 2 is not negative. The 270° angle causes the point to lie below the origin.

EXAMPLE 21.2.   Plot the points $(3, 0°)$, $(4, 90°)$, $(1, 180°)$, and $\left(\dfrac{3}{2}, 270°\right)$.

Solution.   These points lie on the axes:

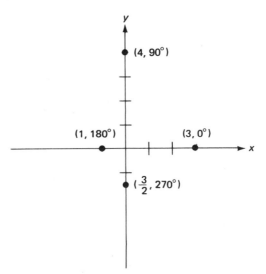

If $r = 0$, the point is the pole itself, regardless of the value of θ, since the distance from the origin is 0.

EXAMPLE 21.3.   Plot the point $(0, 30°)$.

Solution.   We draw a 30° angle, but measure 0 units from the origin:

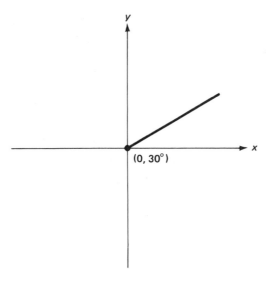

We can also interpret negative values of r. We draw the angle θ in standard position as before. Then, if r is negative, we measure r units from the origin in the direction *opposite* to the terminal side of θ.

EXAMPLE 21.4.   Plot the point $(-2, 210°)$.

Solution.   We draw a 210° angle in standard position. Then, we measure 2 units from the origin in the opposite direction:

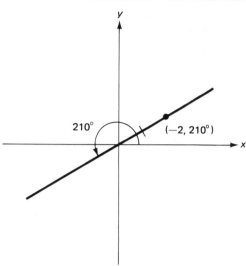

Observe that this point is again the same as the point with coordinates (2, 30°).

EXAMPLE 21.5.   Plot the point $\left(-\dfrac{4}{3}, 0°\right)$.

Solution.   The angle 0° is on the positive x-axis. We measure $\dfrac{4}{3}$ units in the opposite direction:

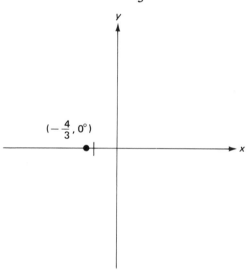

Exercise 21.1

Plot the points in the polar coordinate system:

1. $(1, 45°)$ 2. $(4, 210°)$ 3. $\left(\dfrac{1}{2}, 330°\right)$ 4. $(\sqrt{2}, 135°)$

5. $(1, 90°)$ 6. $(4, 270°)$ 7. $(\sqrt{3}, 180°)$ 8. $\left(\dfrac{5}{2}, 0°\right)$

9. $(-2, 30°)$ 10. $(-\sqrt{2}, 315°)$ 11. $(-1, 120°)$ 12. $\left(-\dfrac{3}{2}, 225°\right)$

13. $(-1, 90°)$ 14. $(-5, 0°)$ 15. $(-3, 180°)$ 16. $(-4, 270°)$

Section 21.2   Rectangular and Polar Coordinates

Consider any point (x, y) in the rectangular coordinate system:

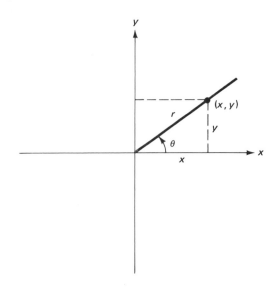

We know that

$$\cos \theta = \frac{x}{r}$$

or

$$x = r \cos \theta,$$

and

$$\sin \theta = \frac{y}{r}$$

or

$$y = r \sin \theta.$$

These formulas hold regardless of the quadrant θ is in. Also,

$$r^2 = x^2 + y^2,$$

where we now allow r to be positive or negative. We can use these formulas to write coordinates given in rectangular form as polar coordinates.

EXAMPLE 21.6.   Write $(-1, \sqrt{3})$ in polar coordinates.

Solution.   We first find r, using

$$r^2 = x^2 + y^2$$

$$r^2 = (-1)^2 + (\sqrt{3})^2$$

$$r^2 = 4.$$

We may choose either $r = 2$ or $r = -2$. For simplicity, we will choose $r = 2$. Then,

$$\cos \theta = \frac{x}{r}$$

$$= -\frac{1}{2}$$

and

$$\sin \theta = \frac{y}{r}$$

$$= \frac{\sqrt{3}}{2}.$$

Therefore, $\theta = 120°$. A pair of polar coordinates is $(2, 120°)$. Of course, there are infinitely many other pairs of polar coordinates associated with the point $(-1, \sqrt{3}\,)$. We may use $(2, 480°)$ or $(2, -240°)$. If we take $r = -2$, we must use an angle in the direction opposite to $120°$ and so $\theta = 300°$. Another pair of polar coordinates is $(-2, 300°)$, and another is $(-2, -60°)$. However, if we always take r to be positive and $0° \leqslant \theta < 360°$, we will have a unique pair of polar coordinates.

EXAMPLE 21.7. Write $(-2, -2)$ in polar coordinates.

Solution. To find r,

$$r^2 = x^2 + y^2$$

$$r^2 = (-2)^2 + (-2)^2$$

$$r^2 = 8.$$

Therefore,

$$r = \sqrt{8}$$

$$= 2\sqrt{2}\,.$$

Then,

$$\cos \theta = \frac{x}{r}$$

$$= \frac{-2}{2\sqrt{2}}$$

$$= -\frac{1}{\sqrt{2}}\,,$$

and also,

$$\sin \theta = \frac{y}{r}$$

$$= -\frac{1}{\sqrt{2}}\,.$$

Therefore, $\theta = 225°$. The polar coordinates are $(2\sqrt{2}\,, 225°)$.

We can also find polar coordinates for points which are not associated with special or quadrantal angles.

EXAMPLE 21.8.   Write $(2, -3)$ in polar coordinates.

Solution.   Again using

$$r^2 = x^2 + y^2$$

$$r^2 = 2^2 + (-3)^2$$

$$r^2 = 13,$$

we find $r \approx 3.6$ using Table I or a calculator. Then,

$$\cos \theta = \frac{x}{r}$$

$$= \frac{2}{\sqrt{13}}$$

$$\approx .5547$$

and

$$\sin \theta = \frac{y}{r}$$

$$= \frac{-3}{\sqrt{13}}$$

$$\approx -.8321.$$

Now, using Table II or a scientific calculator, $\theta \approx 303.7°$. The polar coordinates are $(3.6, 303.7°)$. We have used approximations to the nearest tenth of a unit because we cannot plot a point more accurately. Also, observe that we have used the nearest tenth of a degree instead of minutes. The point is

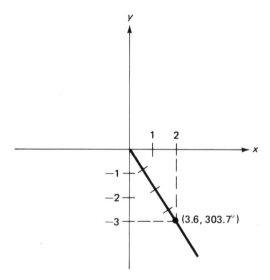

We can also change from polar coordinates to rectangular coordinates using the formulas in the form $x = r \cos \theta$ and $y = r \sin \theta$.

EXAMPLE 21.9.   Write $(2, 270°)$ in rectangular coordinates.

Solution.   Using the formulas above,

$$x = r \cos \theta$$
$$= 2 \cos 270°$$
$$= 2(0)$$
$$= 0$$

and

$$y = r \sin \theta$$
$$= 2 \sin 270°$$
$$= 2(-1)$$
$$= -2.$$

The rectangular coordinates are $(0, -2)$. This result agrees with the conclusion in the preceding section concerning points on the coordinate axes.

EXAMPLE 21.10.   Write $(-2, 210°)$ in rectangular coordinates.

Solution.

$$x = r \cos \theta$$
$$= -2 \cos 210°$$
$$= -2\left(-\frac{\sqrt{3}}{2}\right)$$
$$= \sqrt{3}$$

and

$$y = r \sin \theta$$
$$= -2 \sin 210°$$
$$= -2\left(-\frac{1}{2}\right)$$
$$= 1.$$

The rectangular coordinates are $(\sqrt{3}, 1)$. This point coincides with the point $(2, 30°)$ in polar coordinates, as expected from the preceding section.

EXAMPLE 21.11.   Write $(4.5, 70°)$ in rectangular coordinates.

Solution.   Using Table II or a scientific calculator,

$$x = r \cos \theta$$
$$= 4.5 \cos 70°$$
$$\approx 1.5$$

and

$$y = r \sin \theta$$

$$= 4.5 \sin 70°$$

$$\approx 4.2.$$

The rectangular coordinates are (1.5, 4.2) to the nearest tenth of a unit:

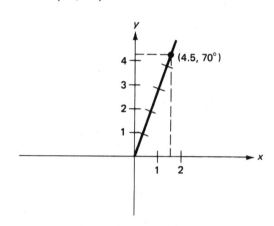

Exercise 21.2

Write in polar coordinates:

1. $(1, \sqrt{3})$ 2. $(1, -1)$ 3. $\left(-\dfrac{1}{2}, -\dfrac{\sqrt{3}}{2}\right)$ 4. $(-\sqrt{2}, \sqrt{2})$

5. $(-2, 0)$ 6. $(\sqrt{2}, 0)$ 7. $\left(0, -\dfrac{1}{2}\right)$ 8. $\left(0, \dfrac{5}{3}\right)$

9. $(3, 4)$ 10. $(-2, 4)$ 11. $(4, -2)$ 12. $(-4, -3)$

Write in rectangular coordinates:

13. $(1, 180°)$ 14. $(3, 270°)$ 15. $(-2, 0°)$ 16. $(-1, 270°)$

17. $(2, 240°)$ 18. $(2\sqrt{2}, 135°)$ 19. $(-1, 45°)$ 20. $(-4, -30°)$

21. $(5, 50°)$ 22. $(4, 335°)$ 23. $(3.5, 115°)$ 24. $(2.5, 255.5°)$

**Section
21.3**   # Graphs in Polar Coordinates

An equation in polar coordinates is an equation in the two variables r and θ. We have special cases when either r or θ is a constant.

EXAMPLE 21.12.   Draw the graph of $r = 3$.

Solution.   We have $r = 3$ for any value of θ:

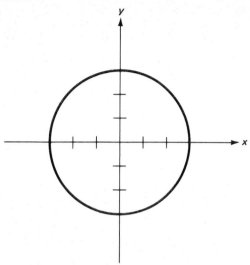

The graph is a circle with its center at the origin and radius 3. In general, the graph of the equation $r = a$ is a circle with its center at the origin and radius a.

EXAMPLE 21.13.   Draw the graph of $\theta = 110°$.

Solution. θ is always 110° while r can have any value:

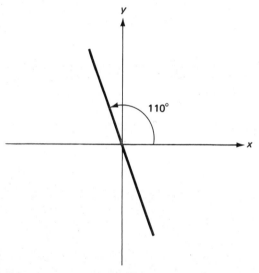

The graph is a line through the origin forming an angle of 110° in standard position. In general, the graph of the equation $\theta = a°$ is a line through the origin forming an angle of $a°$ in standard position.

The equations $r = a \sin \theta$ and $r = a \cos \theta$ are interesting examples of equations in θ and r.

EXAMPLE 21.14.   Draw the graph of $r = 2 \sin \theta$.

Solution.   We make a chart of values for θ and r, approximating r to the nearest tenth of a unit:

θ	0°	30°	45°	60°	90°	120°	135°	150°	180°
r	0	1	1.4	1.7	2	1.7	1.4	1	0

Observe that, if we continue to $\theta = 210°$, we have

$$r = 2\sin 210°$$

$$= 2\left(-\frac{1}{2}\right)$$

$$= -1,$$

which gives the point $(-1, 210°)$. However, this point is identical to the point $(1, 30°)$, which is already on our chart. Similarly, the remaining angles up to 360° will give points identical to those for angles between 0° and 180°. Plotting the points on the chart, we have the graph

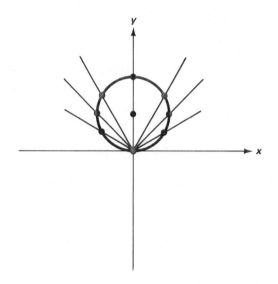

The graph is again a circle, but with its center at $(0, 1)$ (in rectangular coordinates) and radius 1. In general, the graph of $r = a\sin\theta$, where a is positive, is a circle with its center at $\left(0, \frac{a}{2}\right)$ and radius $\frac{a}{2}$. If a is negative, the circle lies below the x-axis. Similarly, the graph of $r = a\cos\theta$, where a is positive, is a circle with its center at $\left(\frac{a}{2}, 0\right)$ and radius $\frac{a}{2}$. If a is negative, the circle lies to the left of the y-axis.

We now show several types of graphs which have simple equations in polar coordinates, but would have very complicated equations in rectangular coordinates.

EXAMPLE 21.15.   Draw the graph of $r = 1 + \sin\theta$.

Solution.   We make a chart of values for θ and r:

θ	0°	45°	90°	135°	180°	225°	270°	315°
r	1	1.7	2	1.7	1	.3	0	.3

Plotting these points, the graph is

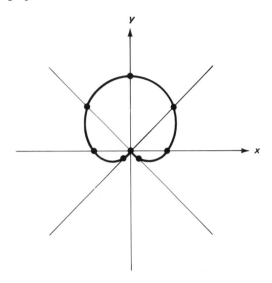

You should fill in the points for the special values related to 30° and 60°. The graph is called a **cardioid** because of its heartlike shape. Any equation of the form $r = a \pm a \sin \theta$ or $r = a \pm a \cos \theta$ has a graph which is a cardioid. The four cases take four positions with respect to the axes.

EXAMPLE 21.16.   Draw the graph of $r = 1 + 2 \sin \theta$.

Solution.   The chart is

θ	0°	45°	90°	135°	180°	225°	270°	315°
r	1	2.4	3	2.4	1	$-.4$	-1	$-.4$

Recall that the points $(-.4, 225°)$, $(-1, 270°)$, and $(-.4, 315°)$ are identical to the points $(.4, 45°)$, $(1, 90°)$, and $(.4, 135°)$. Also observe that $r = 0$ when $\theta = 210°$ and $\theta = 330°$. The graph is

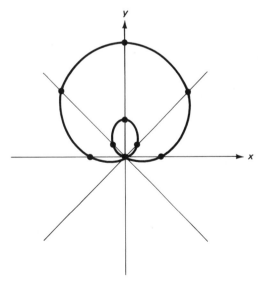

The graph of an equation of the form $r = a \pm b \sin \theta$ or $r = a \pm b \cos \theta$, where a and b are positive and $a < b$, is called a **limaçon**. Again, the four cases take four positions with respect to the coordinate axes. There is an interesting variation when $a > b$.

EXAMPLE 21.17.   Draw the graph of $r = 3 - 2 \cos \theta$.

Solution.   The chart is

θ	0°	45°	90°	135°	180°	225°	270°	315°
r	1	1.6	3	4.4	5	4.4	3	1.6

The graph is

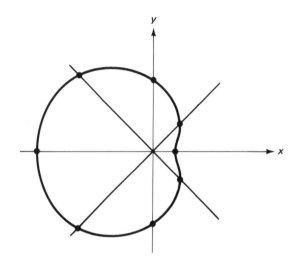

You should supply the points for the special values related to 30° and 60°.

EXAMPLE 21.18.   Draw the graph of $r = 3 \sin 2\theta$.

Solution.   Because we are dealing with 2θ, we need a more extensive chart than was necessary for the preceding examples. Consider this chart from 0° to 90°:

θ	0°	15°	30°	45°	60°	75°	90°
r	0	1.5	2.6	3	2.6	1.5	0

Plotting these points, we have

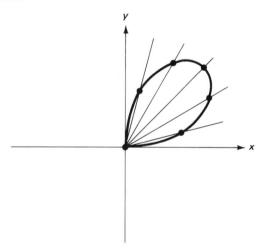

This graph forms one "petal" of a **rose**. To form another "petal," we continue from 90° to 180°:

θ	90°	105°	120°	135°	150°	165°	180°
r	0	-1.5	-2.6	-3	-2.6	-1.5	0

Since r is negative, the next "petal" is in the fourth quadrant:

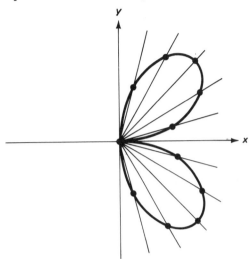

The next set of values for θ, from 180° to 270°, gives positive values for r, and the next "petal" is in the third quadrant. The final set of values for θ, from 270° to 360°, gives negative values for r and a "petal" in the second quadrant. The completed rose is

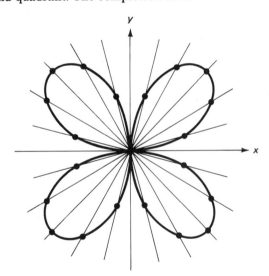

The graph of any equation of the form $r = a \sin n\theta$ or $r = a \cos n\theta$ is a rose. The rose has $2n$ "petals" if n is even and n "petals" if n is odd. The "petals" may be centered on the axes as well as in the quadrants. They should be evenly spaced around the coordinate system.

EXAMPLE 21.19.   Draw the graph of $r = \cos 3\theta$.

Solution.   We choose values for θ which will give special values for 3θ:

θ	0°	10°	15°	20°	30°
r	1	.9	.7	.5	0

Plotting these points, we see that we have half a "petal," centered on the *x*-axis:

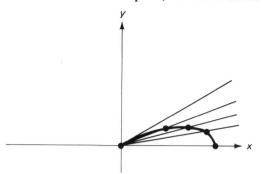

The next set of values is

θ	30°	40°	45°	50°	60°	70°	75°	80°	90°
r	0	− .5	− .7	− .9	− 1	− .9	− .7	− .5	0

Since *r* is negative, these points are identical to points in the third quadrant where *r* is positive and θ goes from 210° to 270°, which forms a "petal" in the third quadrant:

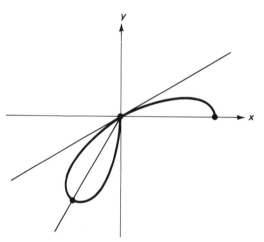

Another set of values for θ, from 90° to 150°, gives positive values for *r* and a "petal" in the second quadrant:

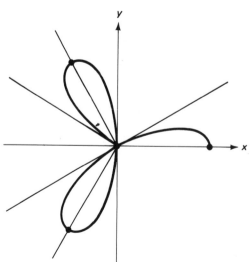

Finally, values for θ from 150° to 180° give negative values for r, and they complete the "petal" centered on the x-axis:

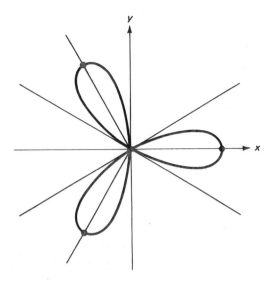

Values for θ between 180° and 360° will repeat the points we already have. Therefore, the rose is complete.

EXAMPLE 21.20.   Draw the graph of $r^2 = 4 \sin 2\theta$.

Solution.   Using half-angles as in Example 21.18, we have the chart

θ	0°	15°	30°	45°	60°	75°	90°
r	0	± 1.4	± 1.9	± 2	± 1.9	± 1.4	0

Taking into account that the negative values for r give points in the third quadrant, these points form the graph

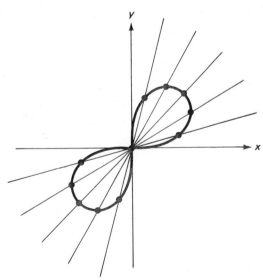

We cannot use values for θ between 90° and 180° because sin 2θ will be negative and r^2 cannot equal a negative number. Values for θ from 180° to 270° will repeat the points above. Values for θ from 270° to 360° again cause r^2 to be equal to a negative number so we cannot use them.

Therefore, the graph is complete. The graph is called a **lemniscate**. The graph of any equation of the form $r^2 = a \sin 2\theta$ or $r^2 = a \cos 2\theta$ is a lemniscate.

Exercise 21.3

Draw the graph:

1. $r = 1$

2. $r = \sqrt{2}$

3. $\theta = 45°$

4. $\theta = 290°$

5. $r = 2 \cos \theta$

6. $r = 4 \sin \theta$

7. $r = -2 \sin \theta$

8. $r = -3 \cos \theta$

9. $r = 2 + 2 \sin \theta$

10. $r = 1 + \cos \theta$

11. $r = 1 - \cos \theta$

12. $r = 2 - 2 \sin \theta$

13. $r = 2 + 3 \sin \theta$

14. $r = 1 - 2 \sin \theta$

15. $r = 1 - 2 \cos \theta$

16. $r = 1 + 3 \cos \theta$

17. $r = 2 - \cos \theta$

18. $r = 2 - \sin \theta$

19. $r = 3 + \sin \theta$

20. $r = 3 + 2 \cos \theta$

21. $r = 2 \sin 2\theta$

22. $r = 2 \cos 2\theta$

23. $r = \sin 3\theta$

24. $r = 4 \sin 4\theta$

25. $r^2 = 9 \sin 2\theta$

26. $r^2 = 9 \cos 2\theta$

27. $r^2 = -4 \sin 2\theta$

28. $r^2 = -9 \cos 2\theta$

Self-test

1. Plot the points:
    a. $(3, 270°)$
    b. $(-3, 300°)$

2. Write $(\sqrt{2}, -\sqrt{2})$ in polar coordinates.

    2._____

3. Write $(4, 135°)$ in rectangular coordinates.

    3._____

Draw the graph:

4. $\theta = 200°$

5. $r = \cos 2\theta$

Answers

Unit 1

Exercise 1.1

1.

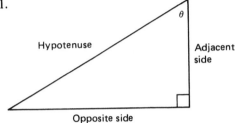

2.

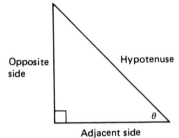

3.

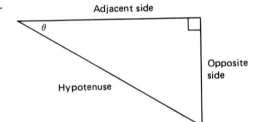

4.

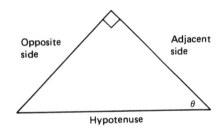

Exercise 1.3

1. $\dfrac{1}{2}$	2. $\dfrac{1}{\sqrt{2}}$	3. 1	4. 2	5. $\dfrac{2}{\sqrt{3}}$
6. $\sqrt{3}$	7. $\sqrt{2}$	8. $\sqrt{3}$	9. $\dfrac{\sqrt{3}}{2}$	10. $\dfrac{1}{2}$

Exercise 1.4

1. .9659	2. 9.514	3. .5324	4. 1.444	5. 1.144
6. .0670	7. 4.745	8. 1.062	9. .9914	10. .9993
11. 13°	12. 41°40′	13. 49°	14. 36°10′	15. 67°50′
16. 19°30′	17. 31°	18. 36°50′	19. 48°20′	20. 82°50′

Exercise 1.5

1. .4310	2. .7485	3. .9548	4. .7668	5. 1.456
6. 1.391	7. .6420	8. .3650	9. 24.39	10. 1.249
11. 23°44′	12. 31°32′	13. 43°38′	14. 16°16′	15. 42°37′
16. 23°06′	17. 73°46′	18. 54°49′	19. 48°16′	20. 84°23′

Self-test

1. 21°29′   (Objective 5)   2. 4.232   (Objective 4)   3.

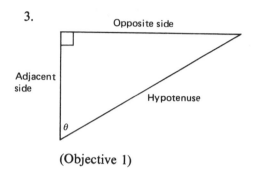

4. $\dfrac{\text{opposite side}}{\text{adjacent side}}$ (Objective 2)

5a. $\dfrac{1}{\sqrt{2}}$ 5b. $\dfrac{1}{\sqrt{3}}$ (Objective 3)

(Objective 1)

Unit 2

These answers were found using a scientific calculator. Your answers may vary by one digit or 10′ if you are using Table II, or if you make a different choice of trigonometric ratio.

Exercise 2.1

1. $\beta = 53°30'$, $b = 13.5$, $c = 16.8$ 2. $\beta = 17°40'$, $a = 13.2$, $c = 13.8$

3. $\alpha = 36°30'$, $\beta = 53°30'$, $b = 12.3$ 4. $\alpha = 51°10'$, $\beta = 38°50'$, $a = 3.97$

5. $\alpha = 77°50'$, $a = 11.1$, $c = 11.4$ 6. $\alpha = 47°10'$, $b = 16.9$, $c = 24.8$

7. $\alpha = 23°30'$, $a = 3.99$, $b = 9.17$ 8. $\beta = 30°10'$, $a = 19.5$, $b = 11.3$

9. $\alpha = 35°20'$, $\beta = 54°40'$, $c = 17.6$ 10. $\alpha = 40°20'$, $\beta = 49°40'$, $a = 13.6$

Exercise 2.2

1. 8.53 feet   2. 48°10′   3. 9.89 feet   4. 76°40′

5. 62°20′   6. 15 feet   7. 10 meters   8. 72°20′

9. 5.87 meters   10. 221 feet   11. 90.5 meters   12. 660 feet

Self-test

1. 16.8   (Objective 1)   2. 36°50′   (Objective 1)   3. 43.6   (Objective 1)

4. 112 feet   (Objective 2)   5. 70°20′   (Objective 2)

Unit 3

Exercise 3.1

1. Yes
2. Yes
3. No
4. No
5. No
6. No
7. Yes
8. No
9. No
10. Yes

Exercise 3.3

1. Positive
2. Positive
3. Positive
4. Negative
5. Negative
6. Positive
7. Negative
8. Negative
9. Positive
10. Negative
11. Negative
12. Negative
13. Positive
14. Positive
15. Positive
16. Negative
17. Negative
18. Positive
19. Positive
20. Positive
21. Negative
22. Negative
23. Negative
24. Positive

Exercise 3.4

1. 1
2. 0
3. 1
4. 0
5. Undefined
6. Undefined
7. Undefined
8. 0
9. 0
10. Undefined
11. -1
12. -1
13. 0
14. -1
15. Undefined
16. 0

Self-test

1. $\dfrac{r}{x}$ (Objective 2)   2. $\dfrac{1}{y}$ (Objective 2)   3a. 1 3b. Undefined (Objective 4)

4a. Negative 4b. Positive (Objective 3)   5. c (Objective 1)

Unit 4

Exercise 4.1

1. 30°
2. 79°
3. 41°
4. 69°
5. 38°
6. 13°
7. 35°40′
8. 13°50′

9. 44°20′   10. 35°40′   11. 1°10′   12. 88°10′

13. 60°, 60°   14. 225°, 45°   15. 134°20′, 45°40′   16. 299°50′, 60°10′

17. 211°40′, 31°40′   18. 140°40′, 39°20′   19. 11°30′, 11°30′   20. 278°30′, 81°30′

21. 326°10′, 33°50′   22. 75°50′, 75°50′   23. 253°40′, 73°40′   24. 144°20′, 35°40′

25. 39°, 39°   26. 250°, 70°   27. 115°, 65°   28. 300°, 60°

Exercise 4.2

1. $\dfrac{\sqrt{3}}{2}$ 2. $-\dfrac{1}{2}$ 3. $\dfrac{1}{\sqrt{3}}$ 4. $\sqrt{3}$

5. -1 6. $\dfrac{1}{\sqrt{2}}$ 7. $\dfrac{2}{\sqrt{3}}$ 8. -2

9. $-\sqrt{3}$ 10. -2 11. $-\dfrac{1}{2}$ 12. $\dfrac{1}{\sqrt{2}}$

13. 1   14. $-\dfrac{1}{\sqrt{3}}$ 15. $-\dfrac{2}{\sqrt{3}}$ 16. $\sqrt{2}$

17. 0   18. 0   19. Undefined   20. Undefined

21. -1 22. 1

Exercise 4.3

1. .6018   2. 1.150   3. $-.6248$ 4. .4514

5. 1.446   6. -1.050 7. $-.4669$ 8. $-.2401$

9. $-.5025$ 10. $-.8923$ 11. .9588   12. 4.843

13. -3.305 14. 1.124   15. $-.9793$ 16. $-.4142$

17. 1.766   18. -2.850 19. $-.8098$ 20. $-.9925$

Exercise 4.4

1. 60°, 120°   2. 30°, 210°   3. 150°, 210°   4. 210°, 330°

5. 45°, 315°   6. 135°, 315°   7. 0°   8. 270°

9. 0°, 180°   10. 90°, 270°   11. 180°   12. 0°, 180°

13. 50°, 130°   14. 35°, 325°   15. 162°, 198°   16. 133°, 313°

17. 67°20′, 247°20′   18. 146°50′, 326°50′   19. 239°10′, 300°50′   20. 53°50′, 126°10′

21. 94°50′, 265°10′   22. 77°50′, 282°10′   23. 164°40′, 344°40′   24. 188°40′, 351°20′

Self-test

1. 236°20′, 56°20′ (Objective 1)   2. $\dfrac{1}{\sqrt{2}}$ (Objective 2)   3. -3.906 (Objective 3)

4. $-.9367$ (Objective 3)   5. 162°20′, 342°20′ (Objective 4)

Unit 5

Self-test

1. $\dfrac{\text{adjacent side}}{\text{hypotenuse}}$ (Unit 1)   2. $\dfrac{1}{x}$ (Unit 3)

3a. 1   3b. 2 (Unit 1)   4a. 0   4b. Undefined (Unit 3)

5. 0° (Unit 4)   6. 142°20′, 37°40′ (Unit 4)

7. -1.263 (Unit 4)   8. 248°, 292° (Unit 4)

9. 26°30′ (Unit 2)   10. 27 feet (Unit 2)

Unit 6

These answers were found using a scientific calculator. Your answers may differ by one digit or 10′ if you are using Table II, or by more if you make a different choice of method where possible.

Exercise 6.1

1. $\gamma = 59°30′, a = 19.2, b = 21.3$ 2. $\gamma = 70°50′, a = 78.9, b = 107$

3. $\gamma = 43°40′, b = 12.2, c = 8.46$ 4. $\gamma = 83°20′, a = 104, c = 108$

5. $\gamma = 26°40′, a = 62.8, b = 45.9$ 6. $\gamma = 18°20′, a = 227, c = 80.2$

7. $\beta = 105°, a = 9.09, c = 10.4$ 8. $\beta = 69°20′, a = 219, b = 207$

9. $\alpha = 98°30′, b = 1.52, c = 1.85$ 10. $\alpha = 85°20′, a = 67.2, c = 49.2$

Exercise 6.2

1. $c = 8.36, \alpha = 40°10′, \beta = 61°30′$ 2. $c = 12.0, \alpha = 61°50′, \beta = 80°20′$

3. $a = 3.48, \beta = 46°30′, \gamma = 79°20′$ 4. $b = 8.65, \alpha = 57°30′, \gamma = 59°50′$

5. $a = 51.0, \beta = 17°50', \gamma = 116°50'$ 6. $c = 7.32, \alpha = 97°40', \beta = 46°50'$

7. $a = 16.4, \beta = 39°, \gamma = 31°$ 8. $c = 14.7, \alpha = 43°20', \beta = 44°30'$

9. $b = 6.76, \alpha = 125°20', \gamma = 13°$ 10. $b = 30.3, \alpha = 51°50', \gamma = 31°40'$

Exercise 6.3

1. $\alpha = 41°40', \beta = 85°30', \gamma = 52°50'$ 2. $\alpha = 46°30', \beta = 54°10', \gamma = 79°10'$

3. $\alpha = 46°, \beta = 98°10', \gamma = 35°50'$ 4. $\alpha = 17°10', \beta = 30°30', \gamma = 132°20'$

5. $\alpha = 145°10', \beta = 20°, \gamma = 14°50'$ 6. $\alpha = 91°10', \beta = 44°10', \gamma = 44°40'$

7. $78°30'$ 8. $89°$

9. $93°10'$ 10. $104°30'$

Exercise 6.4

1. No triangle   2. No triangle

3. $\beta = 86°10', \gamma = 35°50', c = 7.04$ 4. $\gamma = 79°10', \beta = 55°50', b = 42.1$
    $\beta' = 93°50', \gamma' = 28°10', c' = 5.68$ $\gamma' = 100°50', \beta' = 34°10', b' = 28.6$

5. $\beta = 45°, \gamma = 77°, c = 13.8$ 6. $\alpha = 44°30', \beta = 65°30', b = 14.5$

7. $\gamma = 37°, \alpha = 31°, a = 13.9$ 8. No triangle

9. No triangle   10. $\alpha = 52°10', \beta = 84°30', b = 22.0$

    $\alpha' = 127°50', \beta' = 8°50', b' = 3.40$

11. No triangle   12. $\gamma = 17°50', \alpha = 138°40', a = 15.6$

13. $\alpha = 80°10', \gamma = 46°30', c = 3.17$ 14. $\beta = 64°30', \alpha = 48°40', a = 4.57$
    $\alpha' = 99°50', \gamma' = 26°50', c' = 1.97$

    16. $\alpha = 85°, \gamma = 66°30', c = 17.5$
15. $\beta = 28°50', \alpha = 76°30', a = 17.1$ $\alpha' = 95°, \gamma' = 56°30', c' = 15.9$

Self-test

1. $83°20'$ (Objective 3)   2. 16.2, (Objective 2)   3. 182   (Objective 1)

4. $57°10', 122°50'$ (Objective 4)   5. No angles   (Objective 4)

Unit 7

The method you should have used to solve the triangle is given in parentheses.

Exercise 7.1

1. 73° (trigonometric ratio)

2. 1.19 feet (trigonometric ratio)

3. 7.15 feet (law of cosines)

4. 76°50′ (law of sines)

5. 6.33 feet (trigonometric ratio)

6. No solution (law of sines)

Exercise 7.2

1. 14.0 meters (trigonometric ratio)

2. 14.0 meters (law of sines)

3. No solution (law of sines)

4. 7.67 feet (law of sines)

5. 32.3 miles (law of sines)

6. 13.2 kilometers (law of sines)

Exercise 7.3

1. 19.3 inches (law of cosines)

2. 60.8 pounds (law of cosines)

3. 48°30′ (law of cosines)

4. 87°10′ (law of cosines)

5. 45.3 pounds, 55.2 pounds (law of sines)

6. No solution (law of sines)

7. 26.2, 18.4 (trigonometric ratios)

8. 2.06, 16.4 (trigonometric ratios)

9. 5 pounds (trigonometric ratio)

10. 15 pounds (trigonometric ratio)

Exercise 7.4

1. 1.30 nautical miles (trigonometric ratio)

2. 1.54 nautical miles (trigonometric ratio)

3. N 83°30′ W (trigonometric ratio)

4. S 16°20′ W (trigonometric ratio)

5. No solution (law of sines)

6. 4.10 nautical miles (law of cosines)

7. 15.6 nautical miles (law of cosines)

8. N 85°50′ E (law of sines)

9. 6.90 nautical miles (law of cosines)

10. 18.0 nautical miles (law of cosines)

Self-test

1. 12.5 nautical miles (Objective 4, law of cosines)

2. 191 meters (Objective 2, law of sines)

3. No solution (Objective 2, law of sines)

4. 7.2 inches (Objective 1, trigonometric ratio)

5. 60° (Objective 3, law of cosines)

Unit 8

Exercise 8.1

1.
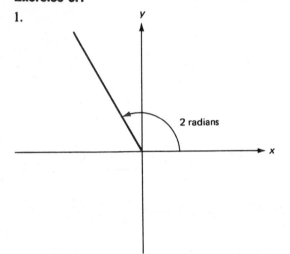

2.
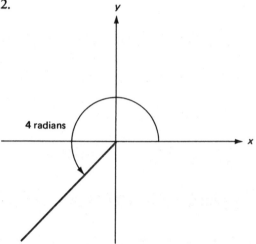

3.
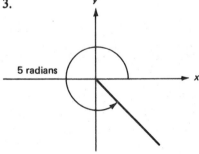

4.
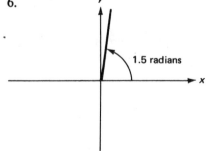

5.
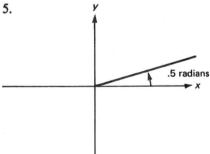

6.
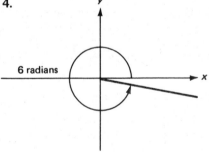

7.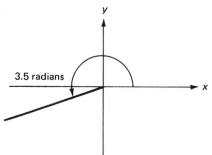

8.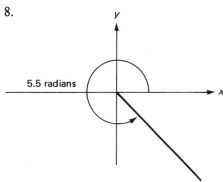

9.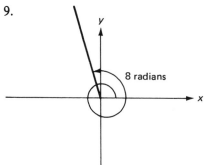

10.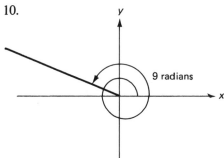

11.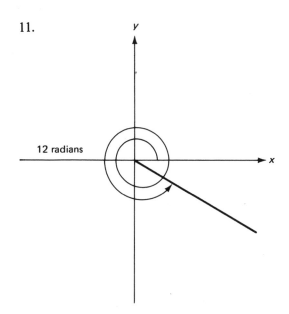

12.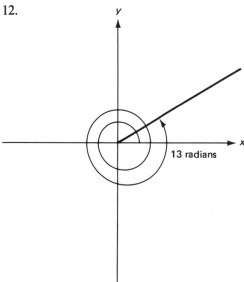

Exercise 8.2

1. $\dfrac{5\pi}{6}$

2. $\dfrac{4\pi}{3}$

3. $\dfrac{7\pi}{4}$

4. $\dfrac{11\pi}{6}$

5. $\dfrac{5\pi}{18}$

6. $\dfrac{13\pi}{18}$

7. $\dfrac{11\pi}{9}$

8. $\dfrac{17\pi}{9}$

9. .94	10. 2.13	11. 4.52	12. 5.43
13. .47	14. 1.55	15. 1.11	16. .74
17. 2.09	18. 4.31	19. 5.45	20. 3.29
21. 135°	22. 210°	23. 300°	24. 270°
25. 80°	26. 200°	27. 54°	28. 170°
29. 74°	30. 195°	31. 281°	32. 126°
33. 38°20′	34. 51°50′	35. 69°50′	36. 80°50′
37. 113°30′	38. 89°20′	39. 122°	40. 102°

Exercise 8.3

1. 2.86 feet

2. 43.6 centimeters

3. 3.32 inches

4. 9°30′

5. 698 miles

6. 63°

7. 1.75 meters per second

8. 101 inches per second

9. 1.47 inches per second

10. 5.24 centimeters per second

11. 25.1 inches per minute

12. .628 inches per minute

13. 5.60 revolutions per second

14. 2.10 revolutions per second

15. 1047 miles per hour

16. 772 miles per hour

Exercise 8.4

1. $\dfrac{1}{2}$	2. 1	3. $\dfrac{1}{2}$	4. $-\dfrac{1}{2}$
5. $-\sqrt{2}$	6. $\dfrac{1}{\sqrt{2}}$	7. $\dfrac{1}{\sqrt{3}}$	8. -2
9. $-\dfrac{1}{2}$	10. $-\dfrac{1}{\sqrt{3}}$	11. $\dfrac{\sqrt{3}}{2}$	12. -1
13. -1	14. 1	15. 0	16. Undefined
17. .8434	18. $-.9881$		

Self-test

1.   (Objective 1)

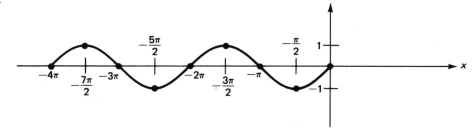

2. 96°50′ (Objective 2)

3. 2.69 (Objective 2)

4. $-\dfrac{1}{\sqrt{3}}$ (Objective 4)

5. 2.62 feet per second (Objective 3)

Unit 9

Exercise 9.1

1.

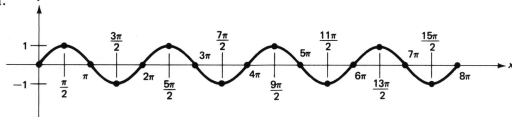

2.

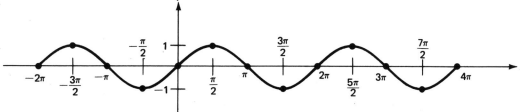

3.

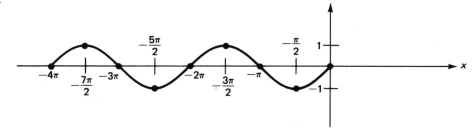

4.

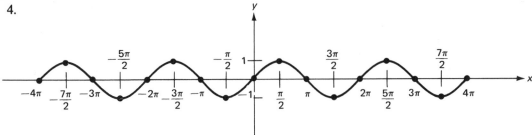

5.

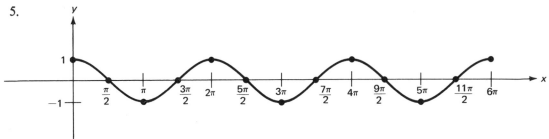

6.

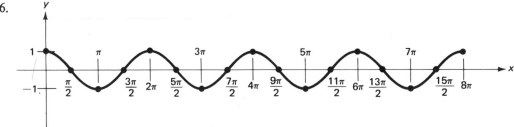

7.

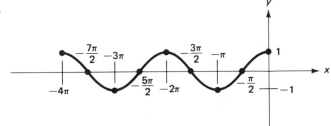

8.

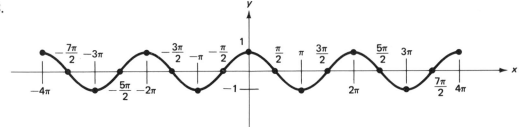

9. All real numbers   10. All real numbers   11. $-1 \leqslant y \leqslant 1$ 12. $-1 \leqslant y \leqslant 1$

Exercise 9.2

1.

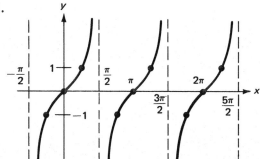

2.

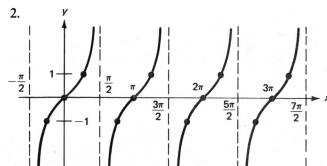

3.

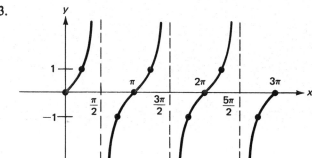

4.

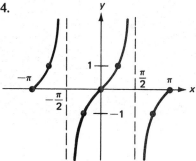

5.

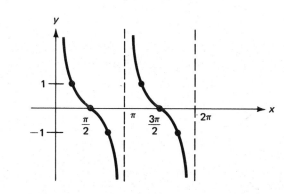

6.

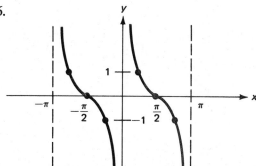

7.

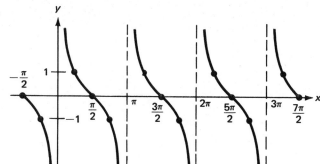

8.

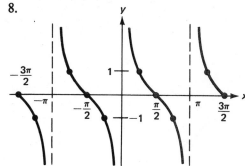

9. $x \neq \pm \dfrac{\pi}{2}, \pm \dfrac{3\pi}{2}, \pm \dfrac{5\pi}{2}, \ldots$ 10. $x \neq 0, \pm\pi, \pm 2\pi, \ldots$

11. All real numbers   12. All real numbers

Exercise 9.3

1.

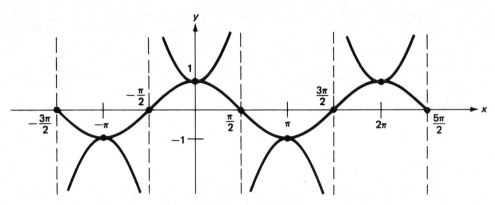

2.

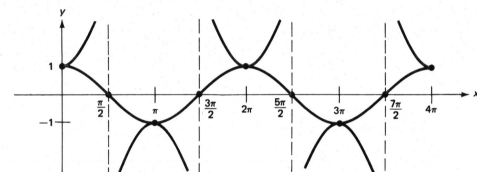

3.

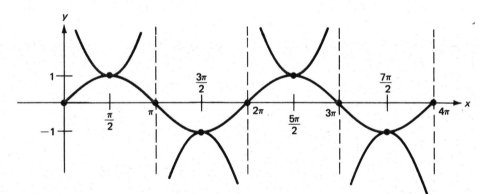

4.

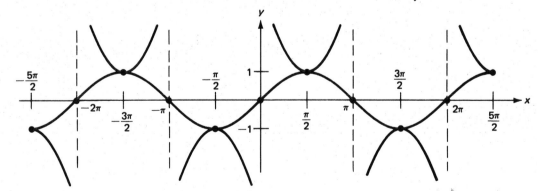

5. $x \neq \pm \dfrac{\pi}{2}, \pm \dfrac{3\pi}{2}, \pm \dfrac{5\pi}{2}, \ldots$

6. $x \neq 0, \pm \pi, \pm 2\pi, \ldots$

7. $y \leqslant -1$ or $y \geqslant 1$

8. $y \leqslant -1$ or $y \geqslant 1$

Self-test

1.   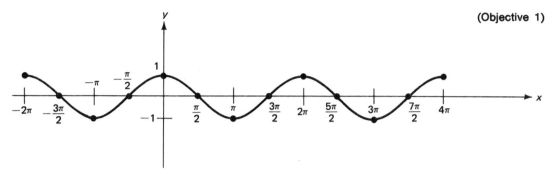   (Objective 1)

2.   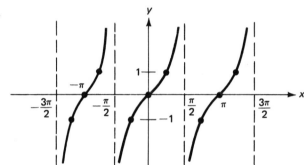   (Objective 2)

3.   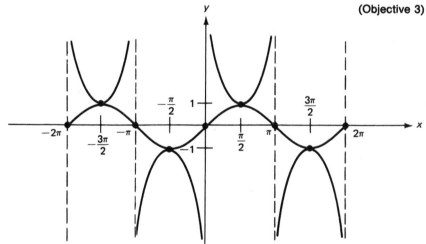   (Objective 3)

4. $y \leqslant -1$ or $y \geqslant 1$ (Objective 3)

5. $x \neq \pm \dfrac{\pi}{2}, \pm \dfrac{3\pi}{2}, \pm \dfrac{5\pi}{2}, \ldots$ (Objective 2)

Unit 10

Exercise 10.1

1.

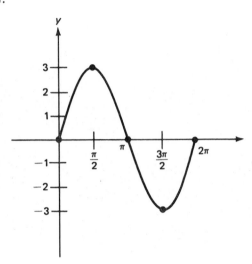

2.

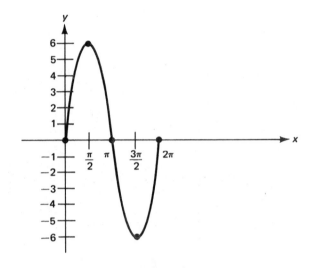

3.

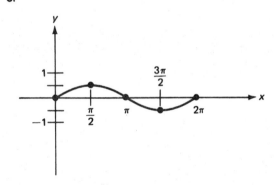

4.

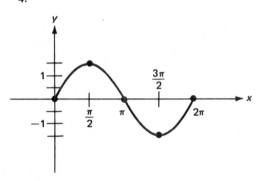

5.

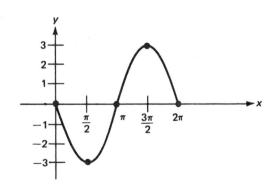

6.

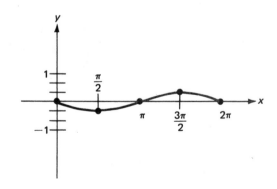

7.

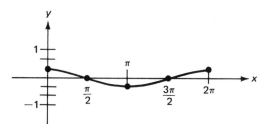

8.

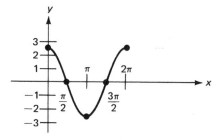

9.

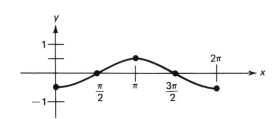

10.

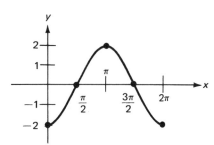

11.

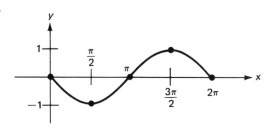

12.

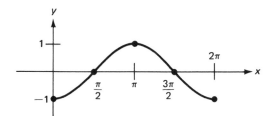

Exercise 10.2

1.

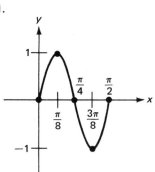

2.

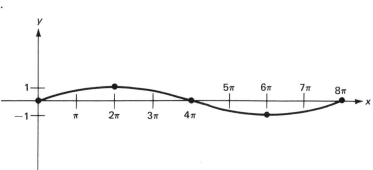

3.

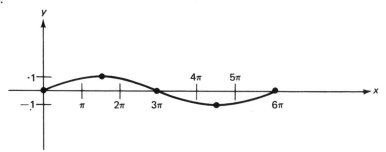

4.

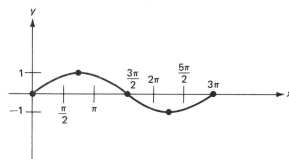

5.

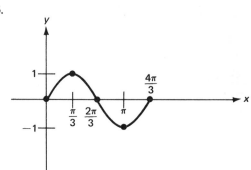

6.

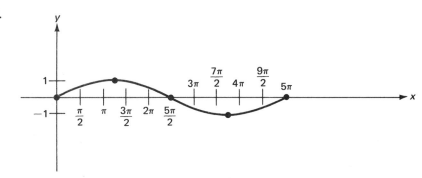

7.

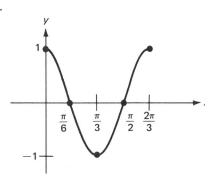

8.

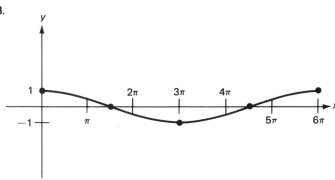

9.

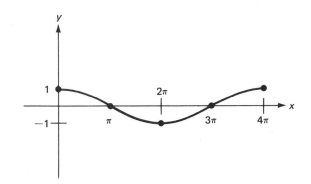

10.

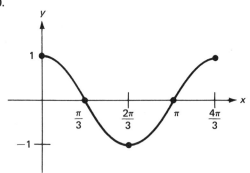

11.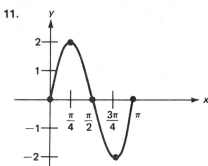

12.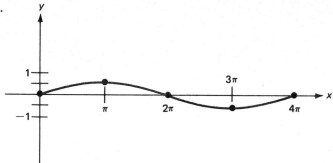

13.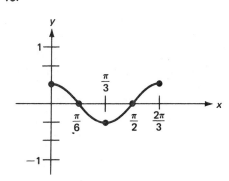

14.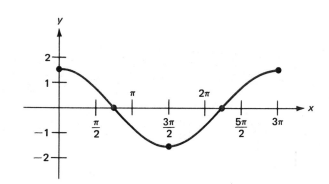

15.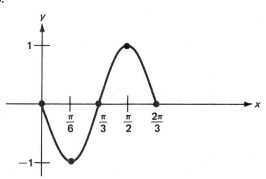

16.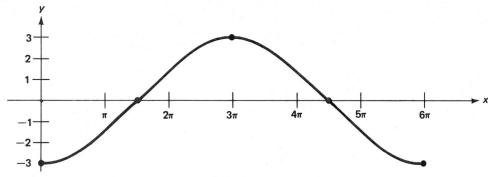

Exercise 10.3

1.

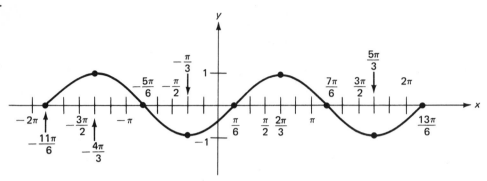

2.

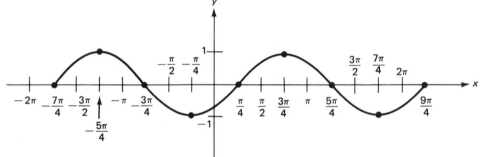

3.

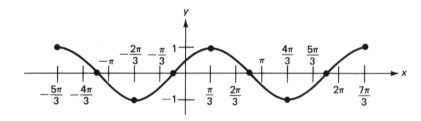

4.

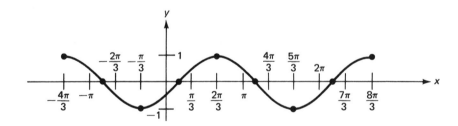

5.

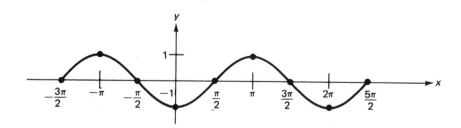

6.

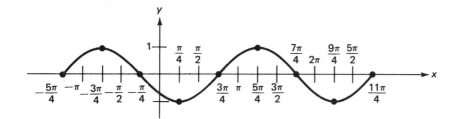

7.

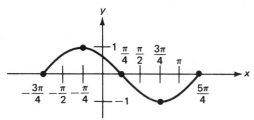

8.

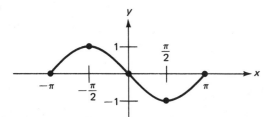

9.

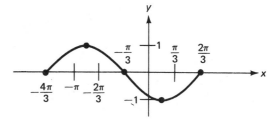

10.

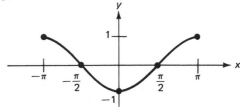

Exercise 10.4

1.

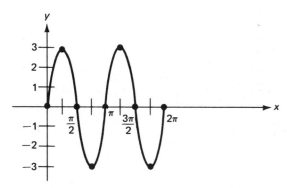

2.

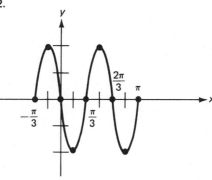

3.

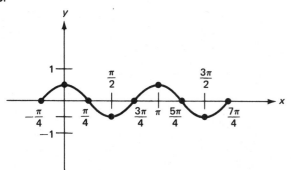

4.

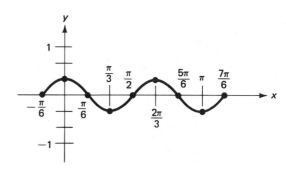

5.

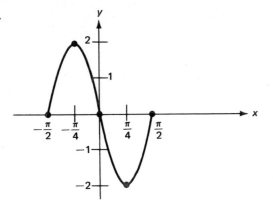

6.

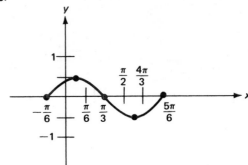

7.

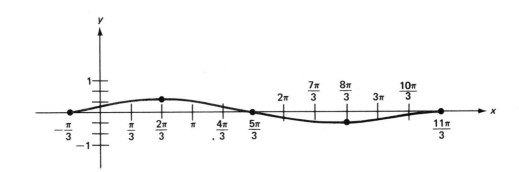

8.

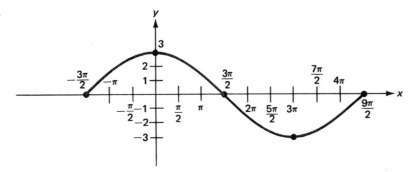

9.

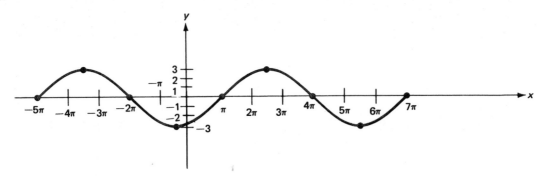

10.

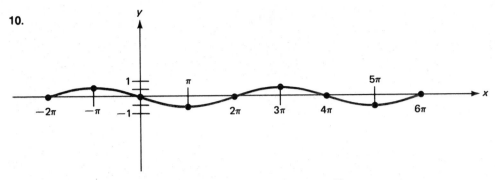

11.   **12.**

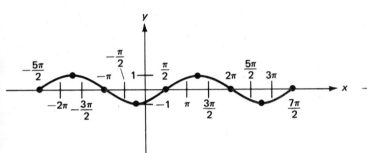 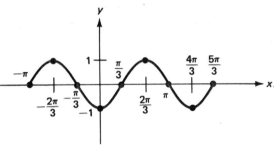

Self-test

1.   **(Objective 3)**

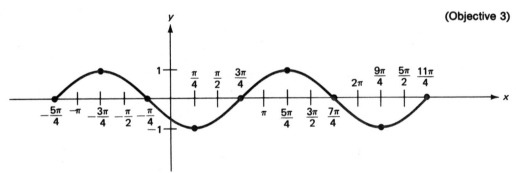

2.

                                                           (Objective 2)

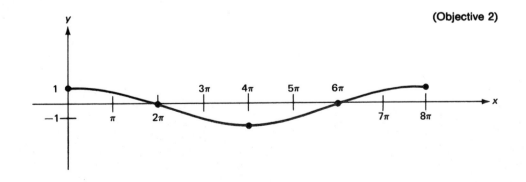

3. (Objective 1)

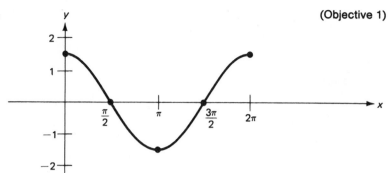

4. Amplitude: 3; Period: $\dfrac{2\pi}{3}$; Phase shift: $-\dfrac{\pi}{3}$ (Objective 4)

5. (Objective 4)

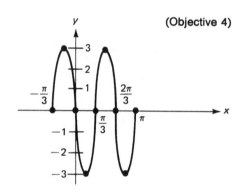

Unit 11

Self-test

1a. 2   (Unit 1)   1b. $\sqrt{3}$ (Unit 4)

2a. Undefined   (Unit 3)   2b. -1 (Unit 4)

3. 29°20′, 330°40′   (Unit 4)

4. 125°10′   (Unit 6)

5. 31°30′   (Unit 6)

6. 66°40′   (Units 2 and 7, trigonometric ratio)

7. No solution   (Unit 7, law of sines)

8. 8.44 inches   (Unit 8)

9.

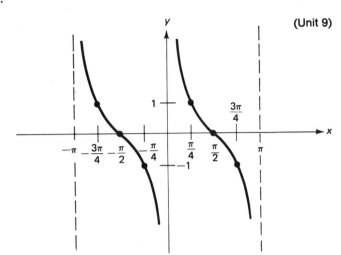

(Unit 9)

10.

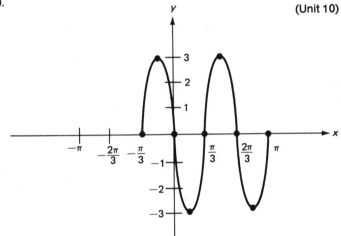

(Unit 10)

Unit 12

Exercise 12.1

1. $\dfrac{\sin x}{\cos x}$

2. $\dfrac{\cos x}{\sin x}$

3. $\dfrac{\cos x}{\sin x}$

4. $\cos^2 x$

5. $\dfrac{1}{\cos x}$

6. $\dfrac{1}{\sin x}$

7. 1

8. 1

9. $\dfrac{1}{\cos x}$

10. $\dfrac{1}{\sin^2 x}$

11. $\sin x$

12. 1

13. $\dfrac{1}{\sin^2 x \cos^2 x}$

14. $\dfrac{\sin x + \cos x}{\sin x \cos x}$

15. $\dfrac{\cos^2 x}{\sin^2 x}$

16. $\dfrac{1}{\cos^2 x}$

17. $\dfrac{1}{\cos x}$ 18. 1   19. $\dfrac{1}{\sin x + \cos x}$ 20. 1

21. $\dfrac{1 - \cos x}{1 + \cos x}$ 22. $\dfrac{1 - \cos x}{1 + \cos x}$ 23. $\dfrac{\sin x}{1 + \sin x}$ 24. $\dfrac{\cos x}{1 - \cos x}$

Self-test

1. $\cos x$ (Objective 1)   2. $\sin^2 x$ (Objective 1)

3. $-5.$ Proofs may vary (Objective 2)

Unit 13

Exercise 13.1

1. 60°, 300°   2. 240°, 300°   3. 135°, 315°   4. 30°, 330°

5. No solution   6. No solution   7. 30°, 150°   8. No solution

9. 270°   10. 90°, 270°   11. 63°, 297°   12. 152°40′, 207°20′

13. 41°50′, 138°10′   14. 139°20′, 319°20′   15. 153°30′, 333°30′   16. No solution

Exercise 13.2

1. 90°, 270°   2. 0°, 180°, 45°, 225°   3. 0°, 180°, 199°30′, 340°30′

4. 48°10′, 311°50′   5. 270°, 30°, 150°   6. 0°, 120°, 240°

7. 0°   8. 90°   9. 60°, 300°

10. 60°, 240°, 30°, 210°   11. No solution   12. No solution

13. 45°, 315°, 135°, 225°   14. 60°, 120°, 240°, 300°   15. 38°10′, 141°50′

16. 51°50′, 308°10′   17. 22°30′, 202°30′, 112°30′, 292°30′   18. 145°20′, 214°40′

19. No solution   20. No solution

Exercise 13.3

1. 270°   2. 180°, 60°, 300°   3. 0°

4. 45°, 225°   5. 90°, 270°   6. 0°, 180°

7. 60°, 300°, 120°, 240°   8. 30°, 150°, 210°, 330°   9. 90°, 270°, 180°

10. 0°, 180°, 30°, 150°   11. 180°   12. 0°, 120°, 240°

13. 90°   14. 180°, 60°, 300°   15. 0°, 120°, 240°   16. 270°

17. 38°10′, 141°50′   18. 141°20′, 218°40′   19. 214°10′, 325°50′   20. No solution

21. 0°, 90°   22. 0°, 270°   23. 0°, 233°10′   24. No solution

Self-test

1. 120°, 240°   (Objective 1)   2. 45°, 315°, 135°, 225°   (Objective 3)

3. 270°, 30°, 150°   (Objective 3)   4. 38°40′, 321°20′   (Objective 2)

5. 90°, 270°, 180°   (Objective 2)

Unit 14

Exercise 14.2

21. 0°, 180°   22. 90°, 270°   23. 90°, 270°, 120°, 240°

24. 0°, 180°, 30°, 150°   25. 270°, 30°, 150°   26. 0°, 120°, 240°

27. 0°, 180°   28. 60°, 120°, 240°, 300°   29. 38°10′, 141°50′

30. 128°10′, 231°50′

Exercise 14.3

9. 180°   10. 0°, 120°, 240°   11. 0°   12. 60°, 300°

Self-test

1. Proofs may vary   (Objective 1)

2. Proofs may vary   (Objective 2)

3. Proofs may vary   (Objective 3)

4. 0°, 180°, 60°, 300°   (Objective 2)

5. 180°   (Objective 2)

Unit 15

Exercise 15.1

1. $y = x + 2$
   function

2. $y = \dfrac{1}{2}x - \dfrac{1}{2}$
   function

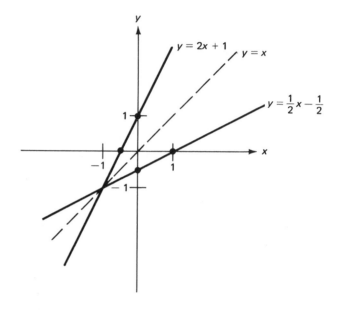

3. $y + 2x = 4$
   function

4. $3y - 4x = 12$
   function

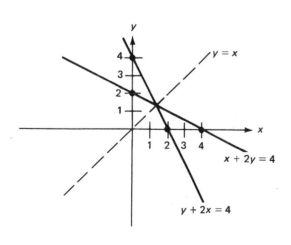

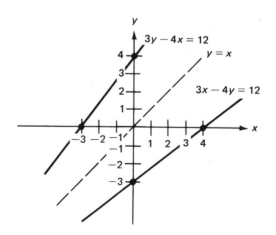

5. $x = 3$

    not a function

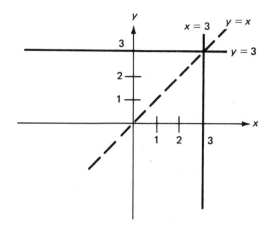

6. $x = -1$

    not a function

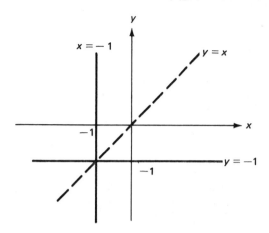

7. $y = \pm \sqrt{x + 1}$

    not a function

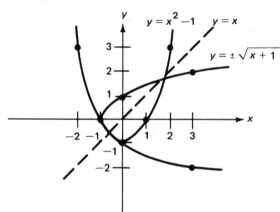

8. $y = \pm \sqrt{x - 2}$

    not a function

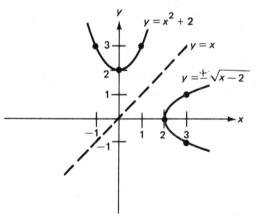

9. $y = \sqrt{x + 1}$

    function

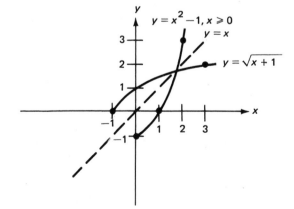

10. $y = \sqrt{x - 2}$

    function

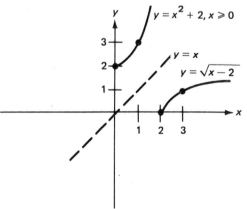

11. $y = -\sqrt{x+1}$

    function

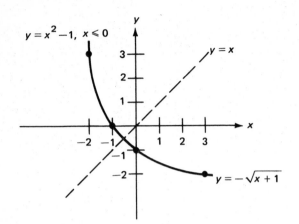

12. $y = -\sqrt{x-2}$

    function

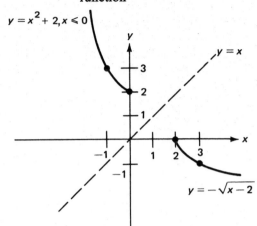

Exercise 15.2

1. $-\dfrac{\pi}{2} \leqslant y \leqslant \dfrac{\pi}{2}$

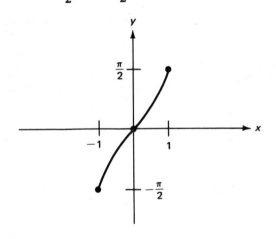

2. $0 \leqslant y \leqslant \pi$

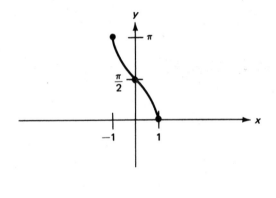

3. $-\dfrac{\pi}{2} < y < \dfrac{\pi}{2}$

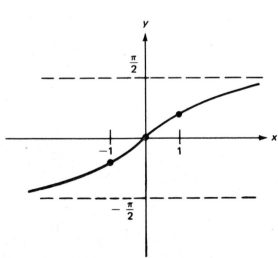

4.

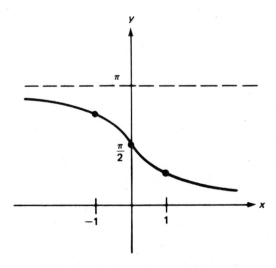

Exercise 15.3

1. $\dfrac{\pi}{4}$

2. $\dfrac{\pi}{3}$

3. $\dfrac{\pi}{3}$

4. $\dfrac{\pi}{4}$

5. $\dfrac{\pi}{3}$

6. $\dfrac{\pi}{6}$

7. $\dfrac{\pi}{2}$

8. 0

9. $-\dfrac{\pi}{3}$

10. $-\dfrac{\pi}{4}$

11. $\dfrac{5\pi}{6}$

12. $-\dfrac{\pi}{4}$

13. π

14. $-\dfrac{\pi}{2}$

15. 0

16. 0

Self-test

1. $y = \dfrac{1}{3}x + 2$ (Objective 1)

2. $0 \leqslant y \leqslant \pi$ (Objective 2)

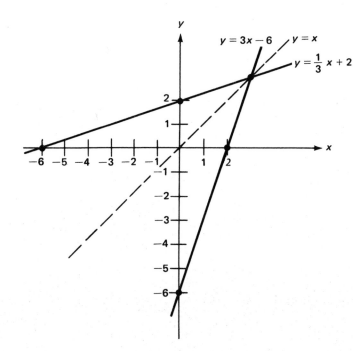

3.   (Objective 2)

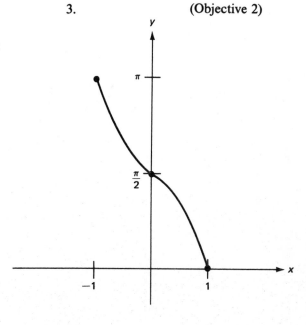

4. $-\dfrac{\pi}{3}$ (Objective 3)

5. $\dfrac{2\pi}{3}$ (Objective 3)

Unit 16

Self-test

1a. -1 (Unit 4)   1b. Undefined   (Unit 8)

2.   171   (Unit 6)

3.   15.1 inches   (Units 2 and 7, trigonometric ratio)

4.   9.12 meters   (Unit 7, law of sines)

5.   (Unit 10)

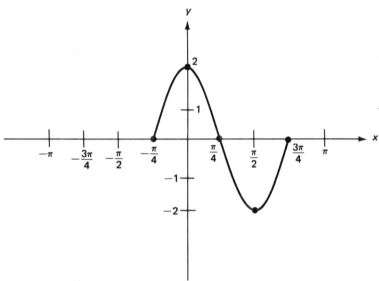

6.   Proofs may vary   (Units 12 and 14)

7.   Proofs may vary   (Unit 14)

8.   180°, 60°, 300°   (Unit 13)

9.   270°, 30°, 150°   (Unit 14)

10a. $\dfrac{5\pi}{6}$ 10b. $-\dfrac{\pi}{2}$ (Unit 15)

Unit 17

Exercise 17.1

1. $\log_4 64 = 3$

2. $\log_{10} .01 = -2$

3. $\log_{64} 8 = \dfrac{1}{2}$

4. $\log_{64} 4 = \dfrac{1}{3}$

5. $\log_4 \dfrac{1}{64} = -3$

6. $\log_4 2 = \dfrac{1}{2}$

7. $\log_{10} 10,000 = 4$ 8. $\log_{10} .0001 = -4$ 9. $2^6 = 64$

10. $27^{\frac{1}{3}} = 3$ 11. $10^2 = 100$ 12. $10^0 = 1$

13. $36^{\frac{1}{2}} = 6$ 14. $2^{-6} = \frac{1}{64}$ 15. $10^{-1} = .1$

16. $10^{-4} = .0001$

Exercise 17.2

1. 3   2. 3   3. 2   4. -2 5. $\frac{1}{2}$

6. $\frac{1}{4}$ 7. 3   8. $\frac{1}{3}$ 9. -3 10. -4

11. -3 12. $\frac{1}{3}$ 13. 3   14. 4   15. 5

16. -3 17. -4 18. -5 19. 1   20. 0

Exercise 17.3

1.

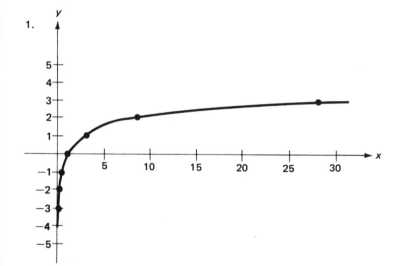

2

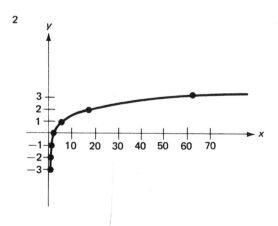

3.

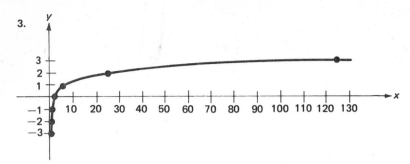

4.

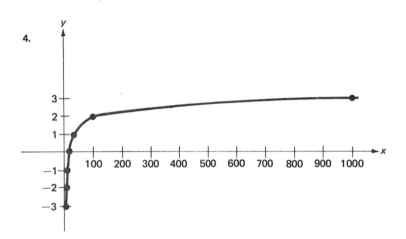

Exercise 17.4

1. $\log_b x + \log_b y + \log_b z$

2. $2\log_b x + 2\log_b y$

3. $\log_b x + \log_b y - \log_b z$

4. $\log_b x - \log_b y - \log_b z$

5. $2\log_b k + \log_b m + 3\log_b n$

6. $2\log_b k - \log_b m - 3\log_b n$

7. $\dfrac{1}{2}\log_b s - 3\log_b t$

8. $2\log_b u - \dfrac{1}{2}\log_b v - \dfrac{1}{2}\log_b w$

9. $\dfrac{1}{3}\log_b p + \dfrac{2}{3}\log_b q - \dfrac{1}{3}\log_b r$

10. $\dfrac{1}{2}\log_b p - \dfrac{1}{4}\log_b q - \dfrac{3}{4}\log_b r$

Exercise 17.5

1. .3404

2. .9191

3. 1.9274

4. .5092 − 1

5. .7404 − 3

6. 4.6693

7. 3.9685

8. 6.6263

9. .8096 − 3

10. .6010 − 5

11. 3.44

12. 1.60

13. .678

14. 545

15. 23,600

16. 590,000

17. .00519

18. .00089

19. .0000827

20. 9,620,000,000

Self-test

1. $2\log_b x + \frac{1}{2}\log_b y - \log_b z$ (Objective 4)   2a. -4 2b. $\frac{1}{4}$ (Objective 2)

3. $\log_3 9 = 2$ (Objective 1)   4a. $.8021 - 1$ 4b. 505 (Objective 5)

5.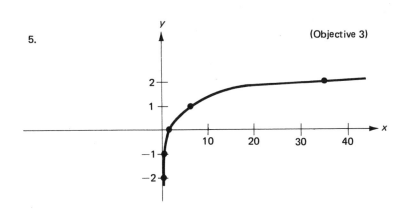

(Objective 3)

$r\theta = arclength(s)$

$\text{×vel.,}\to\omega = \frac{\theta}{t}$

Unit 18

LINEAR VEL. $V = r\omega$

$1 + tan^2\theta = sec^2\theta$

$1 + cot^2\theta = csc^2\theta$

These answers were found using a scientific calculator. Your answers may differ by one digit if you are using Tables III and IV.

Exercise 18.1

1. 1.63   2. 1.96   3. 2.73   4 .699

5. 1.23   6. $-.468$ 7. .753   8. 1.84

Exercise 18.2

1. $4620   2. $650   3. 11.7, or about 12 years

4. 3.63 or about $3\frac{3}{4}$ years   5. 5.79 years   6. $21,200

7. 3310 bacteria   8. 34.7 hours   9. 6.77 grams

10. 576 hours   11. 5570 years   12. 12,900 years

Exercise 18.3

1. 1 hour   2. 9.96 seconds   3. $2110

4. 10.7 years   5. 5.76 milligrams

6. Superior: 431 years; Michigan: 71.4 years; Erie: 6.05 years; Ontario: 17.6 years

Self-test

1. .861 (Objective 1)

2. $743 (Objective 2)

3. 1.73 years (Objective 2 or 3)

4. 485 bacteria (Objective 2 or 3)

5. 4.61 seconds (Objective 2 or 3)

Unit 19

Exercise 19.1

1. $6i$

2. $-3i$

3. $i\sqrt{7}$

4. $-i\sqrt{11}$

5. $3i\sqrt{2}$

6. $3i\sqrt{3}$

7. $4i\sqrt{5}$

8. $7i\sqrt{2}$

9. $-5i\sqrt{2}$

10. $-3i\sqrt{5}$

11. $\frac{1}{2}i$

12. $\frac{3}{5}i$

13. $-\frac{4}{3}i$

14. $-\frac{1}{4}i$

15. $\frac{i\sqrt{5}}{2}$

16. $-\frac{2i\sqrt{2}}{3}$

17. $9+9i$

18. $4-8i$

19. $10-2i\sqrt{5}$

20. $\sqrt{3}+6i\sqrt{2}$

21. $\frac{3}{4}-\frac{3}{2}i$

22. $\frac{2}{3}+\frac{i\sqrt{2}}{3}$

Exercise 19.2

1. $5i$

2. $3i$

3. $3+6i$

4. $7-3i$

5. $3+4i$

6. $-1+2i$

7. $6-12i$

8. $5-3i$

9. $-2+4i$

10. $-5-10i$

11. $4i$

12. $-i$

13. 4

14. 8

15. $5-3i$

16. $-7i$

17. $8-14i$

18. $9+i$

19. $2\sqrt{2}+5i\sqrt{2}$

20. $3\sqrt{3}+(2\sqrt{5}-2\sqrt{3})i$

Exercise 19.3

1. -14

2. -4

3. -4

4. -6

5. -4

6. $-3\sqrt{2}$

7. $24i-16$

8. $8+12i$

9. $1-4i$

10. $-10-25i$

11. $5+14i$

12. $19i-9$

13. $26+18i$

14. $-3-36i$

15. 41

16. 5

17. $13i$

18. $-20i$

19. $4 + 5i\sqrt{2}$

20. $8 + 3i\sqrt{3}$

21. $3 + \sqrt{5} + (\sqrt{3} - \sqrt{15})i$

22. 8

23. $6 - \sqrt{15} + (2\sqrt{5} + 3\sqrt{3})i$

24. $9i$

Exercise 19.4

1. $-\dfrac{5}{6}i$

2. $\dfrac{2}{3}i$

3. $-\dfrac{3}{4}i$

4. $-5i$

5. $-i$

6. $-\dfrac{i\sqrt{2}}{4}$

7. $\dfrac{8 - 5i}{2}$

8. $-\dfrac{4 + 3i}{9}$

9. $\dfrac{8 + 2i}{5}$

10. $1 - i$

11. $\dfrac{2 + i}{2}$

12. $1 - 2i$

13. $\dfrac{9 + 12i}{5}$

14. $\dfrac{22i - 7}{13}$

15. $\dfrac{14 + 5i}{17}$

16. $\dfrac{9 - 13i}{25}$

17. $\dfrac{5 - 5i}{4}$

18. $\dfrac{1 - 6i}{3}$

19. $\dfrac{4 + 3i}{5}$

20. $-\dfrac{3 + 4i}{5}$

21. i

22. $-i$

23. $\dfrac{\sqrt{5} - 3i}{2}$

24. $1 + i\sqrt{2}$

Exercise 19.5

1. $\pm i$

2. $\pm 5i$

3. $\pm 3i\sqrt{5}$

4. $\pm 2i\sqrt{3}$

5. $\dfrac{-1 \pm i\sqrt{3}}{2}$

6. $\dfrac{5 \pm i\sqrt{15}}{4}$

7. $1 \pm i$

8. $-1 \pm 5i$

9. $\dfrac{1 \pm i\sqrt{17}}{2}$

10. $\dfrac{3 \pm i\sqrt{39}}{6}$

Self-test

1. $-3i\sqrt{10}$ (Objective 1)

2. $14 - 34i$ (Objective 3)

3. $2 + 6i$ (Objective 2)

4. $\dfrac{3 - 4i}{5}$ (Objective 4)

5. $-1 \pm i\sqrt{14}$ (Objective 5)

Unit 20

Exercise 20.1

1.

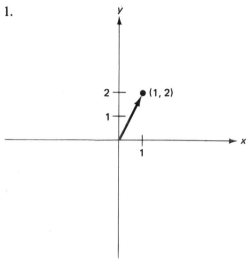

2.

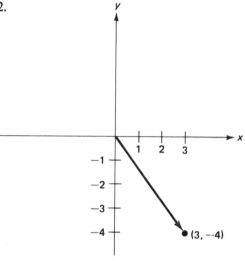

3.

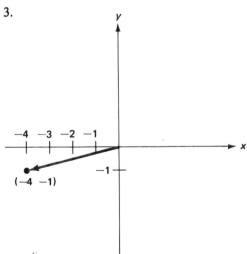

4.

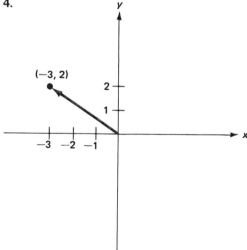

5.

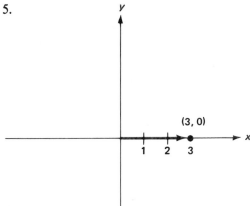

6.

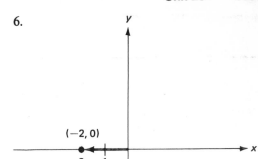

7.

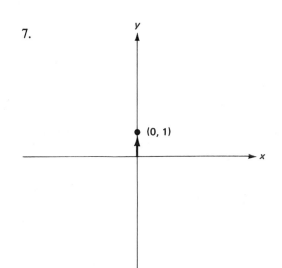

8.

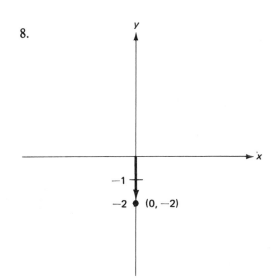

9. $2 \operatorname{cis} 30°$

10. $\sqrt{2} \operatorname{cis} 45°$

11. $\sqrt{2} \operatorname{cis} 225°$

12. $2 \operatorname{cis} 150°$

13. $2 \operatorname{cis} 45°$

14. $2\sqrt{2} \operatorname{cis} 315°$

15. $2\sqrt{3} \operatorname{cis} 120°$

16. $2\sqrt{3} \operatorname{cis} 210°$

17. $\operatorname{cis} 0°$

18. $2 \operatorname{cis} 180°$

19. $3 \operatorname{cis} 270°$

20. $\operatorname{cis} 90°$

21. $1 + i\sqrt{3}$

22. $1 + i$

23. $-1 + i$

24. $\sqrt{3} - i$

25. $\sqrt{2} - i\sqrt{2}$

26. $-2 + 2i$

27. $-3\sqrt{3} + 3i$

28. $-\sqrt{3} - 3i$

29. $2i$

30. -3

31. $-i$

32. 1

33. 0

34. 0

Exercise 20.2

1. cis 90°

2. 8 cis 0°

3. 2 cis 225°

4. 4√3 cis 30°

5. 8 cis 60°

6. 4 cis 180°

7. 4 cis 90°

8. 12 cis 90°

9. √2 cis 135°

10. 2 cis 30°

11. 4 cis 270°

12. 8 cis 180°

13. 2√2 cis 165°

14. 2√2 cis 75°

15. cis 90°

16. 2 cis 180°

17. √2 cis 45°

18. 2 cis 60°

19. $\frac{1}{3}$ cis 300°

20. cis 180°

21. 2 cis 270°

22. $\frac{1}{2}$ cis 270°

23. √2 cis 135°

24. 2 cis 30°

25. $\frac{3}{4}$ cis 60°

26. $\frac{1}{2}$ cis 45°

27. √2 cis 90°

28. cis 150°

29. √2 cis 195°

30. √2 cis 15°

Exercise 20.3

1. −4

2. 2 − 2i

3. −16√3 − 16i

4. 64

5. i

6. 1

7. $\frac{\sqrt{2}}{2} + \frac{i\sqrt{6}}{2}, -\frac{\sqrt{2}}{2} - \frac{i\sqrt{6}}{2}$

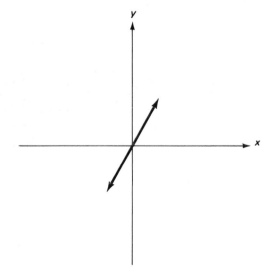

8. $-\frac{\sqrt{3}}{2} + \frac{i}{2}, \frac{\sqrt{3}}{2} - \frac{i}{2}$

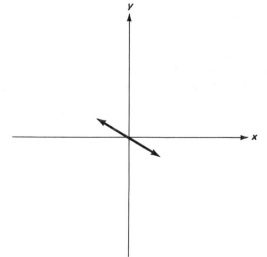

9. $1 + i\sqrt{3}, -2, 1 - i\sqrt{3}$

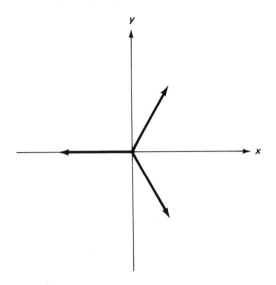

10. $i, -\dfrac{\sqrt{3}}{2} - \dfrac{i}{2}, \dfrac{\sqrt{3}}{2} - \dfrac{i}{2}$

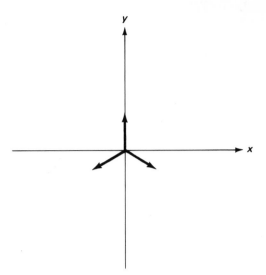

11. $\dfrac{3}{\sqrt{2}} + \dfrac{3i}{\sqrt{2}}, -\dfrac{3}{\sqrt{2}} + \dfrac{3i}{\sqrt{2}},$

$-\dfrac{3}{\sqrt{2}} - \dfrac{3i}{\sqrt{2}}, \dfrac{3}{\sqrt{2}} - \dfrac{3i}{\sqrt{2}}$

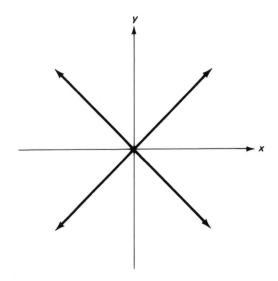

12. $\dfrac{\sqrt{3}}{2} + \dfrac{i}{2}, -\dfrac{1}{2} + \dfrac{i\sqrt{3}}{2},$

$-\dfrac{\sqrt{3}}{2} - \dfrac{i}{2}, \dfrac{1}{2} - \dfrac{i\sqrt{3}}{2}$

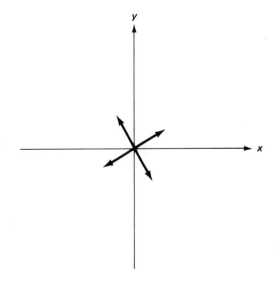

13. $1, i, -1, -i$

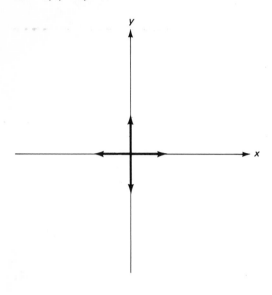

14. $1, \dfrac{1}{2} + \dfrac{i\sqrt{3}}{2}, -\dfrac{1}{2} + \dfrac{i\sqrt{3}}{2}, -1,$

$-\dfrac{1}{2} - \dfrac{i\sqrt{3}}{2}, \dfrac{1}{2} - \dfrac{i\sqrt{3}}{2}$

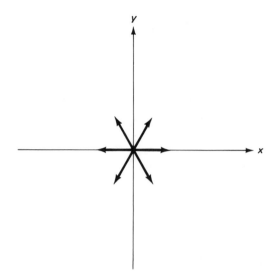

Self-test

1. $1 - i$ (Objective 1)

2. $24 \operatorname{cis} 90°$ (Objective 2)

3. $2 \operatorname{cis} 315°$ (Objective 2)

4. $-i$ (Objective 3)

5. $\sqrt{2} + i\sqrt{2}, -\sqrt{2} + i\sqrt{2},$

   $-\sqrt{2} - i\sqrt{2}, \sqrt{2} - i\sqrt{2}$ (Objective 3)

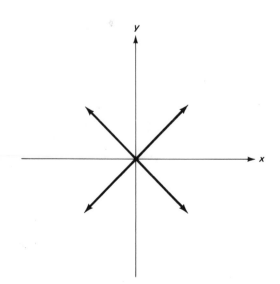

Unit 21

Exercise 21.1

1.

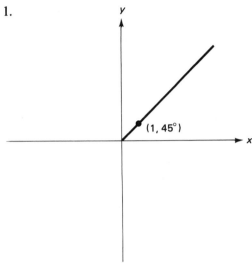

2.

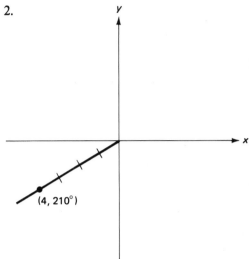

3.

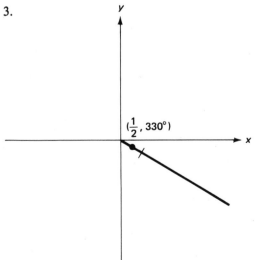

4.

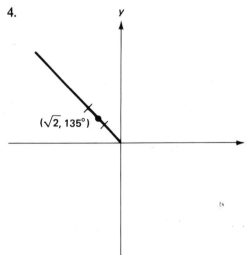

5.

● (1, 90°)

6.

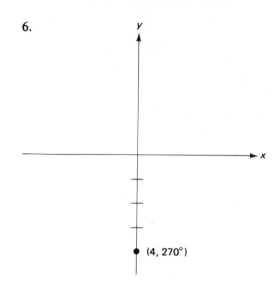

● (4, 270°)

7.

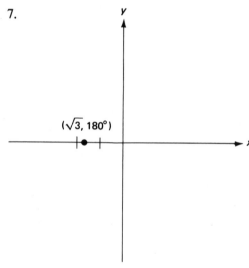

(√3, 180°)

8.

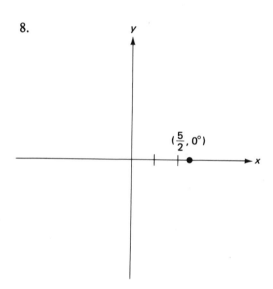

($\frac{5}{2}$, 0°)

9.

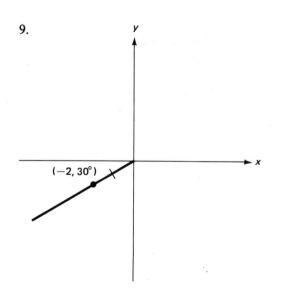

(−2, 30°)

10.

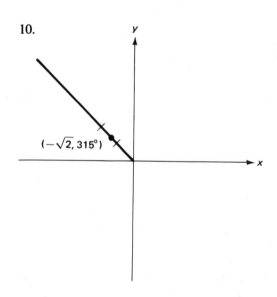

(−√2, 315°)

11.

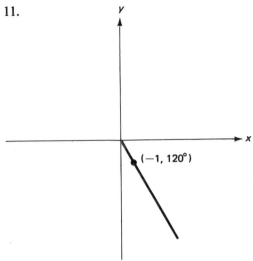

$(-1, 120°)$

12.

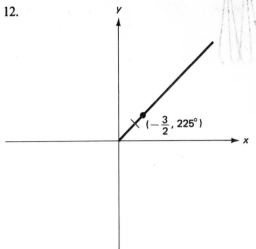

$(-\frac{3}{2}, 225°)$

13.

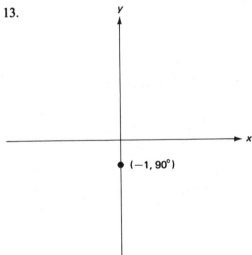

$(-1, 90°)$

14.

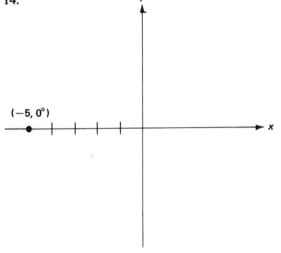

$(-5, 0°)$

15.

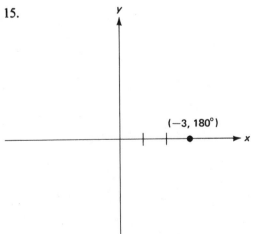

$(-3, 180°)$

16.

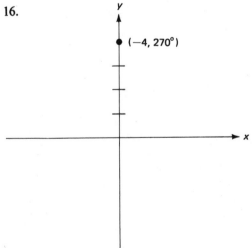

$(-4, 270°)$

Exercise 21.2

1. $(2, 60°)$

2. $(\sqrt{2}, 315°)$

3. $(1, 240°)$

4. $(2, 135°)$

5. $(2, 180°)$

6. $(\sqrt{2}, 0°)$

7. $\left(\frac{1}{2}, 270°\right)$

8. $\left(\frac{5}{3}, 90°\right)$

9. $(5, 53.1°)$

10. $(2\sqrt{5}, 116.6°)$

11. $(2\sqrt{5}, 333.4°)$

12. $(5, 216.9°)$

13. $(-1, 0)$

14. $(0, -3)$

15. $(-2, 0)$

16. $(0, 1)$

17. $(-1, -\sqrt{3})$

18. $(-2, 2)$

19. $\left(-\frac{1}{\sqrt{2}}, -\frac{1}{\sqrt{2}}\right)$

20. $(-2\sqrt{3}, 2)$

21. $(3.2, 3.8)$

22. $(3.6, -1.7)$

23. $(-1.5, 3.2)$

24. $(-.6, -2.4)$

Exercise 21.3

1.

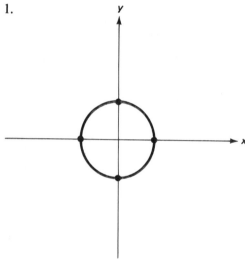

2.

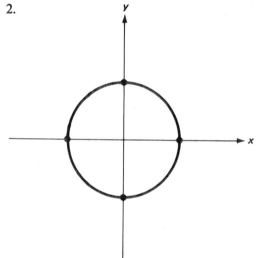

3.

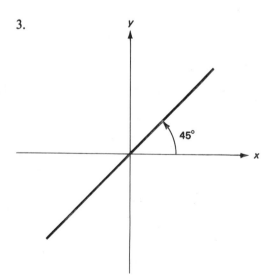

4.

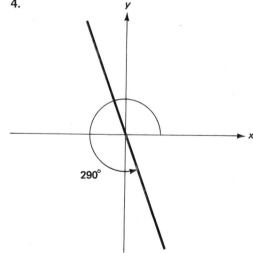

5.

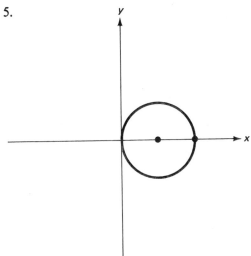

6.

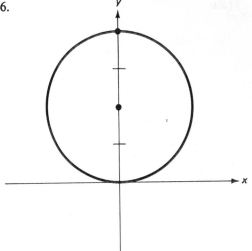

7.

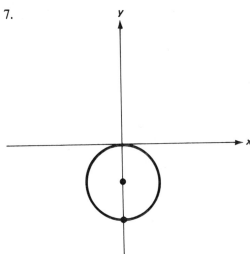

8.

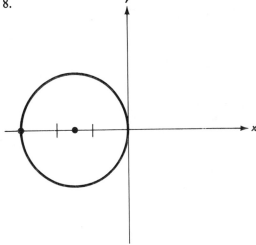

9.

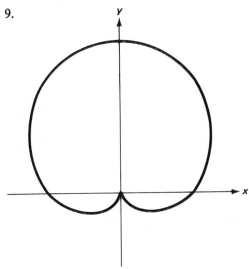

10.

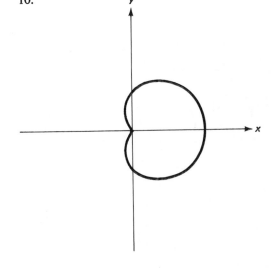

11.

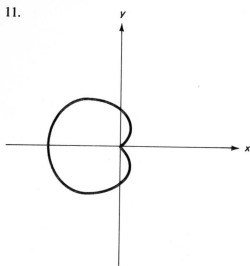

12.

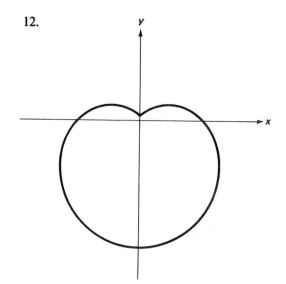

13.

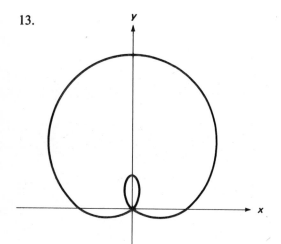

14.

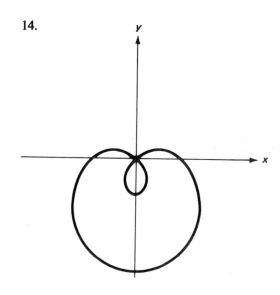

15.

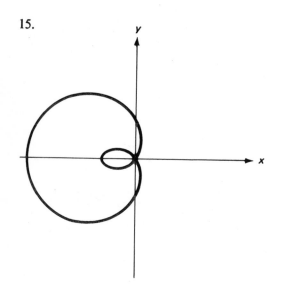

16.

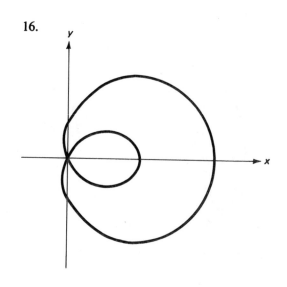

17.

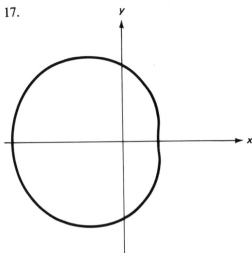

18.

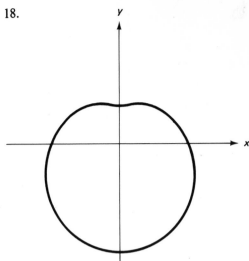

19.

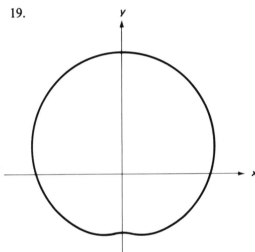

20.

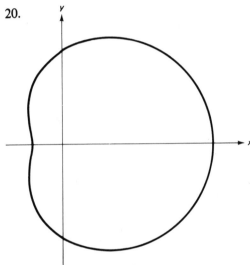

21.

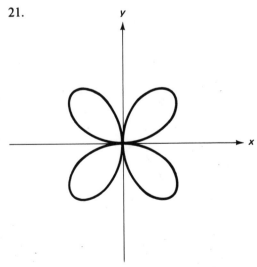

22.

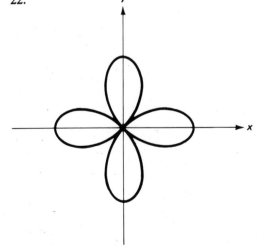

23.

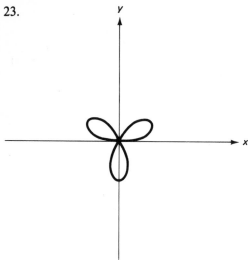

24.

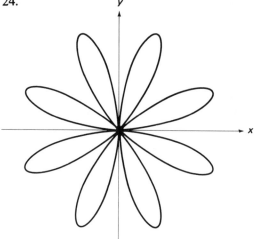

25.

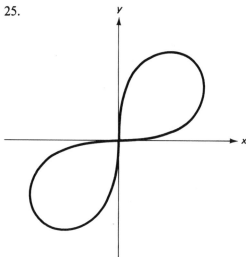

26.

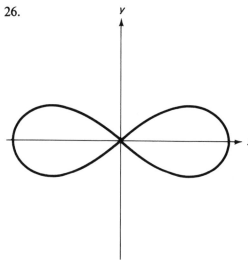

27.

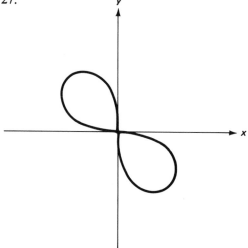

28.

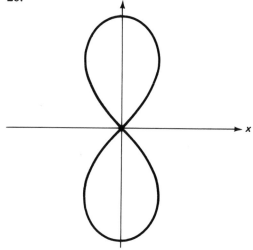

Self-test

1a.

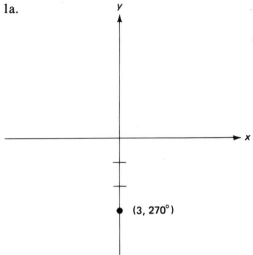

1b.

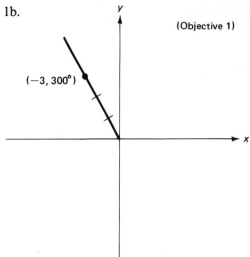

(Objective 1)

2. $(2, 315°)$ (Objective 2)

3. $(-2\sqrt{2}, 2\sqrt{2})$ (Objective 2)

4.

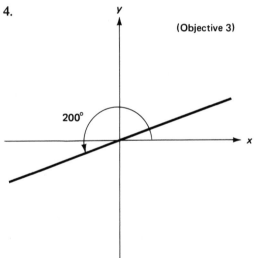

(Objective 3)

5.

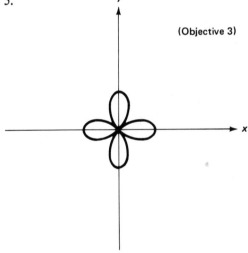

(Objective 3)

Appendix

Table I

Squares and Square Roots

N	N²	√N	√10N	N	N²	√N	√10N
1	1	1.0 00	3.1 62	51	2 601	7.1 41	22.5 83
2	4	1.4 14	4.4 72	52	2 704	7.2 11	22.8 04
3	9	1.7 32	5.4 77	53	2 809	7.2 80	23.0 22
4	16	2.0 00	6.3 25	54	2 916	7.3 48	23.2 38
5	25	2.2 36	7.0 71	55	3 025	7.4 16	23.4 52
6	36	2.4 49	7.7 46	56	3 136	7.4 83	23.6 64
7	49	2.6 46	8.3 67	57	3 249	7.5 50	23.8 75
8	64	2.8 28	8.9 44	58	3 364	7.6 16	24.0 83
9	81	3.0 00	9.4 87	59	3 481	7.6 81	24.2 90
10	100	3.1 62	10.0 00	60	3 600	7.7 46	24.4 95
11	121	3.3 17	10.4 88	61	3 721	7.8 10	24.6 98
12	144	3.4 64	10.9 54	62	3 844	7.8 74	24.9 00
13	169	3.6 06	11.4 02	63	3 969	7.9 37	25.1 00
14	196	3.7 42	11.8 32	64	4 096	8.0 00	25.2 98
15	225	3.8 73	12.2 47	65	4 225	8.0 62	25.4 95
16	256	4.0 00	12.6 49	66	4 356	8.1 24	25.6 90
17	289	4.1 23	13.0 38	67	4 489	8.1 85	25.8 84
18	324	4.2 43	13.4 16	68	4 624	8.2 46	26.0 77
19	361	4.3 59	13.7 84	69	4 761	8.3 07	26.2 68
20	400	4.4 72	14.1 42	70	4 900	8.3 67	26.4 58
21	441	4.5 83	14.4 91	71	5 041	8.4 26	26.6 46
22	484	4.6 90	14.8 32	72	5 184	8.4 85	26.8 33
23	529	4.7 96	15.1 66	73	5 329	8.5 44	27.0 19
24	576	4.8 99	15.4 92	74	5 476	8.6 02	27.2 03
25	625	5.0 00	15.8 11	75	5 625	8.6 60	27.3 86
26	676	5.0 99	16.1 25	76	5 776	8.7 18	27.5 68
27	729	5.1 96	16.4 32	77	5 929	8.7 75	27.7 49
28	784	5.2 92	16.7 33	78	6 084	8.8 32	27.9 28
29	841	5.3 85	17.0 29	79	6 241	8.8 88	28.1 07
30	900	5.4 77	17.3 21	80	6 400	8.9 44	28.2 84
31	961	5.5 68	17.6 07	81	6 561	9.0 00	28.4 60
32	1 024	5.6 57	17.8 89	82	6 724	9.0 55	28.6 36
33	1 089	5.7 45	18.1 66	83	6 889	9.1 10	28.8 10
34	1 156	5.8 31	18.4 39	84	7 056	9.1 65	28.9 83

TABLE I (Continued)

N	N²	√N	√10N	N	N²	√N	√10N
35	1 225	5.9 16	18.7 08	85	7 225	9.2 20	29.1 55
36	1 296	6.0 00	18.9 74	86	7 396	9.2 74	29.3 26
37	1 369	6.0 83	19.2 35	87	7 569	9.3 27	29.4 96
38	1 444	6.1 64	19.4 94	88	7 744	9.3 81	29.6 65
39	1 521	6.2 45	19.7 48	89	7 921	9.4 34	29.8 33
40	1 600	6.3 25	20.0 00	90	8 100	9.4 87	30.0 00
41	1 681	6.4 03	20.2 48	91	8 281	9.5 39	30.1 66
42	1 764	6.4 81	20.4 94	92	8 464	9.5 92	30.3 32
43	1 849	6.5 57	20.7 36	93	8 649	9.6 44	30.4 96
44	1 936	6.6 33	20.9 76	94	8 836	9.6 95	30.6 59
45	2 025	6.7 08	21.2 13	95	9 025	9.7 47	30.8 22
46	2 116	6.7 82	21.4 48	96	9 216	9.7 98	30.9 84
47	2 209	6.8 56	21.6 79	97	9 409	9.8 49	31.1 45
48	2 304	6.9 28	21.9 09	98	9 604	9.8 99	31.3 05
49	2 401	7.0 00	22.1 36	99	9 801	9.9 50	31.4 64
50	2 500	7.0 71	22.3 61	100	10 000	10.0 00	31.6 23
N	N²	√N	√10N	N	N²	√N	√10N

Table II

Values of the Trigonometric Ratios

Degrees	Radians	Sin	Cos	Tan	Cot	Sec	Csc		
0° 00′	.0000	.0000	1.0000	.0000	——	1.000	——	1.5708	90° 00′
10	029	029	000	029	343.8	000	343.8	679	50
20	058	058	000	058	171.9	000	171.9	650	40
30	.0087	.0087	1.0000	.0087	114.6	1.000	114.6	1.5621	30
40	116	116	.9999	116	85.94	000	85.95	592	20
50	145	145	999	145	68.75	000	68.76	563	10
1° 00′	.0175	.0175	.9998	.0175	57.29	1.000	57.30	1.5533	89° 00′
10	204	204	998	204	49.10	000	49.11	504	50
20	233	233	997	233	42.96	000	42.98	475	40
30	.0262	.0262	.9997	.0262	38.19	1.000	38.20	1.5446	30
40	291	291	996	291	34.37	000	34.38	417	20
50	320	320	995	320	31.24	001	31.26	388	10
2° 00′	.0349	.0349	.9994	.0349	28.64	1.001	28.65	1.5359	88° 00′
10	378	378	993	378	26.43	001	26.45	330	50
20	407	407	992	407	24.54	001	24.56	301	40
30	.0436	.0436	.9990	.0437	22.90	1.001	22.93	1.5272	30
40	465	465	989	466	21.47	001	21.49	243	20
50	495	494	988	495	20.21	001	20.23	213	10
3° 00′	.0524	.0523	.9986	.0524	19.08	1.001	19.11	1.5184	87° 00′
10	553	552	985	553	18.07	002	18.10	155	50
20	582	581	983	582	17.17	002	17.20	126	40
30	.0611	.0610	.9981	.0612	16.35	1.002	16.38	1.5097	30
40	640	640	980	641	15.60	002	15.64	068	20
50	669	669	978	670	14.92	002	14.96	039	10
4° 00′	.0698	.0698	.9976	.0699	14.30	1.002	14.34	1.5010	86° 00′
10	727	727	974	729	13.73	003	13.76	981	50
20	756	756	971	758	13.20	003	13.23	952	40
30	.0785	.0785	.9969	.0787	12.71	1.003	12.75	1.4923	30
40	814	814	967	816	12.25	003	12.29	893	20
50	844	843	964	846	11.83	004	11.87	864	10
5° 00′	.0873	.0872	.9962	.0875	11.43	1.004	11.47	1.4835	85° 00′
10	902	901	959	904	11.06	004	11.10	806	50
20	931	929	957	934	10.71	004	10.76	777	40
30	.0960	.0958	.9954	.0963	10.39	1.005	10.43	1.4748	30
40	989	987	951	992	10.08	005	10.13	719	20
50	.1018	.1016	948	.1022	9.788	005	9.839	690	10
6° 00′	.1047	.1045	.9945	.1051	9.514	1.006	9.567	1.4661	84° 00′
10	076	074	942	080	9.255	006	9.309	632	50
20	105	103	939	110	9.010	006	9.065	603	40
30	.1134	.1132	.9936	.1139	8.777	1.006	8.834	1.4573	30
40	164	161	932	169	8.556	007	8.614	544	20
50	193	190	929	198	8.345	007	8.405	515	10
7° 00′	.1222	.1219	.9925	.1228	8.144	1.008	8.206	1.4486	83° 00′
10	251	248	922	257	7.953	008	8.016	457	50
20	280	276	918	287	7.770	008	7.834	428	40
30	.1309	.1305	.9914	.1317	7.596	1.009	7.661	1.4399	30
40	338	334	911	346	7.429	009	7.496	370	20
50	367	363	907	376	7.269	009	7.337	341	10
8° 00′	.1396	.1392	.9903	.1405	7.115	1.010	7.185	1.4312	82° 00′
10	425	421	899	435	6.968	010	7.040	283	50
20	454	449	894	465	6.827	011	6.900	254	40
30	.1484	.1478	.9890	.1495	6.691	1.011	6.765	1.4224	30
40	513	507	886	524	6.561	012	6.636	195	20
50	542	536	881	554	6.435	012	6.512	166	10
9° 00′	.1571	.1564	.9877	.1584	6.314	1.012	6.392	1.4137	81° 00′
		Cos	Sin	Cot	Tan	Csc	Sec	Radians	Degrees

$y = A \sin(Bx - D)$
amp $= |A|$
period $= \dfrac{2\pi}{B}$
PHASE SHIFT
$Bx - D = 0$
$Bx = D$
$y = \dfrac{D}{B}$

$\dfrac{2\pi}{B}$

$\dfrac{D}{B}$

TABLE II   (Continued)

Degrees	Radians	Sin	Cos	Tan	Cot	Sec	Csc	·	
9° 00′	.1571	.1564	.9877	.1584	6.134	1.012	6.392	1.4137	81° 00′
10	600	593	872	614	197	013	277	108	50
20	629	622	868	644	084	013	166	079	40
30	.1658	.1650	.9863	.1673	5.976	1.014	6.059	1.4050	30
40	687	679	858	703	871	014	5.955	1.4021	20
50	716	708	853	733	769	015	855	992	10
10° 00′	.1745	.1736	.9848	.1763	5.671	1.015	5.759	1.3963	80° 00′
10	774	765	843	793	576	016	665	934	50
20	804	794	838	823	485	016	575	904	40
30	.1883	.1822	.9833	.1853	5.396	1.017	5.487	1.3875	30
40	862	851	827	883	309	018	403	846	20
50	891	880	822	914	226	018	320	817	10
11° 00′	.1920	.1908	.9816	.1944	5.145	1.019	5.241	1.3788	79° 00′
10	949	937	811	974	066	019	164	759	50
20	978	965	805	.2004	4.989	020	089	730	40
30	.2007	.1994	.9799	.2035	4.915	1.020	5.106	1.3701	30
40	036	.2022	793	065	843	021	4.945	672	20
50	065	051	787	095	773	022	876	643	10
12° 00′	.2094	.2079	.9781	.2126	4.705	1.022	4.810	1.3614	78° 00′
10	123	108	775	156	638	023	745	584	50
20	153	136	769	186	574	024	682	555	40
30	.2182	.2164	.9763	.2217	4.511	1.024	4.620	1.3526	30
40	211	193	757	247	449	025	560	497	20
50	240	221	750	278	390	026	502	468	10
13° 00′	.2269	.2250	.9744	.2309	4.331	1.026	4.445	1.3439	77° 00′
10	298	278	737	339	275	027	390	410	50
20	327	306	730	370	219	028	336	381	40
30	.2356	.2334	.9724	.2401	4.165	1.028	4.284	1.3352	30
40	385	363	717	432	113	029	232	323	20
50	414	391	710	462	061	030	182	294	10
14° 00′	.2443	.2419	.9703	.2493	4.011	1.031	4.134	1.3265	76° 00′
10	473	447	696	524	3.962	031	086	235	50
20	502	476	689	555	914	032	039	206	40
30	.2531	.2504	.9681	.2586	3.867	1.033	3.994	1.3177	30
40	560	532	674	617	821	034	950	148	20
50	589	560	667	648	776	034	906	119	10
15° 00′	.2618	.2588	.9659	.2679	3.732	1.035	3.864	1.3090	75° 00′
10	647	616	652	711	689	036	822	061	50
20	676	644	644	742	647	037	782	032	40
30	.2705	.2672	.9636	.2773	3.606	1.038	3.742	1.3003	30
40	734	700	628	805	566	039	703	974	20
50	763	728	621	836	526	039	665	945	10
16° 00′	.2793	.2756	.9613	.2867	3.487	1.040	3.628	1.2915	74° 00′
10	822	784	605	899	450	041	592	886	50
20	851	812	596	931	412	042	556	857	40
30	.2880	.2840	.9588	.2962	3.376	1.043	3.521	1.2828	30
40	909	868	580	994	340	044	487	799	20
50	938	896	572	.3026	.305	045	453	770	10
17° 00′	.2967	.2924	.9563	.3057	3.271	1.046	3.420	1.2741	73° 00′
10	996	952	555	089	237	047	388	712	50
20	.3025	979	546	121	204	048	356	683	40
30	.3054	.3007	.9537	.3153	3.172	1.049	3.326	1.2654	30
40	083	035	528	185	140	049	295	625	20
50	113	062	520	217	108	050	265	595	10
18° 00′	.3142	.3090	.9511	.3249	3.078	1.051	3.236	1.2566	72° 00′
		Cos	Sin	Cot	Tan	Csc	Sec	Radians	Degrees

TABLE II   (Continued)

Degrees	Radians	Sin	Cos	Tan	Cot	Sec	Csc		
18° 00′	.3142	.3090	.9511	.3249	3.078	1.051	3.236	1.2566	72° 00′
10	171	118	502	281	047	052	207	537	50
20	200	145	492	314	.018	053	179	508	40
30	.3229	.3173	.9483	.3346	2.989	1.054	3.152	1.2479	30
40	258	201	474	378	960	056	124	450	20
50	287	228	465	411	932	057	098	421	10
19° 00′	.3316	.3256	.9455	.3443	2.904	1.058	3.072	1.2392	71° 00′
10	345	283	446	476	877	059	046	363	50
20	374	311	436	508	850	060	021	334	40
30	.3403	.3338	.9426	.3541	2.824	1.061	2.996	1.2305	30
40	432	365	417	574	798	062	971	275	20
50	462	393	407	607	773	063	947	·246	10
20° 00′	.3491	.3420	.9397	.3640	2.747	1.064	2.924	1.2217	70° 00′
10	520	448	387	673	723	065	901	188	50
20	549	475	377	706	699	066	878	159	40
30	.3578	.3502	.9367	.3739	2.675	1.068	2.855	1.2130	30
40	607	529	356	772	651	069	833	101	20
50	636	557	346	805	628	070	812	072	10
21° 00′	.3665	.3584	.9336	.3839	2.605	1.071	2.790	1.2043	69° 00′
10	694	611	325	872	583	072	769	1.2014	50
20	723	638	315	906	560	074	749	985	40
30	.3752	.3665	.9304	.3939	2.539	1.075	2.729	1.1956	30
40	782	692	293	973	517	076	709	926	20
50	811	719	283	.4006	496	077	689	897	10
22° 00′	.3840	.3746	.9272	.4040	2.475	1.079	2.669	1.1868	68° 00′
10	869	773	261	074	455	080	650	839	50
20	898	800	250	108	434	081	632	810	40
30	.3927	.3827	.9239	.4142	2.414	1.082	2.613	1.1781	30
40	956	854	228	176	394	084	595	752	20
50	985	881	216	210	375	085	577	723	10
23° 00′	.4014	.3907	.9205	.4245	2.356	1.086	2.559	1.1694	67° 00′
10	043	934	194	279	337	088	542	665	50
20	072	961	182	314	318	089	525	636	40
30	.4102	.3987	.9171	.4348	2.300	1.090	2.508	1.1606	30
40	131	.4014	159	383	282	092	491	577	20
50	160	041	147	417	264	093	475	548	10
24° 00′	.4189	.4067	.9135	.4452	2.246	1.095	2.459	1.1519	66° 00′
10	218	094	124	487	229	096	443	490	50
20	247	120	112	522	211	097	427	461	40
30	.4276	.4147	.9100	.4557	2.194	1.099	2.411	1.1432	30
40	305	173	088	592	177	100	396	403	20
50	334	200	075	628	161	102	381	374	10
25° 00′	.4363	.4226	.9063	.4663	2.145	1.103	2.366	1.1345	65° 00′
10	392	253	051	699	128	105	352	316	50
20	422	279	038	734	112	106	337	286	40
30	.4451	.4305	.9026	.4770	2.097	1.108	2.323	1.1257	30
40	480	331	013	806	081	109	309	228	20
50	509	358	001	841	066	111	295	199	10
26° 00′	.4538	.4384	.8988	.4877	2.050	1.113	2.281	1.1170	64° 00′
10	567	410	975	913	035	114	268	141	50
20	596	436	962	950	020	116	254	112	40
30	.4625	.4462	.8949	.4986	2.006	1.117	2.241	1.1083	30
40	654	488	936	.5022	1.991	119	228	054	20
50	683	514	923	059	977	121	215	1.1025	10
27° 00′	.4712	.4540	.8910	.5095	1.963	1.122	2.203	1.0996	63° 00′
		Cos	Sin	Cot	Tan	Csc	Sec	Radians	Degrees

TABLE II   **(Continued)**

Degrees	Radians	Sin	Cos	Tan	Cot	Sec	Csc		
27° 00′	.4712	.4540	.8910	.5095	1.963	1.122	2.203	1.0996	63° 00′
10	741	566	897	132	949	124	190	966	50
20	771	592	884	169	935	126	178	937	40
30	.4800	.4617	.8870	.5206	1.921	1.127	2.166	1.0908	30
40	829	643	857	243	907	129	154	879	20
50	858	669	843	280	894	131	142	850	10
28° 00′	.4887	.4695	.8829	.5317	1.881	1.133	2.130	1.0821	62° 00′
10	916	720	816	354	868	134	118	792	50
20	945	746	802	392	855	136	107	763	40
30	.4974	.4772	.8788	.5430	1.842	1.138	2.096	1.0734	30
40	.5003	797	774	467	829	140	085	705	20
50	032	823	760	505	816	142	074	676	10
29° 00′	.5061	.4848	.8746	.5543	1.804	1.143	2.063	1.0647	61° 00′
10	091	874	732	581	792	145	052	617	50
20	120	899	718	619	780	147	041	588	40
30	.5149	.4924	.8704	.5658	1.767	1.149	2.031	1.0559	30
40	178	950	689	696	756	151	020	530	20
50	207	975	675	735	744	153	010	501	10
30° 00′	.5236	.5000	.8660	.5774	1.732	1.155	2.000	1.0472	60° 00′
10	265	025	646	812	720	157	1.990	443	50
20	294	050	631	851	709	159	980	414	40
30	.5323	.5075	.8616	.5890	1.698	1.161	1.970	1.0385	30
40	352	100	601	930	686	163	961	356	20
50	381	125	587	969	675	165	951	327	10
31° 00′	.5411	.5150	.8572	.6009	1.664	1.167	1.942	1.0297	59° 00′
10	440	175	557	048	653	169	932	268	50
20	469	200	542	088	643	171	923	239	40
30	.5498	.5225	.8526	.6128	1.632	1.173	1.914	1.0210	30
40	527	250	511	168	621	175	905	181	20
50	556	275	496	208	611	177	896	152	10
32° 00′	.5585	.5299	.8480	.6249	1.600	1.179	1.887	1.0123	58° 00′
10	614	324	465	289	590	181	878	094	50
20	643	348	450	330	580	184	870	065	40
30	.5672	.5373	.8434	.6371	1.570	1.186	1.861	1.0036	30
40	701	398	418	412	560	188	853	1.0007	20
50	730	422	403	453	550	190	844	977	10
33° 00′	.5760	.5446	.8387	.6494	1.540	1.192	1.836	.9948	57° 00′
10	789	471	371	536	530	195	828	919	50
20	818	495	355	577	520	197	820	890	40
30	.5847	.5519	.8339	.6619	1.511	1.199	1.812	.9861	30
40	876	544	323	661	501	202	804	832	20
50	905	568	307	703	1.492	204	796	803	10
34° 00′	.5934	.5592	.8290	.6745	1.483	1.206	1.788	.9774	56° 00′
10	963	616	274	787	473	209	781	745	50
20	992	640	258	830	464	211	773	716	40
30	.6021	.5664	.8241	.6873	1.455	1.213	1.766	.9687	30
40	050	688	225	916	446	216	758	657	20
50	080	712	208	959	437	218	751	628	10
35° 00′	.6109	.5736	.8192	.7002	1.428	1.221	1.743	.9599	55° 00′
10	138	760	175	046	419	223	736	570	50
20	167	783	158	089	411	226	729	541	40
30	.6196	.5807	.8141	.7133	1.402	1.228	1.722	.9512	30
40	225	831	124	177	393	231	715	483	20
50	254	854	107	221	385	233	708	454	10
36° 00′	.6283	.5878	.8090	.7265	1.376	1.236	1.701	.9425	54° 00′
		Cos	Sin	Cot	Tan	Csc	Sec	Radians	Degrees

TABLE II (Continued)

Degrees	Radlans	Sin	Cos	Tan	Cot	Sec	Csc		
36° 00′	.6283	.5878	.8090	.7265	1.376	1.236	1.701	.9425	54° 00′
10	312	901	073	310	368	239	695	396	50
20	341	925	056	355	360	241	688	367	40
30	.6370	.5948	.8039	.7400	1.351	1.244	1.681	.9338	30
40	400	972	021	445	343	247	675	308	20
50	429	995	004	490	335	249	668	279	10
37° 00′	.6458	.6018	.7986	.7536	1.327	1.252	1.662	.9250	53° 00′
10	487	041	969	581	319	255	655	221	50
20	516	065	951	627	311	258	649	192	40
30	.6545	.6088	.7934	.7673	1.303	1.260	1.643	.9163	30
40	574	111	916	720	295	263	636	134	20
50	603	134	898	766	288	266	630	105	10
38° 00′	.6632	.6157	.7880	.7813	1.280	1.269	1.624	.9076	52° 00′
10	661	180	862	860	272	272	618	047	50
20	690	202	844	907	265	275	612	.9018	40
30	.6720	.6225	.7826	.7954	1.257	1.278	1.606	.8988	30
40	749	248	808	.8002	250	281	601	959	20
50	778	271	790	050	242	284	595	930	10
39° 00′	.6807	.6293	.7771	.8098	1.235	1.287	1.589	.8901	51° 00′
10	836	316	753	146	228	290	583	872	50
20	865	338	735	195	220	293	578	843	40
30	.6894	.6361	.7716	.8243	1.213	1.296	1.572	.8814	30
40	923	383	698	292	206	299	567	785	20
50	952	406	679	342	199	302	561	756	10
40° 00′	.6981	.6428	.7660	.8391	1.192	1.305	1.556	.8727	50° 00′
10	.7010	450	642	441	185	309	550	698	50
20	039	472	623	491	178	312	545	668	40
30	.7069	.6494	.7604	.8541	1.171	1.315	1.540	.8639	30
40	098	517	585	591	164	318	535	610	20
50	127	539	566	642	157	322	529	581	10
41° 00′	.7156	.6561	.7547	.8693	1.150	1.325	1.524	.8552	49° 00′
10	185	583	528	744	144	328	519	523	50
20	214	604	509	796	137	332	514	494	40
30	.7243	.6626	.7490	.8847	1.130	1.335	1.509	.8465	30
40	272	648	470	899	124	339	504	436	20
50	301	670	451	952	117	342	499	407	10
42° 00′	.7330	.6691	.7431	.9004	1.111	1.346	1.494	.8378	48° 00′
10	359	713	412	057	104	349	490	348	50
20	389	734	392	110	098	353	485	319	40
30	.7418	.6756	.7373	.9163	1.091	1.356	1.480	.8290	30
40	447	777	353	217	085	360	476	261	20
50	476	799	333	271	079	364	471	232	10
43° 00′	.7505	.6820	.7314	.9325	1.072	1.367	1.466	.8203	47° 00′
10	534	841	294	380	066	371	462	174	50
20	563	862	274	435	060	375	457	145	40
30	.7592	.6884	.7254	.9490	1.054	1.379	1.453	.8116	30
40	621	905	234	545	048	382	448	087	20
50	650	926	214	601	042	386	444	058	10
44° 00′	.7679	.6947	.7193	.9657	1.036	1.390	1.440	.8029	46° 00′
10	709	967	173	713	030	394	435	999	50
20	738	988	153	770	024	398	431	970	40
30	.7767	.7009	.7133	.9827	1.018	1.402	1.427	.7941	30
40	796	030	112	884	012	406	423	912	20
50	825	050	092	942	006	410	418	883	10
45° 00′	.7854	.7071	.7071	1.000	1.000	1.414	1.414	.7854	45° 00′
		Cos	Sin	Cot	Tan	Csc	Sec	Radians	Degrees

Table III

Common Logarithms

N	0	1	2	3	4	5	6	7	8	9
1.0	.0000	.0043	.0086	.0128	.0170	.0212	.0253	.0294	.0334	.0374
1.1	.0414	.0453	.0492	.0531	.0569	.0607	.0645	.0682	.0719	.0755
1.2	.0792	.0828	.0864	.0899	.0934	.0969	.1004	.1038	.1072	.1106
1.3	.1139	.1173	.1206	.1239	.1271	.1303	.1335	.1367	.1399	.1430
1.4	.1461	.1492	.1523	.1553	.1584	.1614	.1644	.1673	.1703	.1732
1.5	.1761	.1790	.1818	.1847	.1875	.1903	.1931	.1959	.1987	.2014
1.6	.2041	.2068	.2095	.2122	.2148	.2175	.2201	.2227	.2253	.2279
1.7	.2304	.2330	.2355	.2380	.2405	.2430	.2455	.2480	.2504	.2529
1.8	.2553	.2577	.2601	.2625	.2648	.2672	.2695	.2718	.2742	.2765
1.9	.2788	.2810	.2833	.2856	.2878	.2900	.2923	.2945	.2967	.2989
2.0	.3010	.3032	.3054	.3075	.3096	.3118	.3139	.3160	.3181	.3201
2.1	.3222	.3243	.3263	.3284	.3304	.3324	.3345	.3365	.3385	.3404
2.2	.3424	.3444	.3464	.3483	.3502	.3522	.3541	.3560	.3579	.3598
2.3	.3617	.3636	.3655	.3674	.3692	.3711	.3729	.3747	.3766	.3784
2.4	.3802	.3820	.3838	.3856	.3874	.3892	.3909	.3927	.3945	.3962
2.5	.3979	.3997	.4014	.4031	.4048	.4065	.4082	.4099	.4116	.4133
2.6	.4150	.4166	.4183	.4200	.4216	.4232	.4249	.4265	.4281	.4298
2.7	.4314	.4330	.4346	.4362	.4378	.4393	.4409	.4425	.4440	.4456
2.8	.4472	.4487	.4502	.4518	.4533	.4548	.4564	.4579	.4594	.4609
2.9	.4624	.4639	.4654	.4669	.4683	.4698	.4713	.4728	.4742	.4757
3.0	.4771	.4786	.4800	.4814	.4829	.4843	.4857	.4871	.4886	.4900
3.1	.4914	.4928	.4942	.4955	.4969	.4983	.4997	.5011	.5024	.5038
3.2	.5051	.5065	.5079	.5092	.5105	.5119	.5132	.5145	.5159	.5172
3.3	.5185	.5198	.5211	.5224	.5237	.5250	.5263	.5276	.5289	.5302
3.4	.5315	.5328	.5340	.5353	.5366	.5378	.5391	.5403	.5416	.5428
3.5	.5441	.5453	.5465	.5478	.5490	.5502	.5514	.5527	.5539	.5551
3.6	.5563	.5575	.5587	.5599	.5611	.5623	.5635	.5647	.5658	.5670
3.7	.5682	.5694	.5705	.5717	.5729	.5740	.5752	.5763	.5775	.5786
3.8	.5798	.5809	.5821	.5832	.5843	.5855	.5866	.5877	.5888	.5899
3.9	.5911	.5922	.5933	.5944	.5955	.5966	.5977	.5988	.5999	.6010
4.0	.6021	.6031	.6042	.6053	.6064	.6075	.6085	.6096	.6107	.6117
4.1	.6128	.6138	.6149	.6160	.6170	.6180	.6191	.6201	.6212	.6222
4.2	.6232	.6243	.6253	.6263	.6274	.6284	.6294	.6304	.6314	.6325
4.3	.6335	.6345	.6355	.6365	.6375	.6385	.6395	.6405	.6415	.6425
4.4	.6435	.6444	.6454	.6464	.6474	.6484	.6493	.6503	.6513	.6522
4.5	.6532	.6542	.6551	.6561	.6571	.6580	.6590	.6599	.6609	.6618
4.6	.6628	.6637	.6646	.6656	.6665	.6675	.6684	.6693	.6702	.6712
4.7	.6721	.6730	.6739	.6749	.6758	.6767	.6776	.6785	.6794	.6803
4.8	.6812	.6821	.6830	.6839	.6848	.6857	.6866	.6875	.6884	.6893
4.9	.6902	.6911	.6920	.6928	.6937	.6946	.6955	.6964	.6972	.6981
5.0	.6990	.6998	.7007	.7016	.7024	.7033	.7042	.7050	.7059	.7067
5.1	.7076	.7084	.7093	.7101	.7110	.7118	.7126	.7135	.7143	.7152
5.2	.7160	.7168	.7177	.7185	.7193	.7202	.7210	.7218	.7226	.7235
5.3	.7243	.7251	.7259	.7267	.7275	.7284	.7292	.7300	.7308	.7316
5.4	.7324	.7332	.7340	.7348	.7356	.7364	.7372	.7380	.7388	.7396
N	0	1	2	3	4	5	6	7	8	9

TABLE III (Continued)

N	0	1	2	3	4	5	6	7	8	9
5.5	.7404	.7412	.7419	.7427	.7435	.7443	.7451	.7459	.7466	.7474
5.6	.7482	.7490	.7497	.7505	.7513	.7520	.7528	.7536	.7543	.7551
5.7	.7559	.7566	.7574	.7582	.7589	.7597	.7604	.7612	.7619	.7627
5.8	.7634	.7642	.7649	.7657	.7664	.7672	.7679	.7686	.7694	.7701
5.9	.7709	.7716	.7723	.7731	.7738	.7745	.7752	.7760	.7767	.7774
6.0	.7782	.7789	.7796	.7803	.7810	.7818	.7825	.7832	.7839	.7846
6.1	.7853	.7860	.7868	.7875	.7882	.7889	.7896	.7903	.7910	.7917
6.2	.7924	.7931	.7938	.7945	.7952	.7959	.7966	.7973	.7980	.7987
6.3	.7993	.8000	.8007	.8014	.8021	.8028	.8035	.8041	.8048	.8055
6.4	.8062	.8069	.8075	.8082	.8089	.8096	.8102	.8109	.8116	.8122
6.5	.8129	.8136	.8142	.8149	.8156	.8162	.8169	.8176	.8182	.8189
6.6	.8195	.8202	.8209	.8215	.8222	.8228	.8235	.8241	.8248	.8254
6.7	.8261	.8267	.8274	.8280	.8287	.8293	.8299	.8306	.8312	.8319
6.8	.8325	.8331	.8338	.8344	.8351	.8357	.8363	.8370	.8376	.8382
6.9	.8388	.8395	.8401	.8407	.8414	.8420	.8426	.8432	.8439	.8445
7.0	.8451	.8457	.8463	.8470	.8476	.8482	.8488	.8494	.8500	.8506
7.1	.8513	.8519	.8525	.8531	.8537	.8543	.8549	.8555	.8561	.8567
7.2	.8573	.8579	.8585	.8591	.8597	.8603	.8609	.8615	.8621	.8627
7.3	.8633	.8639	.8645	.8651	.8657	.8663	.8669	.8675	.8681	.8686
7.4	.8692	.8698	.8704	.8710	.8716	.8722	.8727	.8733	.8739	.8745
7.5	.8751	.8756	.8762	.8768	.8774	.8779	.8785	.8791	.8797	.8802
7.6	.8808	.8814	.8820	.8825	.8831	.8837	.8842	.8848	.8854	.8859
7.7	.8865	.8871	.8876	.8882	.8887	.8893	.8899	.8904	.8910	.8915
7.8	.8921	.8927	.8932	.8938	.8943	.8949	.8954	.8960	.8965	.8971
7.9	.8976	.8982	.8987	.8993	.8998	.9004	.9009	.9015	.9020	.9025
8.0	.9031	.9036	.9042	.9047	.9053	.9058	.9063	.9069	.9074	.9079
8.1	.9085	.9090	.9096	.9101	.9106	.9112	.9117	.9122	.9128	.9133
8.2	.9138	.9143	.9149	.9154	.9159	.9165	.9170	.9175	.9180	.9186
8.3	.9191	.9196	.9201	.9206	.9212	.9217	.9222	.9227	.9232	.9238
8.4	.9243	.9248	.9253	.9258	.9263	.9269	.9274	.9279	.9284	.9289
8.5	.9294	.9299	.9304	.9309	.9315	.9320	.9325	.9330	.9335	.9340
8.6	.9345	.9350	.9355	.9360	.9365	.9370	.9375	.9380	.9385	.9390
8.7	.9395	.9400	.9405	.9410	.9415	.9420	.9425	.9430	.9435	.9440
8.8	.9445	.9450	.9455	.9460	.9465	.9469	.9474	.9479	.9484	.9489
8.9	.9494	.9499	.9504	.9509	.9513	.9518	.9523	.9528	.9533	.9538
9.0	.9542	.9547	.9552	.9557	.9562	.9566	.9571	.9576	.9581	.9586
9.1	.9590	.9595	.9600	.9605	.9609	.9614	.9619	.9624	.9628	.9633
9.2	.9638	.9643	.9647	.9652	.9657	.9661	.9666	.9671	.9675	.9680
9.3	.9685	.9689	.9694	.9699	.9703	.9708	.9713	.9717	.9722	.9727
9.4	.9731	.9736	.9741	.9745	.9750	.9754	.9759	.9763	.9768	.9773
9.5	.9777	.9782	.9786	.9791	.9795	.9800	.9805	.9809	.9814	.9818
9.6	.9823	.9827	.9832	.9836	.9841	.9845	.9850	.9854	.9859	.9863
9.7	.9868	.9872	.9877	.9881	.9886	.9890	.9894	.9899	.9903	.9908
9.8	.9912	.9917	.9921	.9926	.9930	.9934	.9939	.9943	.9948	.9952
9.9	.9956	.9961	.9965	.9969	.9974	.9978	.9983	.9987	.9991	.9996
N	0	1	2	3	4	5	6	7	8	9

Table IV

Natural Logarithms

N	0	1	2	3	4	5	6	7	8	9
1.0	0000	0100	0198	0296	0392	0488	0583	0677	0770	0862
1.1	0953	1044	1133	1222	1310	1398	1484	1570	1655	1740
1.2	1823	1906	1989	2070	2151	2231	2311	2390	2469	2546
1.3	2624	2700	2776	2852	2927	3001	3075	3148	3221	3293
1.4	3365	3436	3507	3577	3646	3716	3874	3853	3920	3988
1.5	4055	4121	4187	4253	4318	4383	4447	4511	4574	4637
1.6	4700	4762	4824	4886	4947	5008	5068	5128	5188	5247
1.7	5306	5365	5423	5481	5539	5596	5653	5710	5766	5822
1.8	5878	5933	5988	6043	6098	6152	6206	6259	6313	6366
1.9	6419	6471	6523	6575	6627	6678	6729	6780	6831	6881
2.0	6931	6981	7031	7080	7129	7178	7227	7275	7324	7372
2.1	7419	7467	7514	7561	7608	7655	7701	7747	7793	7839
2.2	7885	7930	7975	8020	8065	8109	8154	8198	8242	8286
2.3	8329	8372	8416	8459	8502	8544	8587	8629	8671	8713
2.4	8755	8796	8838	8879	8920	8961	9002	9042	9083	9123
2.5	9163	9203	9243	9282	9322	9361	9400	9439	9478	9517
2.6	9555	9594	9632	9670	9708	9746	9783	9821	9858	9895
2.7	9933	9969	*0006	*0043	*0080	*0116	*0152	*0188	*0225	*0260
2.8	1.0296	0332	0367	0403	0438	0473	0508	0543	0578	0613
2.9	0647	0682	0716	0750	0784	0818	0852	0886	0919	0953
3.0	1.0986	1019	1053	1086	1119	1151	1184	1217	1249	1282
3.1	1314	1346	1378	1410	1442	1474	1506	1537	1569	1600
3.2	1632	1663	1694	1725	1756	1787	1817	1848	1878	1909
3.3	1939	1969	2000	2030	2060	2090	2119	2149	2179	2208
3.4	2238	2267	2296	2326	2355	2384	2413	2442	2470	2499
3.5	1.2528	2556	2585	2613	2641	2669	2698	2726	2754	2782
3.6	2809	2837	2865	2892	2920	2947	2975	3002	3029	3056
3.7	3083	3110	3137	3164	3191	3218	3244	3271	3297	3324
3.8	3350	3376	3403	3429	3455	3481	3507	3533	3558	3584
3.9	3610	3635	3661	3686	3712	3737	3762	3788	3813	3838
4.0	1.3863	3888	3913	3938	3962	3987	4012	4036	4061	4085
4.1	4110	4134	4159	4183	4207	4231	4255	4279	4303	4327
4.2	4351	4375	4398	4422	4446	4469	4493	4516	4540	4563
4.3	4586	4609	4633	4656	4679	4702	4725	4748	4770	4793
4.4	4816	4839	4861	4884	4907	4929	4951	4974	4996	5019
4.5	1.5041	5063	5085	5107	5129	5151	5173	5195	5217	5239
4.6	5261	5282	5304	5326	5347	5369	5390	5412	5433	5454
4.7	5476	5497	5518	5539	5560	5581	5602	5623	5644	5665
4.8	5686	5707	5728	5748	5769	5790	5810	5831	5851	5872
4.9	5892	5913	5933	5953	5974	5994	6014	6034	6054	6074
5.0	1.6094	6114	6134	6154	6174	6194	6214	6233	6253	6273
5.1	6292	6312	6332	6351	6371	6390	6409	6429	6448	6467
5.2	6487	6506	6525	6544	6563	6582	6601	6620	6639	6658
5.3	6677	6696	6715	6734	6752	6771	6790	6808	6827	6845
5.4	6864	6882	6901	6919	6938	6956	6974	6993	7011	7029
N	0	1	2	3	4	5	6	7	8	9

TABLE IV (Continued)

N	0	1	2	3	4	5	6	7	8	9
5.5	1.7047	7066	7084	7102	7120	7138	7156	7174	7192	7210
5.6	7228	7246	7263	7281	7299	7317	7334	7352	7370	7387
5.7	7405	7422	7440	7457	7475	7492	7509	7527	7544	7561
5.8	7579	7596	7613	7630	7647	7664	7681	7699	7716	7733
5.9	7750	7766	7783	7800	7817	7834	7851	7867	7884	7901
6.0	1.7918	7934	7951	7967	7984	8001	8017	8034	8050	8066
6.1	8083	8099	8116	8132	8148	8165	8181	8197	8213	8229
6.2	8245	8262	8278	8294	8310	8326	8342	8358	8374	8390
6.3	8405	8421	8437	8453	8469	8485	8500	8516	8532	8547
6.4	8563	8579	8594	8610	8625	8641	8656	8672	8687	8703
6.5	1.8718	8733	8749	8764	8779	8795	8810	8825	8840	8856
6.6	8871	8886	8901	8916	8931	8946	8961	8976	8991	9006
6.7	9021	9036	9051	9066	9081	9095	9110	9125	9140	9155
6.8	9169	9184	9199	9213	9228	9242	9257	9272	9286	9301
6.9	9315	9330	9344	9359	9373	9387	9402	9416	9430	9445
7.0	1.9459	9473	9488	9502	9516	9530	9544	9559	9573	9587
7.1	9601	9615	9629	9643	9657	9671	9685	9699	9713	9727
7.2	9741	9755	9769	9782	9796	9810	9824	9838	9851	9865
7.3	9879	9892	9906	9920	9933	9947	9961	9974	9988	*0001
7.4	2.0015	0028	0042	0055	0069	0082	0096	0109	0122	0136
7.5	2.0149	0162	0176	0189	0202	0215	0229	0242	0255	0268
7.6	0281	0295	0308	0321	0334	0347	0360	0373	0386	0399
7.7	0412	0425	0438	0451	0464	0477	0490	0503	0516	0528
7.8	0541	0554	0567	0580	0592	0605	0618	0630	0643	0656
7.9	0669	0681	0694	0707	0719	0732	0744	0757	0769	0782
8.0	2.0794	0807	0819	0832	0844	0857	0869	0882	0894	0906
8.1	0919	0931	0943	0956	0968	0980	0992	1005	1017	1029
8.2	1041	1054	1066	1078	1090	1102	1114	1126	1138	1150
8.3	1163	1175	1187	1199	1211	1223	1235	1247	1258	1270
8.4	1282	1294	1306	1318	1330	1342	1353	1365	1377	1389
8.5	2.1401	1412	1424	1436	1448	1459	1471	1483	1494	1506
8.6	1518	1529	1541	1552	1564	1576	1587	1599	1610	1622
8.7	1633	1645	1656	1668	1679	1691	1702	1713	1725	1736
8.8	1748	1759	1770	1782	1793	1804	1815	1827	1838	1849
8.9	1861	1872	1883	1894	1905	1917	1928	1939	1950	1961
9.0	2.1972	1983	1994	2006	2017	2028	2039	2050	2061	2072
9.1	2083	2094	2105	2116	2127	2138	2148	2159	2170	2181
9.2	2192	2203	2214	2225	2235	2246	2257	2268	2279	2289
9.3	2300	2311	2322	2332	2343	2354	2364	2375	2386	2396
9.4	2407	2418	2428	2439	2450	2460	2471	2481	2492	2502
9.5	2.2513	2523	2534	2544	2555	2565	2576	2586	2597	2607
9.6	2618	2628	2638	2649	2659	2670	2680	2690	2701	2711
9.7	2721	2732	2742	2752	2762	2773	2783	2793	2803	2814
9.8	2824	2834	2844	2854	2865	2875	2885	2895	2905	2915
9.9	2925	2935	2946	2956	2966	2976	2986	2996	3006	3016
N	0	1	2	3	4	5	6	7	8	9

Index

$c^2 = a^2 + b^2 - 2ab \cos \gamma$